# Laser Capture Microdissection

# METHODS IN MOLECULAR BIOLOGY™

## John M. Walker, SERIES EDITOR

309. **RNA Silencing:** *Methods and Protocols*, edited by *Gordon Carmichael, 2005*

308. **Therapeutic Proteins:** *Methods and Protocols*, edited by *C. Mark Smales and David C. James, 2005*

307. **Phosphodiesterase Mehtods and Protocols,** edited by *Claire Lugnier, 2005*

306. **Receptor Binding Techniques:** *Second Edition*, edited by *Anthony P. Davenport, 2005*

305. **Protein–Ligand Interactions:** *Methods and Protocols*, edited by *G. Ulrich Nienhaus, 2005*

304. **Human Retrovirus Protocols:** *Virology and Molecular Biology*, edited by *Tuofu Zhu, 2005*

303. **NanoBiotechnology Protocols,** edited by *Sandra J. Rosenthal and David W. Wright, 2005*

302. **Handbook of ELISPOT:** *Methods and Protocols*, edited by *Alexander E. Kalyuzhny, 2005*

301. **Ubiquitin–Proteasome Protocols,** edited by *Cam Patterson and Douglas M. Cyr, 2005*

300. **Protein Nanotechnology:** *Protocols, Instrumentation, and Applications*, edited by *Tuan Vo-Dinh, 2005*

299. **Amyloid Proteins:** *Methods and Protocols*, edited by *Einar M. Sigurdsson, 2005*

298. **Peptide Synthesis and Application,** edited by *John Howl, 2005*

297. **Forensic DNA Typing Protocols,** edited by *Angel Carracedo, 2005*

296. **Cell Cycle Protocols:** *Mechanisms and Protocols*, edited by *Tim Humphrey and Gavin Brooks, 2005*

295. **Immunochemical Protocols,** *Third Edition*, edited by *Robert Burns, 2005*

294. **Cell Migration:** *Developmental Methods and Protocols*, edited by *Jun-Lin Guan, 2005*

293. **Laser Capture Microdissection:** *Methods and Protocols*, edited by *Graeme I. Murray and Stephanie Curran, 2005*

292. **DNA Viruses:** *Methods and Protocols*, edited by *Paul M. Lieberman, 2005*

291. **Molecular Toxicology Protocols,** edited by *Phouthone Keohavong and Stephen G. Grant, 2005*

290. **Basic Cell Culture Protocols,** *Third Edition*, edited by *Cheryl D. Helgason and Cindy L. Miller, 2005*

289. **Epidermal Cells,** *Methods and Applications*, edited by *Kursad Turksen, 2005*

288. **Oligonucleotide Synthesis,** *Methods and Applications*, edited by *Piet Herdewijn, 2005*

287. **Epigenetics Protocols,** edited by *Trygve O. Tollefsbol, 2004*

286. **Transgenic Plants:** *Methods and Protocols*, edited by *Leandro Peña, 2005*

285. **Cell Cycle Control and Dysregulation Protocols:** *Cyclins, Cyclin-Dependent Kinases, and Other Factors*, edited by *Antonio Giordano and Gaetano Romano, 2004*

284. **Signal Transduction Protocols,** *Second Edition*, edited by *Robert C. Dickson and Michael D. Mendenhall, 2004*

283. **Bioconjugation Protocols,** edited by *Christof M. Niemeyer, 2004*

282. **Apoptosis Methods and Protocols,** edited by *Hugh J. M. Brady, 2004*

281. **Checkpoint Controls and Cancer, Volume 2:** *Activation and Regulation Protocols*, edited by *Axel H. Schönthal, 2004*

280. **Checkpoint Controls and Cancer, Volume 1:** *Reviews and Model Systems*, edited by *Axel H. Schönthal, 2004*

279. **Nitric Oxide Protocols,** *Second Edition*, edited by *Aviv Hassid, 2004*

278. **Protein NMR Techniques,** *Second Edition*, edited by *A. Kristina Downing, 2004*

277. **Trinucleotide Repeat Protocols,** edited by *Yoshinori Kohwi, 2004*

276. **Capillary Electrophoresis of Proteins and Peptides,** edited by *Mark A. Strege and Avinash L. Lagu, 2004*

275. **Chemoinformatics,** edited by *Jürgen Bajorath, 2004*

274. **Photosynthesis Research Protocols,** edited by *Robert Carpentier, 2004*

273. **Platelets and Megakaryocytes, Volume 2:** *Perspectives and Techniques*, edited by *Jonathan M. Gibbins and Martyn P. Mahaut-Smith, 2004*

272. **Platelets and Megakaryocytes, Volume 1:** *Functional Assays*, edited by *Jonathan M. Gibbins and Martyn P. Mahaut-Smith, 2004*

271. **B Cell Protocols,** edited by *Hua Gu and Klaus Rajewsky, 2004*

270. **Parasite Genomics Protocols,** edited by *Sara E. Melville, 2004*

269. **Vaccina Virus and Poxvirology:** *Methods and Protocols*, edited by *Stuart N. Isaacs, 2004*

268. **Public Health Microbiology:** *Methods and Protocols*, edited by *John F. T. Spencer and Alicia L. Ragout de Spencer, 2004*

267. **Recombinant Gene Expression:** *Reviews and Protocols*, *Second Edition*, edited by *Paulina Balbas and Argelia Johnson, 2004*

266. **Genomics, Proteomics, and Clinical Bacteriology:** *Methods and Reviews*, edited by *Neil Woodford and Alan Johnson, 2004*

265. **RNA Interference, Editing, and Modification:** *Methods and Protocols*, edited by *Jonatha M. Gott, 2004*

METHODS IN MOLECULAR BIOLOGY™

# Laser Capture Microdissection

## *Methods and Protocols*

Edited by

## Graeme I. Murray

and

## Stephanie Curran

*Department of Pathology,*
*University of Aberdeen, Aberdeen, UK*

HUMANA PRESS ✳ TOTOWA, NEW JERSEY

© 2005 Humana Press Inc.
999 Riverview Drive, Suite 208
Totowa, New Jersey 07512

**www.humanapress.com**

This publication is printed on acid-free paper. ∞
ANSI Z39.48-1984 (American Standards Institute)

Permanence of Paper for Printed Library Materials.
Cover illustration: Provided by Graeme I. Murray.

Production Editor: Mark J. Breaugh
Cover design by Patricia F. Cleary

For additional copies, pricing for bulk purchases, and/or information about other Humana titles, contact Humana at the above address or at any of the following numbers: Tel.: 973-256-1699; Fax: 973-256-8341; E-mail: humana@humanapr.com; or visit our Website: www.humanapress.com.

Printed in the United States of America. 10 9 8 7 6 5 4 3 2 1

eISBN: 1-59259-853-6

Library of Congress Cataloging in Publication Data

Laser capture microdissection : methods and protocols / edited by Graeme I. Murray and Stephanie Curran.
   p. ; cm. -- (Methods in molecular biology ; v. 293)
 Includes bibliographical references and index.
 ISBN 1-58829-260-6 (alk. paper)
 1. Molecular biology--Methodology. 2. Microdissection. 3. Lasers.
 [DNLM: 1. Lasers--diagnostic use. 2. Microdissection--methods. 3.
Genetic Techniques. QS 130 L3426 2005] I. Murray, Graeme I. II. Curran,
Stephanie. III. Series: Methods in molecular biology (Clifton, N.J.) ; v.
293.
 QH506.L25 2005
 572.8'028--dc22
                                                              2004021929

# Preface

Laser microdissection techniques have revolutionized the ability of researchers, generally, and pathologists in particular, to carry out molecular analysis on specific normal and diseased cells and fully utilize the power of current molecular technologies, including polymerase chain reaction (PCR), microarrays, and proteomics. The primary purpose of *Laser Capture Microdissection: Methods and Protocols* is to provide readers with practical advice on how to carry out tissue-based laser microdissection successfully in their own laboratories using the microdissection systems available and how best to apply a wide range of molecular technologies. The individual chapters encompass detailed descriptions of each of the laser-based microdissection systems. Applications of the laser microdissected tissue described in the book include PCR in its many different forms and gene expression analysis involving microarrays and proteomics.

The editors are especially grateful to all the contributing authors for the time and effort they have put into writing their chapters.

The series editor, John Walker, has expertly guided us through the editorial process, while Craig Adams of Humana Press has been very helpful in dealing with all the publication related issues.

We are particularly pleased to acknowledge the excellent secretarial support of Ms. Anne McMillan of the Department of Pathology, University of Aberdeen who helped us deal efficiently with all the correspondence relating to this book. We hope the readers will find this volume valuable.

*Graeme I. Murray*
*Stephanie Curran*

# Contents

Preface ........................................................................................................ v

Contributors ............................................................................................ xi

**PART I. INTRODUCTION**

1. An Introduction to Laser-Based Tissue Microdissection Techniques
   *Stephanie Curran and Graeme I. Murray* ........................................... 3

**PART II. MICRODISSECTION AND DNA ANALYSIS**

2. Methacarn Fixation for Genomic DNA Analysis
   in Microdissected Cells
   *Makoto Shibutani and Chikako Uneyama* ........................................ 11

3. Multiplex Quantitative Real-Time PCR
   of Laser Microdissected Tissue
   *Patrick H. Rooney* ........................................................................... 27

4. Comparative Genomic Hybridization Using DNA
   From Laser Capture Microdissected Tissue
   *Grace Callagy, Lucy Jackson, and Carlos Caldas* ............................. 39

5. Detection of Ki-*ras* and *p53* Mutations
   by Laser Capture Microdissection/PCR/SSCP
   *Deborah Dillon, Karl Zheng, Brina Negin, and José Costa* ............... 57

6. Whole-Genome Allelotyping Using Laser Microdissected Tissue
   *Colleen M. Feltmate and Samuel C. Mok* ....................................... 69

7. Microdissection for Detecting Genetic Aberrations
   in Early and Advanced Human Urinary Bladder Cancer
   *Arndt Hartmann, Robert Stoehr, Peter J. Wild,*
   *Wolfgang Dietmaier, and Ruth Knuechel* ....................................... 79

8. Laser Microdissection for Microsatellite Analysis
   in Colon and Breast Cancer
   *Peter J. Wild, Robert Stoehr, Ruth Knuechel, Arndt Hartmann,*
   *and Wolfgang Dietmaier* ................................................................. 93

9. Assessment of RET/PTC Oncogene Activation in Thyroid Nodules
   Utilizing Laser Microdissection Followed by Nested RT-PCR
   *Giovanni Tallini and Guilherme Brandao* ...................................... 103

10. Combined Laser-Assisted Microdissection and Short
    Tandem Repeat Analysis for Detection of *In Situ*
    Microchimerism After Solid Organ Transplantation
    *Ulrich Lehmann, Anne Versmold, and Hans Kreipe* ....................... 113

## PART III.  RNA AND GENE EXPRESSION STUDIES USING MICRODISSECTED CELLS

11.  Laser-Assisted Microdissection of Membrane-Mounted
     Tissue Sections
     ***Lise Mette Gjerdrum and Stephen Hamilton-Dutoit ........................ 127***

12.  Laser-Assisted Microdissection of Membrane-Mounted Sections
     Following Immunohistochemistry and *In Situ* Hybridization
     ***Lise Mette Gjerdrum and Stephen Hamilton-Dutoit ........................ 139***

13.  Laser-Assisted Cell Microdissection Using the PALM System
     ***Patrick Micke, Arne Östman, Joakim Lundeberg,***
     ***and Fredrik Ponten .......................................................... 151***

14.  Laser Microdissection and RNA Analysis
     ***Ludger Fink and Rainer Maria Bohle ....................................... 167***

15.  Gene Expression Profiling of Primary Tumor Cell Populations
     Using Laser Capture Microdissection, RNA Transcript
     Amplification, and GeneChip® Microarrays
     ***Veronica I. Luzzi, Victoria Holtschlag, and Mark A. Watson .......... 187***

16.  Quantification of Gene Expression in Mouse
     and Human Renal Proximal Tubules
     ***Jun-ya Kaimori, Masaru Takenaka, and Kousaku Okubo ................. 209***

17.  Laser Capture Microdissection for Analysis of Macrophage
     Gene Expression From Atherosclerotic Lesions
     ***Eugene Trogan and Edward A. Fisher ...................................... 221***

18.  Analysis of Pituitary Cells by Laser Capture Microdissection
     ***Ricardo V. Lloyd, Xiang Qian, Long Jin, Katharina Ruebel,***
     ***Jill Bayliss, Shuya Zhang, and Ikuo Kobayaski ........................... 233***

## PART IV.  MICRODISSECTION TECHNIQUES AND APPLICATIONS IN PROTEOMICS

19.  Laser Capture Microdissection and Colorectal Cancer Proteomics
     ***Laura C. Lawrie and Stephanie Curran ..................................... 245***

20.  Proteomic Analysis of Human Bladder Tissue
     Using SELDI® Approach Following Microdissection Techniques
     ***Rene C. Krieg, Nadine T. Gaisa, Cloud P. Paweletz,***
     ***and Ruth Knuechel ........................................................... 255***

## PART V.  MICRODISSECTION AND MOLECULAR ANALYSIS OF MICROORGANISMS

21.  Genetic Analysis of HIV by *In Situ* PCR-Directed
     Laser Capture Microscopy of Infected Cells
     ***Daniele Marras ............................................................... 271***

Contents                                                                                          *ix*

22. Use of Laser Capture Microdissection Together
        With *In Situ* Hybridization and Real-Time PCR
        to Study the Distribution of Latent Herpes Simplex
        Virus Genomes in Mouse Trigeminal Ganglion
        **Xiao-Ping Chen, Marina Mata, and David J. Fink** ............................ 285

23. Laser Capture Microdissection and PCR
        for Analysis of Human Papilloma Virus Infection
        **Kheng Chew, Patrick H. Rooney, Margaret E. Cruickshank,
        and Graeme I. Murray** ..................................................................... 295

24. Laser Capture Microdissection of Hepatic Stages of the Human
        Parasite *Plasmodium falciparum* for Molecular Analysis
        **Jean-Philippe Semblat, Olivier Silvie, Jean-François Franetich,
        and Dominique Mazier** .................................................................. 301

Index ............................................................................................................. 309

# Contributors

JILL BAYLISS • *Department of Laboratory Medicine and Pathology, Mayo Clinic, Rochester, MN*

RAINER MARIA BOHLE • *Department of Pathology, Justus-Liebig-University, Giessen, Germany*

GUILHERME BRANDAO • *Division of Pathology, Yale University School of Medicine, New Haven, CT*

CARLOS CALDAS • *Cancer Genomics Program, Department of Oncology, University of Cambridge, Cambridge, UK*

GRACE CALLAGY • *Cancer Genomics Program, Department of Oncology, University of Cambridge, Cambridge, UK*

XIAO-PING CHEN • *Department of Microbiology and Immunology, Tong-Ji University School of Medicine, Shanghai, China*

KHENG CHEW • *Department of Obstetrics and Gynaecology, University of Aberdeen, Aberdeen, UK*

JOSÉ COSTA • *Department of Pathology, Yale University School of Medicine, New Haven, CT*

MARGARET E. CRUICKSHANK • *Department of Obstetrics and Gynaecology, University of Aberdeen, Aberdeen, UK*

STEPHANIE CURRAN • *Department of Pathology, University of Aberdeen, Aberdeen, UK*

WOLFGANG DIETMAIER • *Institute of Pathology, University of Regensburg, Regensburg, Germany*

DEBORAH DILLON • *Department of Pathology, Brigham and Women's Hospital and Harvard Medical School, Boston, MA*

COLLEEN M. FELTMATE • *Department of Obstetrics, Gynecology, and Reproductive Biology, Brigham and Women's Hospital and Harvard Medical School, Boston, MA*

DAVID J. FINK • *Department of Neurology, University of Michigan, Ann Arbor, MI*

LUDGER FINK • *Department of Pathology, University of Giessen, Giessen, Germany*

EDWARD A. FISHER • *Division of Cardiology (Department of Medicine), Department of Cell Biology, NYU School of Medicine, New York, NY*

JEAN-FRANÇOIS FRANETICH • *INSERM U-511, Immunobiologie Cellulaire et Moléculaire des Infections Parasitaires, Centre Hospitalier-Universitaire Pitié-Salpêtriere, Université Pierre et Marie Curie, Paris, France*

*xi*

NADINE T. GAISA • *Institute of Pathology, University of Regensburg, Regensburg, Germany*

LISE METTE GJERDRUM • *Institute of Pathology, Aarhus University Hospital, Aarhus, Denmark*

STEPHEN HAMILTON-DUTOIT • *Institute of Pathology, Aarhus University Hospital, Aarhus, Denmark*

ARNDT HARTMANN • *Institute of Pathology, University of Basel, Basel, Switzerland*

VICTORIA HOLTSCHLAG • *Siteman Cancer Center, Washington University School of Medicine, St. Louis, MO*

LUCY JACKSON • *Cancer Genomics Program, Department of Oncology, University of Cambridge, Cambridge, UK*

LONG JIN • *Department of Laboratory Medicine and Pathology, Mayo Clinic, Rochester, MN*

JUN-YA KAIMORI • *Department of Internal Medicine and Therapeutics, Osaka University Graduate School of Medicine, Osaka, Japan*

RUTH KNUECHEL • *Institute of Pathology, University Hospital Aachen, RWTH, Aachen, Germany*

IKUO KOBAYASKI • *Department of Laboratory Medicine and Pathology, Mayo Clinic, Rochester, MN*

HANS KREIPE • *Institute of Pathology, Medizinische Hochschule Hannover, Hannover, Germany*

RENE C. KRIEG • *Institute of Pathology, University Hospital Aachen, RWTH, Aachen, Germany*

LAURA C. LAWRIE • *Department of Pathology, University of Aberdeen, Aberdeen, UK*

ULRICH LEHMANN • *Institute of Pathology, Medizinische Hochschule Hannover, Hannover, Germany*

RICARDO V. LLOYD • *Department of Laboratory Medicine and Pathology, Mayo Clinic, Rochester, MN*

JOAKIM LUNDEBERG • *Department of Biotechnology, KTH, Stockholm, Sweden*

VERONICA I. LUZZI • *Department of Pathology and Immunology, Washington University School of Medicine, St. Louis, MO*

DANIELE MARRAS • *Division of Nephrology, Department of Medicine, Mount Sinai School of Medicine, New York, NY*

MARINA MATA • *Department of Neurology, University of Michigan, Ann Arbor, MI*

DOMINIQUE MAZIER • *INSERM U-511, Immunobiologie Cellulaire et Moléculaire des Infections Parasitaires, Centre Hospitalier-Universitaire Pitié-Salpêtrière, Université Pierre et Marie Curie, Paris, France*

PATRICK MICKE • *Cancer Center Karolinska, Department of Oncology and Pathology, Karolinska Institute, Stockholm, Sweden*

SAMUEL C. MOK • *Department of Obstetrics, Gynecology, and Reproductive Biology, Brigham and Women's Hospital and Harvard Medical School, Boston, MA*

GRAEME I. MURRAY • *Department of Pathology, University of Aberdeen, Aberdeen, UK*

BRINA NEGIN • *Department of Pathology, Brigham and Women's Hospital, Boston, MA*

KOUSAKU OKUBO • *Center for Information Biology, National Institute of Genetics, Shizuoka, Japan*

ARNE ÖSTMAN • *Cancer Center Karolinska, Department of Oncology and Pathology, Karolinska Institute, Stockholm, Sweden*

CLOUD P. PAWELETZ • *Department of Anatomy, Physiology, and Genetics, Institute for Molecular Medicine, Uniformed Services University of the Health Sciences, Bethesda, MD*

FREDRIK PONTEN • *Institute for Genetics and Pathology, University of Uppsala, Uppsala, Sweden*

XIANG QIAN • *Department of Laboratory Medicine and Pathology, Mayo Clinic, Rochester, MN*

PATRICK H. ROONEY • *Department of Pathology, University of Aberdeen, Aberdeen, UK*

KATHARINA RUEBEL • *Department of Laboratory Medicine and Pathology, Mayo Clinic, Rochester, MN*

JEAN-PHILIPPE SEMBLAT • *INSERM U-511, Immunobiologie Cellulaire et Moléculaire des Infections Parasitaires, Centre Hospitalier-Universitaire Pitié-Salpêtriere, Université Pierre et Marie Curie, Paris, France*

MAKOTO SHIBUTANI • *Division of Pathology, National Institute of Health Sciences, Tokyo, Japan*

OLIVIER SILVIE • *INSERM U-511, Immunobiologie Cellulaire et Moléculaire des Infections Parasitaires, Centre Hospitalier-Universitaire Pitié-Salpêtriere, Université Pierre et Marie Curie, Paris, France*

ROBERT STOEHR • *Institute of Pathology, University of Regensburg, Regensburg, Germany*

MASARU TAKENAKA • *Graduate School of Life Sciences, Kobe Women's University, Kobe, Japan*

GIOVANNI TALLINI • *Division of Pathology, Yale University School of Medicine, New Haven, CT*

EUGENE TROGAN • *Graduate School of Biological Sciences, Mount Sinai School of Medicine, New York, NY*

CHIKAKO UNEYAMA • *Division of Pathology, National Institute of Health Sciences, Tokyo, Japan*

ANNE VERSMOLD • *Institute of Pathology, Medizinische Hochschule Hannover, Hannover, Germany*

MARK A. WATSON • *Department of Pathology and Immunology, Washington University School of Medicine, St. Louis, MO*

PETER J. WILD • *Institute of Pathology, University of Regensburg, Regensburg, Germany*

SHUYA ZHANG • *Department of Laboratory Medicine and Pathology, Mayo Clinic, Rochester, MN*

KARL ZHENG • *Columbia University School of Medicine, New York, NY*

# I

## INTRODUCTION

# 1

# An Introduction to Laser-Based Tissue Microdissection Techniques

## Stephanie Curran and Graeme I. Murray

### Summary

The development and application of laser-based tissue microdissection techniques has provided a major impetus to the sensitive and specific molecular analysis of solid tissues and tumors. This chapter provides an overview of the different laser-based microdissection systems and an introduction to the principles involved in the function and applications of these individual systems.

**Key Words:** Laser capture microdissection; laser microbeam microdissection; molecular analysis.

## 1. Introduction

Tissues, especially diseased tissues, are complex three-dimensional structures composed of heterogeneous mixtures of morphologically and phenotypically distinct cell types. The meaningful molecular analysis of morphologically and/or phenotypically distinct cell types from such tissues requires rapid, efficient, and accurate methods for obtaining specific population of cells.

The molecular investigation of solid tissues, especially tumors, has been revolutionized over the past decade by the development of accurate, rapid, and effective laser-based methods of tissue microdissection *(1,2)*. This has provided an extremely valuable and sophisticated tool to fully utilize the power and sensitivity of modern molecular analytical technologies in the detailed investigation of many different diseases and provided significant new insights into the pathogenesis of these diseases. Many of the investigations using laser microdissected cells have focused on specific types of cancers, where the morphological and phenotypic heterogeneity and complexity of tissues is often the greatest *(1–3)*.

From: *Methods in Molecular Biology, vol. 293: Laser Capture Microdissection: Methods and Protocols*
Edited by: G. I. Murray and S. Curran © Humana Press Inc., Totowa, NJ

Individual studies have usually based cell selection on specific morphological criteria of stained histological sections, but phenotypic characteristics as defined by immunohistochemistry of antigen expression *(4)* or genotypic features as demonstrated by *in situ* hybridization have also been used as selection criteria *(5)* and demonstrate the power of laser-based microdissection techniques. One of the major advantages of using laser microdissection methods to obtain specific cells for molecular analysis, especially from the viewpoint of the pathologist, is that the procedure is carried out under direct-light microscopic visualization of the cells. Whereas other technologies used to isolate specific cell populations for molecular analysis—e.g., fluorescent-activated cell sorting or magnetic bead-based cell separation—are indirect techniques with no microscopic visualization of the cells and require the availability of suitable antibodies to aid cell selection. Moreover, the methods used to prepare single cell suspensions (e.g., proteolytic enzyme digestion) that are necessary for antibody-based cell separation techniques from solid tissues may result in alteration or modification of cellular constituents. There is no doubt that the availability of laser-based microdissection technologies has provided a major impetus to molecular pathology research and this technology is now found in many laboratories worldwide (as represented in the diverse geographic locations of contributors to this volume). The wide availablity of this easy-to-use technology has allowed many questions in a range of research disciplines to be answered that previously could not be asked or answered using manual methods of tissue microdissection because of the imprecise nature of manual methods of microdissection or the time required to obtain tissue.

There are two major systems that have been developed for performing laser-assisted tissue microdissection—namely, laser capture microdissection and laser microbeam microdissection. Both types of systems have now been commercially available for several years. An overview of the principles, advantages, and potential disadvantages of each of the systems will be provided in this introductory chapter, detailed descriptions and applications of the individual systems are given in the relevant chapters in this volume.

## 2. Overview of Laser Microdissection Systems

### 2.1. Laser Capture Microdissection

The laser capture microdissection system was developed in the mid-1990s at the National Institutes of Health by Emmert-Buck and colleagues *(6,7)*, who recognized the need to develop a microscope-based microdissection system for accurately and efficiently microdissecting cells from histological tissue sections to fully exploit emerging molecular analytical technologies. They developed this system primarily to facilitate the molecular analysis of solid tumors.

This system rapidly moved into commercial production by Arcturus Engineering *(8)*, which has further developed the system since then. The laser capture microdissection system is now probably the most widely used laser-based microdissection system worldwide and our own experience of more than 5 yr is with this system, in particular the Arcturus PixCell II laser capture microdissection system.

The basic principle of the laser capture microdissection system is the capture of groups of cells or even individual cells onto a thermoplastic membrane from stained tissue histological sections (frozen sections or fixed wax-embedded sections) or cytological preparations. This plastic membrane that overlies the tissue section is attached to a specially designed "cap" and ensures that the plastic membrane is held in direct contact with the tissue section. The plastic is transiently melted by a low-power narrow-beam infrared laser directed at the cells of interest under microscope control. As the plastic cools and solidifies again, the cells are embedded into the plastic membrane and are removed from the tissue section by lifting the cap along with the plastic membrane off the tissue sections. Multiple groups of cells can be readily captured onto the same membrane. The laser beam in the current model of the laser capture microdissection system, the Arcturus PixCell II system, has three settings of the diameter of the laser beam, and using the laser beam at its narrowest diameter (7.5 μm) setting permits even single cells to be microdissected *(9)*. Subsequently cellular components including DNA, RNA, and protein can all be readily extracted with appropriate procedures and used for an extensive range of molecular analysis including PCR, gene expression studies, and proteomics *(6,7,10)*. To date, laser capture microdissected cells have been found to be compatible with all the molecular analytical techniques that have been used. Histological staining of the tissue, which is generally necessary for laser capture microdissection to allow morphological visualization, does not appear to significantly alter most cellular constituents. Moreover, the process of acquiring the cells onto the thermoplastic membrane does not appear to alter or damage the integrity of DNA, RNA, or protein, nor does capturing the cells onto a thermoplastic membrane appear to prevent a subsequent high rate of recovery of such material.

The procedure of laser capture microdissection is generally rapid, although morphological visualization of the tissue sections and cells is only moderate in comparison with stained and mounted histological slides as the tissue sections are not mounted with a coverslip. The inferior-quality morphology that is observed with this system in our experience only requires a little readjustment by an operator used to observing and interpreting high-quality morphology and we have found that operating this technology is easily learnt. In our laboratory we have also found that even relatively inexperienced operators, often with no background in histology/morphology, can easily be taught how to use the system, including recognition of cell types of interest, and thus rapidly and successfully acquire appropriate cells for downstream molecular analysis.

## 2.2. Laser Microbeam Microdissection

This method of laser microdissection operates on an entirely different principle from that used in laser capture microdissection *(11,12)*. The basic principle is that a pulsed very narrow-beam ultraviolet light laser is used to "draw" around the cell or cells of interest and unwanted tissue is photoablated *(12)*. Three different manufacturers have produced laser microbeam microdissection systems, all of which vary in the precise details of operation, especially with regard to the method of collection and transfer of the microdissected tissue for subsequent molecular analysis. In the PALM system the laser can be used to "catapult" the microdissected cells into a collecting tube, whereas in the Leica AS LMD system the section is inverted so that after microdissection the microdissected cells fall into the collecting tube under the influence of gravity. The advantages of this method is that there is no physical contact between the cells and plastic, unlike in laser capture microdissection, and this method clearly avoids the potential risk of modification of the molecules of interest by especially the heating and cooling of the thermoplastic membrane. In the microcut MMI system transfer of the microdissected cells is by a very fine needle or glass micropipet, this can run the risk of loss of microdissected cells during the transfer process or contamination of the transfer needle.

All laser microbeam microdissection systems use a much finer diameter laser beam (smallest laser diameter is 0.5 μm) compared with the laser capture microdissection system (smallest laser diameter size is 7.5 μm), making these systems ideally suited for the precise microdissection of single cells. However, this makes operating these systems potentially more time-consuming compared with laser capture microdissection, especially when a large number of cells require to be microdissected. For nucleic acid-based molecular analysis this is unlikely to be a significant issue, as analysis can often be performed on very few cells, but the time taken to acquire the large number of cells required for many types of proteomic studies may then become a significant factor in experimental planning and design.

## 3. Conclusions

Both types of laser microdissecting systems greatly facilitate the acquisition of specific cells for a wide range of downstream molecular analysis. For both laser capture microdissection and laser microbeam microdissection, cell selection can be based on morphological features or phenotypic criteria; both systems have their advantages and possible limitations

The further development of these laser-based microdissection systems—especially the integration of sophisticated image analysis software to permit the automatic dissection of pre-defined cells of interest—will greatly facilitate the more widespread application of these technologies. Indeed, Arcturus *(8)* has recently introduced an automated laser capture microdissection system (AutoPix™ Automated Laser Capture Microdissection Instrument). This instrument allows the user to predefine the areas of each tissue section to be microdissected and then automatically microdissects the areas of interest. PALM Microlaser Technologies has also introduced an automated system to its PALM microbeam system allowing highly automated specimen handling (isolation, transportation, and capturing) without mechanical contact for pure contamination-free target cells *(13)*. The development of automated laser microdissection systems will greatly enhance the speed and accuracy at which cells can be microdissected and acquired for molecular analysis and will support higher-throughput and larger-scale molecular analysis of specific cell types.

## References

1. Curran, S., McKay, J. A., McLeod, H. L., and Murray, G. I. (2000) Laser capture microscopy. *J. Clin. Pathol.: Mol. Pathol.* **53,** 64–68.
2. Curran, S. and Murray, G. I. (2002) Tissue microdissection and its applications in pathology. *Curr. Diagn. Pathol.* **8,** 183–192.
3. Dundas, S. R., Curran, S., and Murray, G. I. (2002) Laser capture microscopy: application to urological cancer research. *UroOncology* **2,** 33–35.
4. Fend, F., Emmert-Buck, M. R., Chuaqui, R., et al. (1999) Immuno-LCM: laser capture microdissection of immunostained frozen sections for mRNA analysis. *Am. J. Pathol.* **154,** 61–66.
5. Gjerdrum, L. M., Lielpetere, I., Rasmussen, L. M., Bendix, K., and Hamilton-Dutoit S. (2001) Laser-assisted microdissection of membrane-mounted paraffin sections for polymerase chain reaction analysis: identification of cell populations using immunohistochemistry and in situ hybridization. *J. Mol. Diagn.* **3,** 105–110.
6. Emmert-Buck, M. R., Bonner, R. F., Smith, P. D., et al. (1996) Laser capture microdissection. *Science* **274,** 998–1001.
7. Simone, N. L., Bonner, R. F., Gillespie, J. W., Emmert-Buck, M. R., and Liotta, L. A. (1998) Laser-capture microdissection: opening the microscopic frontier to molecular analysis. *Trends Genet.* **14,** 272–276.
8. http://www.arctur.com, accessed on 13 January 2004.
9. Suarez-Quian, C. A., Goldstein, S. R., Pohida, T., et al. (1999) Laser capture microdissection of single cells from complex tissues. *Biotechniques* **26,** 328–35.
10. Lawrie, L., Curran, S., McLeod, H. L., Fothergill, J. E., and Murray, G. I. (2001) Application of laser capture microdissection and proteomics in colon cancer. *J. Clin. Pathol.: Mol. Pathol.* **54,** 253–258.
11. Bohm, M., Wieland, I., Schutze, K., and Rubben, H. (1997) Microbeam MOMeNT: non-contact laser microdissection of membrane-mounted native tissue. *Am. J. Pathol.* **151,** 63–7.

12. Schutze, K. and Lahr, G. (1998) Identification of expressed genes by laser-mediated manipulation of single cells. *Nat. Biotechnol.* **16,** 737–742.
13. http://www.palm-mikrolaser.com, accessed on 14 January 2004.

# II

## MICRODISSECTION AND DNA ANALYSIS

# 2

# Methacarn Fixation for Genomic DNA Analysis in Microdissected Cells

## Makoto Shibutani and Chikako Uneyama

### Summary

We have found methacarn, a non-crosslinking protein-precipitating fixative, to be useful for the analysis of DNA from microdissected specimens of wax-embedded tissue. In this chapter, we present the procedure regarding genomic DNA analysis in methacarn-fixed wax-embedded microdissected rat tissue. Using nested polymerase chain reaction (PCR), and a rapid extraction procedure, fragments of DNA up to 2.8 kb in size can be amplified from a $1 \times 1$ mm area of a 10-µm-thick tissue section. Target fragments of about 500 bp can be amplified from a single cell, but 10–20 cells are necessary for practical detection by nested PCR. Although tissue staining with hematoxylin and eosin inhibits the PCR, amplification of about 500-bp fragments is successful with 150–270 cells by single-step PCR. Immunostaining results in a substantial decrease of yield and degradation of extracted DNA. However, even after immunostaining, fragments of about 180 bp can be amplified with 150–270 cells by single-step PCR. These features demonstrate the suitability of methacarn-fixed wax-embedded tissue for practical genomic DNA analysis in terms of tissue handling, extraction efficiency, and satisfactory PCR results.

**Key Words:** DNA analysis; methacarn; microdissection; PCR; wax-embedded tissue.

## 1. Introduction

Tissue fixation and subsequent wax embedding are routinely employed for histological assessment because of the ease of handling tissues and subsequent staining as well as the good morphological preservation. Usually, formaldehyde-based fixatives are used for this purpose. However, with such crosslinking agents, there is limited performance in terms of the efficiency of extraction and quality of extracted RNA *(1–3)*, protein *(4,5)*, and genomic DNA *(6–9)*, with consequent difficulty in the analysis of microdissected, histologically defined tissue areas.

From: *Methods in Molecular Biology, vol. 293: Laser Capture Microdissection: Methods and Protocols*
Edited by: G. I. Murray and S. Curran © Humana Press Inc., Totowa, NJ

Extraction efficiency and integrity of DNA are critical for the molecular analysis of microdissected cells. Recently, we have found that methacarn, a non-crosslinking protein-precipitating fixative *(10,11)*, meets critical criteria for analysis of RNA and proteins in wax-embedded tissue sections *(5)*. In the case of DNA extraction from formalin-fixed wax-embedded tissues, the extraction protocol usually requires proteinase K treatment with extended incubation periods *(12–14)*. On the other hand, we found that methacarn fixation allows high yields and amplification of long genomic DNA segments in wax-embedded tissue sections by a simple extraction procedure *(9,15)*.

Tissue staining is essential for cellular identification in practical molecular analysis using microdissection techniques *(13,16–21)*; therefore it is important to assess the effect of tissue staining on the performance of molecular analysis *(13,17,20)*. Furthermore, analysis of gene expression or mutation in immuno-phenotypically defined cells would be a versatile research technique *(19)*.

In this chapter, we detail the procedures for genomic DNA analysis in methacarn-fixed wax-embedded microdissected tissue specimens *(15)*, and illustrate its suitability in terms of target fragment size and the number of microdissected cells required for DNA analysis using cresyl-violet-stained sections. We also assess the effects of tissue staining with hematoxylin and eosin (H&E) or immunohistochemical stains on subsequent analysis of genomic DNA.

## 2. Materials

1. Methacarn, consisting of 60% (v/v) absolute methanol, 30% chloroform, and 10% glacial acetic acid.
2. Ethanol, 99.5% (v/v).
3. Shaker for tissue agitation.
4. Xylene, reagent-grade.
5. Tissue cassettes (Tissue-Tek® Cassette series; Sakura Finetek Japan Co. Ltd., Tokyo, Japan).
6. Tissue-embedding console system (Tissue-Tek® TEC™ 5; Sakura Finetek Japan).
7. Embedding molds (Base Molds for Tissue-Tek® Embedding Rings; Sakura Finetek Japan).
8. Embedding rings (Sakura Finetek Japan).
9. Wax (Sakura Finetek Japan).
10. Microtome.
11. Hematoxylin (Tissue-Tek® Hematoxylin 3G; Sakura Finetek Japan).
12. Eosin (Tissue-Tek® Eosin; Sakura Finetek Japan).
13. 0.1% Cresyl violet solution.
14. Primary antibodies for immunohistochemistry.
15. 1% Periodic acid solution.
16. Immunostaining kit (Vectastain Elite kit; Vector Laboratories Inc., Burlingame, CA).

17. 3,3'-Diaminobenzidine, tetrahydrochloride (DAB; Dojindo Laboratories; Kumamoto, Japan).
18. Hydrogen peroxide, 30% (w/w).
19. Casein (Merck, Darmstadt, Germany).
20. Microdissector (PALM Robot-MicroBeam equipment; Carl Zeiss Co., Ltd., Tokyo, Japan).
21. Polyethylene film for microdissection, 1.35 μm thick (PALM GmbH; Wolfratshausen, Germany).
22. Nail polish.
23. TaKaRa DEXPAT™ (Takara Bio Inc., Shiga, Japan).
24. Hoechst 33258 (Molecular Probe, Eugene, OR).
25. Fluorescence spectrophotometer.
26. Thermal cycler.
27. Oligonucleotide primers.
28. PCR buffer: 20 m$M$ Tris-HCl (pH 8.4), 50 m$M$ KCl, 0.2 m$M$ dNTP, 1.5 m$M$ MgCl$_2$.
29. PLATINUM *Taq* DNA polymerase (Invitrogen Corp, Carlsbad, CA).
30. Agarose gel electrophoresis equipment.
31. Ethidium bromide (10 mg/mL; Invitrogen).
32. Agarose and DNA sequencing equipment.
33. Autoclaved ultrapure water for preparation of solutions.

## 3. Methods

The methods described below outline (1) the preparation of methacarn-fixed wax-embedded tissue specimens, (2) tissue staining, (3) microdissection, (4) DNA extraction from microdissected cells, and (5) polymerase chain reaction (PCR).

### 3.1. Preparation of Methacarn-Fixed Wax-Embedded Tissue Specimens

Methacarn solution, which is easily prepared, should be freshly made and stored at 4°C before fixation *(22)* (*see* **Note 1**).

### 3.1.1. Fixation and Tissue Embedding

1. Trim tissues/organs to 3 mm in thickness if possible.
2. If necessary, each tissue can be placed on a piece of filter paper or into a tissue cassette (Sakura Finetek Japan) to support tissue shape.
3. Fix tissues with methacarn for 2 h at 4°C with gentle agitation using a shaker.
4. Dehydrate tissues three times for 1 h in fresh 99.5% ethanol at 4°C with agitation.
5. Trim tissues for embedding during **step 4** if necessary.
6. Immerse tissue in xylene for 1 h and then three times for 30 min at room temperature.
7. Immerse tissues in hot wax (60°C) three successive 1-h periods.
8. Embed tissue specimens in fresh wax using a tissue-embedding console system (Sakura Finetek Japan).
9. Store wax-embedded tissue blocks at 4°C until sectioning.

**Table 1**
**Comparison of DNA Yields Between Unfixed or Ethanol-Fixed Frozen Tissues and Methacarn-Fixed Wax-Embedded Tissues[a]**

| Tissue condition | No. of sample | DNA yield (μg/mg wet tissue) |
|---|---|---|
| Unfixed frozen | 5 | 1.05 ± 0.10 |
| Ethanol-fixed frozen | 5 | 1.15 ± 0.11 |
| Methacarn-fixed wax-embedded | 4 | 0.76 ± 0.06[b] |

[a]Rat liver tissue was used as described previously in **ref. 15**. After dewaxing, extraction of DNA from methacarn-fixed wax-embedded small liver tissue blocks was performed by digestion with 500 μL of 10 m$M$ Tris-HCl (pH 8.0), 150 m$M$ NaCl, 10 m$M$ ethylenediaminetetraacetic acid, 0.1% sodium dodecyl sulfate, and 1 U of proteinase K at 55°C for 2 h. The film was removed from the tube at this time point. Then 500 μL of Tris buffer-saturated phenol was added, mixed well, and centrifuged at 10,000$g$ for 15 min. The supernatant was further extracted again with 500 μL of Tris-phenol/chloroform (1:1), and the separated aqueous portion after centrifugation was transferred to a new tube and treated with 0.5 U of RNase A at 37°C for 1 h. The solution was extracted with 500 μL of phenol/chloroform (1:1) and then treated with ether. Extracted DNA was precipitated by adding 1 μL of cold 99.5% ethanol, and after storing at –20°C overnight, centrifuged at 5000$g$ for 5 min. The pellet was washed twice with 75% ethanol, dried, and resuspended in 10 μL of water. One milliliter of sample was used to measure DNA concentration by Hoechst 33258 and a fluorescence spectrophotometer. Extraction of DNA from methacarn-fixed wax-embedded small liver tissue blocks was performed after dewaxing. Frozen tissue blocks of unfixed or ethanol-fixed liver were directly subjected to DNA extraction, and dewaxed blocks of methacarn-fixed wax-embedded liver tissue were air-dried before extraction.
[b]Significantly different from unfixed frozen sample ($p < 0.01$ by ANOVA).
(Reproduced with permission from **ref. 15**.)

The yield and quality of extracted DNA are critical for the subsequent analysis of microdissected cells. DNA yield from ethanol-fixed frozen tissues is similar to that from unfixed frozen tissues (*see* **Table 1**) *(15)*. On the other hand, DNA yield from methacarn-fixed wax-embedded tissue is slightly reduced, the mean value being about 70% of that from unfixed frozen tissues (**Table 1**). The integrity of extracted DNA from methacarn-fixed wax-embedded tissue is assessed by electrophoresis on 1.5% agarose gel (*see* **Fig. 1**). Although a slightly greater intensity at the top of the DNA smear is observed in unfixed or ethanol-fixed frozen samples, DNA in every case distributes within the high molecular weight range, suggesting good preservation of extracted DNA.

### 3.1.2. Preparation of Methacarn-Fixed Wax-Embedded Tissue Sections

1. Section at 5–10 μm in thickness using a microtome.
2. Stretch sections with slide warmer (Sakura Finetek Japan).
3. Mount stretched section onto a 1.35-μm thin polyethylene film (PALM GmbH) overlaid on a glass slide.

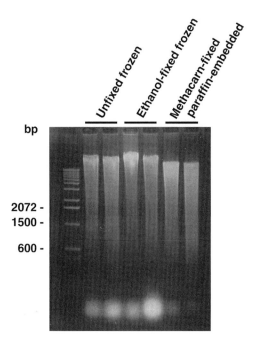

Fig. 1. Comparison of the integrity of DNA extracted from rat liver of unfixed or ethanol-fixed frozen tissue blocks and dewaxed methacarn-fixed tissue blocks. Tissue blocks were directly subjected to DNA extraction as described in the footnote of **Table 1**, and 2.0 µg of extracted DNA was subjected to electrophoresis in 1.5% agarose gel and stained with ethidium bromide as previously described *(15)*. (Reproduced with permission from **ref. 15**.)

4. Dry sections overnight at 37°C in an incubator.
5. Store sections at 4°C until use.

## 3.2. Tissue Staining

Dewax sections by immersing in xylene for 3 × 2 min followed by 99.5% ethanol for 2 × 2 min. Tissue sections can be stained with cresyl violet or H&E, or immunostained (*see* **Note 2**).

### 3.2.1. Cresyl Violet Staining

1. Dissolve 0.5 g cresyl violet in 500 mL water.
2. Add 8 drops of acetic acid.
3. If necessary, boil the solution until completely dissolved.
4. Filter the solution before using.
5. Immerse dewaxed sections briefly in water.

6. Incubate sections in cresyl violet solution for 20 min.
7. Wash sections once with 95% ethanol that contains 0.5% acetic acid, and then with 99.5% ethanol twice.
8. Air-dry.

### 3.2.2. Hematoxylin and Eosin Staining

1. Immerse dewaxed sections briefly in water.
2. Immerse sections in hematoxylin solution (Tissue-Tek® Hematoxylin 3G) for 10 s.
3. Wash sections briefly with water.
4. Immerse sections with eosin solution (Tissue-Tek® Eosin) for 10 s.
5. Wash sections briefly with 99.5% ethanol.
6. Air-dry.

### 3.2.3. Immunostaining

1. Treat dewaxed sections with 1% periodic acid solution for 10 min.
2. Wash sections briefly with water and 1X phosphate-buffered saline (PBS; pH 7.4).
3. Block nonspecific binding sites with 0.5% casein in PBS for 30 min.
4. Incubate with primary antibody of appropriate dilution for 2 h.
5. Wash sections with PBS for 5 min × 3.
6. Incubate sections with biotin-labeled secondary antibody.
7. Repeat **step 5**.
8. Incubate sections with avidin-biotin complex utilizing Vectastain Elite kit (Vector Laboratories).
9. Repeat **step 5**.
10. Visualize immunoreaction using the avidin/biotin system with 0.004% hydrogen peroxide as substrate and DAB as chromogen.
11. Rinse sections with water.
12. Perform nuclear staining with hematoxylin if desired (*see* **Subheading 3.2.2.**).
13. Air-dry.

Tissue staining can affect the yield and quality of extracted DNA with methacarn-fixed wax-embedded tissue sections (*see* **Table 2** and **Fig. 2**) *(15)*. **Table 2** shows the DNA yield from methacarn-fixed wax-embedded tissue sec-

---

Fig. 2. (*opposite page*) Integrity of DNA extracted from stained sections of methacarn-fixed wax-embedded tissue. Liver of a rat treated with thioacetamide at the promotion stage in the two-stage hepatocarcinogenesis model *(23,24)* was used. Tissue blocks were trimmed to obtain sections of 100 mm$^2$ in area before sectioning, and sectioned at 10 µm in thickness. Serial sections were randomized and mounted onto polyethylene film overlaid on a glass slide. Dewaxed sections were either unstained, stained with H&E, or immunostained with glutathione-*S*-transferase placental form

**Table 2**
**Effect of Staining on Yield of Extracted DNA From Methacarn-Fixed Rat Liver Wax-Embedded Tissue Sections[a]**

| Tissue condition | No. of samples | Yield of DNA (ng/100 mm² area)[b] | Ratio of unstained section (%) |
|---|---|---|---|
| Unstained | 10 | 2705.1 ± 853.4 | 100 |
| H&E-stained | 5 | 2687.4 ± 632.4 | 99.3 |
| Immunostained[c] | 13 | 314.5 ± 85.2[d] | 11.6 |

[a]Liver of a rat treated with thioacetamide at the promotion stage in the two-stage hepatocarcinogenesis model *(23,24)* was used as described previously in **ref. 15**.

[b]Tissue blocks were trimmed to 100 mm² before sectioning, and sectioned at 10 µm. DNA extraction and estimation of its concentration were performed according to the methods described in **Table 1**.

[c]Sections were immunostained with GST-P.

[d]Significantly different from the unstained and H&E-stained samples ($p < 0.0001$ by ANOVA).

(Reproduced with permission from **ref. 15**.)

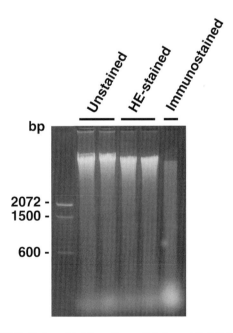

Fig. 2. (*continued*) (GST-P) (*see* **Subheadings 3.2.2.** and **3.2.3.**). One µg of extracted DNA from whole tissue section was subjected to electrophoresis in 1.5% agarose gel and stained with ethidium bromide as described previously *(15)*. (Reproduced with permission from **ref. 15**.)

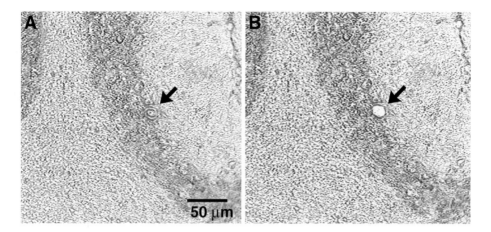

Fig. 3. Microdissection of single Purkinje cells from cresyl violet-stained rat cerebellum section as described previously in **ref. *15***. (**A**) Wax-embedded rat cerebellum sectioned at 10 μm in thickness was mounted on a film, dewaxed, and stained with cresyl violet (*see* **Subheading 3.2.1.**). A single Purkinje cell was selected and microdissected from the surrounding tissue with a laser beam (arrow). (**B**) The Purkinje cell has been cut out and catapulted by laser pressure (arrow). (Reproduced with permission from **ref. *15***.)

tions after tissue staining. The yield recovered from H&E-stained sections is similar to that from unstained sections. Immunostained tissue sections, on the other hand, result in very low DNA yield, values being 12% those of unstained sections. **Fig. 2** shows the integrity of DNA extracted from stained sections as visualized by electrophoresis on 1.5% agarose gel. DNA from unstained sections distributes mainly within the high molecular weight range. Similar to the unstained tissue section, H&E-stained sections show good preservation of the extracted DNA. As compared to unstained and H&E-stained cases, DNA extracted after immunostaining shows significant degradation of the DNA with small DNA fragments of approx 100 bp size (*see* **Note 3**).

### *3.3. Microdissection*

Microdissection is performed with PALM Robot-MicroBeam equipment (Carl Zeiss Co., Ltd.) as described previously (*see* **Note 4**) *(25)*. Briefly, the film with the attached specimen is mounted in reverse (film side up) onto a new cover slip (26 × 76 mm) by adhering the film to the cover slip with nail polish. The specimens are then subjected to Robot-MicroBeam dissection by laser beam and the selected cells are catapulted by laser pressure into mineral oil-coated PCR tube caps (**Fig. 3**) *(15)*. In case of large specimens (circle areas of 150–200 μm in radius or square areas larger than 60 × 60 μm), the excised

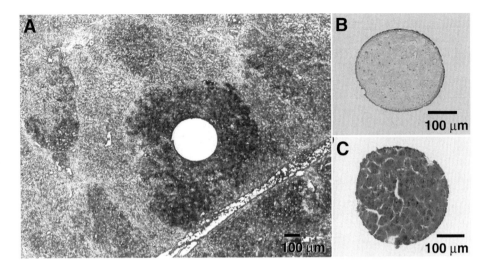

Fig. 4. Microdissection using a liver section immunostained with GST-P from a thioacetamide-treated rat in the two-stage hepatocarcinogenesis model as described previously in **ref. *15***. (**A**) A liver section (10-μm thick) mounted on polyethylene film was immunostained with GST-P (*see* **Subheading 3.2.3.**) and a circle area of 150 μm radius within a GST-P-positive focus was microdissected. (**B**) Removed GST-P-positive cellular area from the section shown in **Fig. 4A**. (**C**) Identical portions of circle area in GST-P-positive foci were microdissected from H&E-stained adjacent section. A mean of 150 cells were contained in the circle area (*n* = 5). (Reproduced with permission from **ref. *15*.**)

cells can be picked up with a thin needle tip (**Fig. 4**) *(15)*. Transfer of microdissected specimen on the cap of a PCR tube should be verified under a microscope.

### 3.4. DNA Extraction From Microdissected Cells

Microdissected cells or tissue areas on PCR tube caps are subjected to extraction with 4 μL of TaKaRa DEXPAT™ (*see* **Note 5**) at 95°C for 10 min, and the entire extracts are used as a template for PCR by adding to the master mix of total 50 μL directly as described in **Subheadings 3.5.1.** and **3.5.2.** *(15)*. In the case of a large cellular area such as 1 × 1 mm area, tissue specimens are extracted with 40 μL of DEXPAT.

### 3.5. PCR

PCR is the major tool for analysis of genomic DNA; cycle numbers should be minimized to avoid amplification-derived DNA-polymerization errors. Hot-start PCR of the genomic sequence of the gene of interest is performed with

PLATINUM *Taq* DNA polymerase in a 50-μL total reaction volume *(15)*. If nested PCR is intended, 1 μL of the first PCR product is used as a template in a 20-μL total volume (*see* **Note 6**).

### 3.5.1. Amplification by Nested PCR

1. Aliquot 4 μL of extracted DNA and mix with PCR reaction mixture contained 20 m$M$ Tris-HCl (pH 8.4), 50 m$M$ KCl, 0.2 m$M$ dNTP, 1.5 m$M$ MgCl$_2$, 0.2 μ$M$ each primer, and 2.5 U of *Taq* DNA polymerase in a 50-μL total volume.
2. Perform first-step PCR of 20–35 cycles.
3. Aliquot 1 μL of the first PCR product and mix with PCR reaction mixture with 1 U of PLATINUM *Taq* DNA polymerase in a 20-μL total volume.
4. Perform second-step PCR of 20–35 cycles.
5. Aliquot 8 μL of the PCR product and run agarose gel electrophoresis to identify the amplified target fragment (*see* **Fig. 5**).
6. Aliquot 10 μL of the PCR product for direct sequencing.

By nested PCR, the 522-bp DNA fragment of the α$_{2u}$-globulin gene is successfully amplified in 20% of the PCR attempts of single Purkinje cells with a total of 70 PCR cycles from cresyl violet-stained rat cerebellum sections (*see* **Table 3**) *(15)*. Similar, but less effective, amplification can be obtained with microdissected areas of hippocampal CA1 region, in which a successful detection is obtained in 15% of 20 × 20 μm samples (corresponding to 2.4 cells). The frequency of PCR detection increases with the area microdissected, but does not reach 100% even in a 60 × 60 μm area.

### 3.5.2. Amplification by Single-Step PCR

1. Aliquot 4 μL of extracted DNA and mix with PCR reaction mixture as described in **Subheading 3.5.1.** and 2.5 U of *Taq* DNA polymerase in a 50-μL total volume.
2. Perform PCR of 35 cycles.
3. Aliquot 8 μL of the PCR product and run agarose gel electrophoresis to identify the amplified target fragment.
4. Aliquot 10 μL of the PCR product for direct sequencing.

In the PALM system, either a rectangle or a circle of any size can be microdissected in automated mode. In H&E-stained rat liver sections as described in **Fig. 4**, a 522-bp fragment can be amplified by single-step PCR of 35 cycles with both 150- and 200-μm-radius samples after DEXPAT extraction, although the amplification of 969-bp fragments is unsuccessful even with 200-μm-radius samples (*see* **Table 4**) *(15)*. Liver samples of 150- and 200-μm-radius areas in this case contain 150 and 270 cells, respectively. In immunostained tissue, a weak 522-bp band can be amplified only with 200-μm-radius samples. In the case of 150-μm-radius samples, only a 184-bp fragment can be amplified.

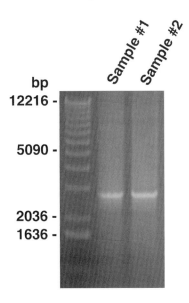

Fig. 5. Nested PCR results for the $\alpha_{2u}$-globulin genomic sequence (Accession no. M24108 in GenBank/EMBL Data Bank), sized 2.8 kb, with DNA extracted from methacarn-fixed wax-embedded rat cerebral cortex as described previously *(15)*. From 10-μm thick, cresyl violet-stained brain sections, 1 × 1 mm areas of cerebral cortex were microdissected and extracted with 40 μL of DEXPAT to extract DNA. Four-μL aliquots of cell extracts were directly applied for the first PCR reaction with upstream-outside primer, 5'-ACGGATCCAG GCTTCAAGTT CCGTATTA-3' and downstream primer for the 2954-bp fragment, 5'-TGAAATCCTG AGACTAAGCT-3'. With 1 μL of the first PCR product, second-step PCR was performed to amplify a 2.8-kb fraction with a combination of upstream-inside primer, 5'-AAAGTTAAAT GGAATCAGAA-3', and the downstream primer used for the 2954-bp fragment in the first PCR. Nested PCR in a 20-μL total volume was performed using 1 μL of the first PCR product as a template. This figure shows results of two different samples. (Reproduced with permission from **ref. *15*.**)

## 4. Notes

1. Exposure of tissues to saline prior to fixation may cause a severe tissue shrinkage artifact *(10)*. Glassware for the preparation of methacarn should be autoclaved before use. Do not use disposable plasticware that can be damaged by chloroform for the preparation and/or storage of methacarn solution. For fixation, the ratio of the fixative volume to tissue volume should be 20:1–30:1. If necessary, tissue processing can be stopped at the step of ethanol dehydration, and tissue blocks can be kept in ethanol at 4°C for several days after fixation.

2. Immersion of the tissue section in aqueous solution for a long time may increase the risk of degradation of DNA *(15)*. If nuclear staining is intended, the staining

**Table 3**
**Detection of 522-bp DNA Fragment From Microdissected Single Cells
or Cellular Areas by Nested PCR[a]**

| Cell or cellular area | Microdissected | No. of samples | PCR-detection (%) |
|---|---|---|---|
| Purkinje cell | single cell | 15 | 20 |
| Hippocampus, | $20 \times 20$ µm (2.4 cells)[b] | 26 | 15 |
| CA1 region | $40 \times 40$ µm (9.5 cells)[b] | 15 | 67 |
| | $60 \times 60$ µm (21.3 cells)[b] | 15 | 87 |
| Cerebral cortex[c] | $1 \times 1$ mm | 24 | 100 |

[a]Rat $\alpha_{2u}$-globulin gene was used for PCR as described previously in **ref. 15**. First-step PCR was performed to amplify a 969-bp fragment with an upstream-outside primer, 5'-ACGGATCCAG GCTTCAAGTT CCGTATTA-3', and a downstream primer, 5'-CGTCATCTGT GGAGGAAATT-3'. With 1 µL of the first PCR product, second-step was performed to amplify a 522-bp fragment with an upstream-inside primer, 5'-AAAGTTAAAT GGAATCAGAA-3', and a downstream-inside primer, 5'-TAAGTCCGTC TCACATGGCT-3'.

[b]Mean cell number in a $60 \times 60$ µm area ($n = 18$) was estimated with the aid of an objective micrometer, and the mean cell number in each square area was calculated.

[c]Extract from $1 \times 1$ mm area of cerebral cortex was further diluted with DEXPAT to adjust the concentration of template to correspond to a $60 \times 60$ µm area for first-step PCR.

(Reproduced with permission from **ref. 15**.)

solution should be autoclaved or filtrated if possible. Methacarn-fixed tissue sections can be stained more quickly than formalin-fixed tissue sections, and therefore the time of the histological staining procedure can be reduced.

3. Immunostaining of methacarn-fixed wax-embedded tissue results in a substantial decrease in DNA yield, in particular the loss of high molecular weight DNA. There is progressive decrease in DNA yield in proportion to the length of the immunostaining process *(15)*.

4. There are two major techniques for microdissection utilizing the precision of lasers. One technique is laser microbeam microdissection as employed in our laboratory; this system is based on a pulsed UV laser with a small beam focus to cut out areas or cells of interest by photoablation of adjacent tissue. Another technique is laser capture microdissection, which uses a low-energy infrared laser pulse to capture the targeted cells by focal melting of the thermoplastic membrane through laser activation. Advantages and disadvantages of these systems are described elsewhere *(26)*.

5. TaKaRa DEXPAT is a reagent originally designed for one-step extraction of DNA from wax-embedded tissue fixed with 10% formalin. DEXPAT is designed to optimize DNA extraction from wax embedded tissue; it utilizes ion exchange resin and detergents, and DNA is extracted in the supernatant. We use only the detergent component.With methacarn-fixed wax-embedded tissues, the time for the preparation of PCR-ready DNA from wax-embedded tissue is dramatically reduced from 2 to 3 d required for a conventional method for formalin-fixed tissues to 25 min.

**Table 4**
**Detection of Genomic DNA Fragment by Single-Step PCR in H&E-Stained or Immunostained Tissue Areas From Methacarn-Fixed Rat Liver Wax-Embedded Tissue Sections**[a]

| Stain | Tissue area (μm in radius) | Fragment size (bp) | Positive detection |
|---|---|---|---|
| H&E-stained | 200 | 184[b] | 8/8 (100%) |
| | 200 | 522[c] | 9/9 (100%) |
| | 200 | 969[c] | 0/8 (0%) |
| | 150 | 184 | 5/5 (100%) |
| | 150 | 522 | 5/5 (100%) |
| Immunostained[d] | 200 | 184 | 5/5 (100%) |
| | 200 | 522 | 5/5 (100%) |
| | 150 | 184 | 10/10 (100%) |
| | 150 | 522 | 0/5 (0%) |

[a]Liver of a rat treated with thioacetamide at the promotion stage in the two-stage hepatocarcinogenesis model was used as described previously in **ref. 15**. Circle areas of 150 or 200 μm in radius were microdissected and solubilized with 4 μL of DEXPAT solution in PCR tubes at 95°C for 10 min and whole extracts were subjected to PCR directly. PCR with 50 μL reaction volume was performed to amplify 184-, 522-, and 969-bp fragments with the same cycle parameters of 95°C for 2 min, 35 cycles of 95°C for 1 min, 55°C for 1 min, and 72°C for 30 s, and final extension at 72°C for 7 min. Eight microliters of PCR product was applied to 2.0% agarose gel electrophoresis.

[b]Rat GST-P gene (Accession no. L29427 in GenBank/EMBL Data Bank) was amplified using upstream primer, 5'-GGAGCAGGAC CCAAAAATGA-3', and downstream primer, 5'-GCA GACGAAT AAAGGCCCCA-3'.

[c]Rat $\alpha_{2u}$-globulin gene was used for amplification. Primer pairs for each DNA fragment were similar to those described in the footnote of **Table 3**.

[d]Sections were immunostained with GST-P.

(Reproduced with permission from **ref. 15**.)

6. To amplify target fragment sizes smaller than 1 kb, PCR was performed with cycle parameters of 95°C for 5 min, 35 cycles of 95°C for 1 min, 55°C for 1 min, 72°C for 30 s. The extension time for 2, 3, and 4 kb is 1.5, 2.5, and 3.5 min respectively. Although the source of cells and the detection system are different from those in the present study, similar performance was obtained when DNA from 25 cells of alcohol-fixed cytology specimens was used in the multiplex PCR (27).

## Acknowledgments

This work was supported in part by Health and Labour Sciences Research Grants (Risk Analysis Research on Food and Pharmaceuticals) from the Ministry of Health, Labour, and Welfare of Japan.

## References

1. Rupp, G. M. and Locker, J. (1988) Purification and analysis of RNA from paraffin-embedded tissues. *Biotechniques* **6,** 56–60.
2. Stanta, G. and Schneider, C. (1991) RNA extracted from paraffin-embedded human tissues is amenable to analysis by PCR amplification. *Biotechniques* **11,** 304–308.
3. Finke, J., Fritzen, R., Ternes, P., Lange, W., and Dolken, G. (1993) An improved strategy and a useful housekeeping gene for RNA analysis from formalin-fixed, paraffin-embedded tissues by PCR. *Biotechniques* **14,** 448–453.
4. Ikeda, K., Monden, T., Kanoh, T., et al. (1998) Extraction and analysis of diagnostically useful proteins from formalin-fixed, paraffin-embedded tissue sections. *J. Histochem. Cytochem.* **46,** 397–403.
5. Shibutani, M., Uneyama, C., Miyazaki, K., Toyoda, K., and Hirose, M. (2000) Methacarn fixation: a novel tool for analysis of gene expressions in paraffin-embedded tissue specimens. *Lab. Invest.* **80,** 199–208.
6. Shibata, D. (1994) Extraction of DNA from paraffin-embedded tissue for analysis by polymerase chain reaction: new tricks from an old friend. *Hum. Pathol.* **25,** 561–563.
7. Frank, T. S., Svoboda-Newman, S. M., and Hsi, E. D. (1996) Comparison of methods for extracting DNA from formalin-fixed paraffin sections for nonisotopic PCR. *Diagn. Mol. Pathol.* **5,** 220–224.
8. Poncin, J., Mulkens, J., Arends, J. W., and de Goeij, A. (1999) Optimizing the APC gene mutation analysis in archival colorectal tumor tissue. *Diagn. Mol. Pathol.* **8,** 11–19.
9. Uneyama, C., Shibutani, M., Nakagawa, K., Masutomi, N., and Hirose, M. (2000) Methacarn, a fixation tool for multipurpose gene expression analysis from paraffin-embedded tissue materials. *Current Topics in Biochem. Res.* **3,** 237–242.
10. Puchtler, H., Waldrop, F. S., Meloan, S. N., Terry, M. S., and Conner, H. M. (1970) Methacarn (Methanol-Carnoy) fixation. Practical and theoretical considerations. *Histochemie* **21,** 97–116.
11. Mitchell, D., Ibrahim, S., and Gusterson, B. A. (1985) Improved immunohistochemical localization of tissue antigens using modified methacarn fixation. *J. Histochem. Cytochem.* **33,** 491–495.
12. Dietmaier, W., Hartmann, A., Wallinger, S., et al. (1999) Multiple mutation analyses in single tumor cells with improved whole genome amplification. *Am. J. Pathol.* **154,** 83–95.
13. Murase, T., Inagaki, H., and Eimoto, T. (2000) Influence of histochemical and immunohistochemical stains on polymerase chain reaction. *Mod. Pathol.* **13,** 147–151.
14. Hirose, Y., Aldape, K., Takahashi, M., Berger, M. S., and Feuerstein, B. G. (2001) Tissue microdissection and degenerate oligonucleotide primed-polymerase chain reaction (DOP-PCR) is an effective method to analyze genetic aberrations in invasive tumors. *J. Mol. Diagn.* **3,** 62–67.

15. Uneyama, C., Shibutani, M., Masutomi, N., Takagi, H., and Hirose, M. (2002) Methacarn fixation for genomic DNA analysis in microdissected paraffin-embedded tissue specimens. *J. Histochem. Cytochem.* **50,** 1237–1245.

16. Pontén, F., Williams, C., Ling, G., et al. (1997) Genomic analysis of single cells from human basal cell cancer using laser-assisted capture microscopy. *Mutat. Res.* **382,** 45–55.

17. Burton, M. P., Schneider, B. G., Brown, R., Escamilla-Ponce, N., and Gulley, M. L. (1998) Comparison of histologic stains for use in PCR analysis of microdissected, paraffin-embedded tissues. *Biotechniques* **24,** 86–92.

18. Alcock, H. E., Stephenson, T. J., Royds, J. A., and Hammond, D. W. (1999) A simple method for PCR based analyses of immunohistochemically stained, microdissected, formalin fixed, paraffin wax embedded material. *Mol. Pathol.* **52,** 160–163.

19. Fend, F., Emmert-Buck, M. R., Chuaqui, R., et al. (1999) Immuno-LCM: laser capture microdissection of immunostained frozen sections for mRNA analysis. *Am. J. Pathol.* **154,** 61–66.

20. Serth, J., Kuczyk, M. A., Paeslack, U., Lichtinghagen, R., and Jonas, U. (2000) Quantitation of DNA extracted after micropreparation of cells from frozen and formalin-fixed tissue sections. *Am. J. Pathol.* **156,** 1189–1196.

21. Gjerdrum, L. M., Lielpetere, I., Rasmussen, L. M., Bendix, K., and Hamilton-Dutoit, S. (2001) Laser-assisted microdissection of membrane-mounted paraffin sections for polymerase chain reaction analysis: identification of cell populations using immunohistochemistry and *in situ* hybridization. *J. Mol. Diagn.* **3,** 105–110.

22. Shibutani, M. and Uneyama, C. (2002) Methacarn, a fixation tool for multipurpose genetic analysis from paraffin-embedded tissues, in *Methods in Enzymology* (Conn, M., ed.), Academic Press, New York, vol. 356, pp. 114–125.

23. Shirai, T. (1997) A medium-term rat liver bioassay as a rapid in vivo test for carcinogenic potential: a historical review of model development and summary of results from 291 tests. *Toxicol. Pathol.* **25,** 453–460.

24. Ito, N., Imaida, K., Asamoto, M., and Shirai, T. (2000) Early detection of carcinogenic substances and modifiers in rats. *Mutat. Res.* **462,** 209–217.

25. Schütze, K. and Lahr, G. (1998) Identification of expressed genes by laser-mediated manipulation of single cells. *Nature Biotechnol.* **16,** 737–742.

26. Fend, F. and Raffeld, M. (2000) Laser capture microdissection in pathology. *J. Clin. Pathol.* **53,** 666–672.

27. Euhus, D. M., Maitra, A., Wistuba, I. I., et al. (1999) Use of archival fine-needle aspirates for the allelotyping of tumors. *Cancer* **87,** 372–379.

# 3

# Multiplex Quantitative Real-Time PCR of Laser Microdissected Tissue

## Patrick H. Rooney

### Summary

This chapter describes a method for the rapid assessment of gene copy number in laser microdissected material using multiplex real-time polymerase chain reaction (PCR). Here a putative oncogene (*ZNF217*) was evaluated in a series of colon tumors, but the method is applicable to any locus for which a nucleic acid sequence is available. The preparation, laser microdissection, and optimum storage of snap-frozen tumor material from freshly resected tissue is described. A set of guidelines specific for real-time PCR assays is included to assist with optimum primer and probe design. In this assay multiplex real-time PCR was performed and our experience has demonstrated that a multiplex reaction allows for a more accurate assessment of gene copy number than a "singleplex" assay because it removes the need for an external control.

**Key Words:** Colon cancer; quantitative real-time PCR; *ZNF217*; gene copy number; laser microdissection; multiplex PCR; ABI7700 sequence detector.

## 1. Introduction

Detection of gene amplification is a recognized process through which oncogenes can be identified. Traditionally, gene copy number has been assessed through labor-intensive methods requiring several micrograms of test DNA, such as Southern blotting *(1)* and fluorescent *in situ* hybridization (FISH) *(2)*. With the advent of laser microdissection *(3)* and quantitative real-time polymerase chain reaction (PCR) *(4)*, the accurate quantification of gene copy in samples containing quantities of DNA several orders of magnitude less than the more traditional techniques has become possible. Nanograms of starting material rather than micrograms allows an investigator more freedom to precisely assess gene copy number in a particular region of tissue they have previ-

From: *Methods in Molecular Biology, vol. 293: Laser Capture Microdissection: Methods and Protocols*
Edited by: G. I. Murray and S. Curran © Humana Press Inc., Totowa, NJ

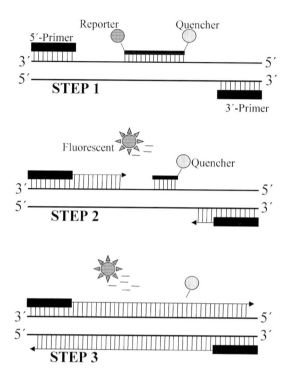

Fig. 1. Real-time PCR. **Step 1:** A fluorescent reporter and a quencher, are attached to the 5' and 3' ends of a probe specific for the locus of interest (when both molecules are attached to the probe, reporter dye emission is quenched). **Step 2:** During each extension cycle, polymerase cleaves the reporter from the probe. **Step 3:** The quencher is also released and this allows the reporter to emit fluorescence. The intensity of the fluorescent signal generated is directly proportional to the cleavage of probe, which is directly proportional to the amount of starting template.

ously selected (e.g., tumor). The application of laser microdissection coupled with quantitative real-time PCR removes the need for large quantities of DNA that can often be contaminated with surrounding material. With no dilution of gene amplification from surrounding diploid tissue, a more accurate measure of gene amplification can be obtained.

Quantitative real-time PCR determines the copy number of a specific gene by measuring the accumulation of this candidate at every point in a polymerase chain reaction *(5,6)*. This is achieved with conventional forward and reverse primers and by a fluorogenic probe *(7)* specific for the sequence of interest that is incorporated into the PCR product by a polymerase enzyme containing 5' nuclease activity (**Fig. 1**). Detection of the fluorescent signal from the resulting PCR product allows quantification of the starting template. The fewer

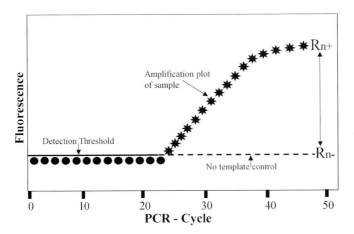

Fig. 2. Amplification plot of a singleplex assay. During the first cycles of the PCR reaction the product is below the detection threshold of the ABI PRISM. As the reaction enters the exponential phase the fluorescent signal ($R_n$) from the newly forming PCR product is detectable. The threshold cycle (or $C_t$ value) is the cycle at which a statistically significant increase is first detected in $R_n$.

cycles it takes to reach a detectable level of fluorescence ($C_t$), the greater the initial copy number (**Fig. 2**).

This chapter describes how multiplex quantitative real-time PCR was used to determine the precise copy number of a candidate gene (*ZNF217*) from laser-microdissected tumor material using a reference gene ($\beta$ *globin*) for comparison purposes. The difference between the $C_t$ of the control gene and the $C_t$ of the candidate gene varies depending on the gene copy number of the candidate gene (**Eq. 1**) *(8)*. Exact gene copy number can then be calculated from the $\Delta C_t$ value.

$$C_t\ ZNF217 - C_t\ \beta\ globin = \Delta C_t \qquad \textbf{(Eq. 1)}$$

This approach can be used to measure the gene copy of any candidate gene for which a gene sequence is available.

## 2. Materials

1. Quantitative real-time PCR system (ABI7700 PRISM sequence detector from PE Applied Biosystems, Warrington, Cheshire, UK).
2. Laser capture microdissection system (PixCell II from Arcturus Engineering, Mountain View, CA).
3. Laser microdissection caps (CapSure™ LCM transfer film; Arcturus Engineering).
4. Primer and probe design program (Primer Express program from PE Applied Biosystems).

5. Notepad (Microsoft, Windows Millennium Edition).
6. Gene sequence databases (http://www.ncbi.nlm.nih.gov/).
7. Oligonucleotide primers.
8. Taqman Universal PCR master mix (PE Applied Biosystems).
9. Fluorescently labeled oligonucleotide probes.
10. Proteinase K.
11. –80°C Storage facility.
12. –20°C Cryotome.
13. Ethanol.
14. Xylene.
15. Toluidine blue: 1% toluidine blue in 50% isopropanol.
16. Digestion buffer: 50 m$M$ Tris-HCl, 1 m$M$ EDTA, 0.5% Tween-20 (pH 8.5).
17. Water bath.
18. Liquid nitrogen.

## 3. Methods

The methods described below outline (1) the preparation of snap-frozen tumorous material for laser microdissection, (2) laser microdissection and DNA extraction, (3) design of primer and probe from gene sequence using specific parameters for quantitative real-time PCR, (4) quantitative real-time PCR, and (5) calculation of gene copy number using $C_t$ values generated from quantitative real-time PCR.

### 3.1. Preparation of Snap-Frozen Colon Tumor Material

This section describes the steps involved in preparing the selected samples for laser microdissection.

1. Take the selected tissue samples from –80°C storage and transport in liquid nitrogen to a suitable cutting area containing a –20°C cryotome.
2. Using the cryotome cut six sections 10 µm in thickness per case and adhere this tissue to glass microscope slides (three sections per slide).
3. Allow each slide to briefly air-dry at room temperature.
4. Fix each slide by placing it in 70% ethanol for 1 min at room temperature.
5. Stain each section in toluidine blue using a rapid staining method (*9*) (*see* **Note 1**). Staining with toluidine blue is performed by immersing the sections in 0.25% toluidine blue (pH 4.5) for 5 s at room temperature, washing briefly in 100% ethanol, and then dehydrating the sections sequentially in 100% ethanol and xylene.
6. Allow the xylene to evaporate completely from each slide and then perform laser microdissection.

### 3.2. Laser Microdissection and DNA Extraction

The next steps in this process involve isolation (in this case, of tumor cells) and DNA retrieval from the cells of interest within the specimen using laser microdissection and protease digestion, respectively.

### 3.2.1. Laser Microdissection of Tumor Material From Surrounding Normal Cells

Tumor microdissection was performed using a PixCell II laser capture microdissection system. The laser capture system was equipped with PixCell II image archiving software. Laser settings were as follows: spot diameter set at 15 µm, pulse duration 5 ms and power 100 mW. Tumor tissue was identified and removed by the laser to a microdissection cap. Approximately 500 laser pulses were taken per cap per tumor. One cap was used for each case.

1. Place the section for laser microdissection on the microscope stage.
2. Identify regions of the tumor within the tissue section.
3. Move the laser site over the region selected for microdissection and activate the laser.
4. After 500 laser pulses remove the microdissection cap from the section and place on a labeled microfuge tube containing 140 µL of digestion buffer.

### 3.2.2. DNA Extraction

1. Add 3.5 µL of 20 mg/mL proteinase K to each microfuge tube and incubate in a water bath overnight at 37°C.
2. The following day, stop the digestion by placing the microfuge tube in a preheated thermal block for 10 min at 90°C.
3. Store the samples at 4°C until required.

## 3.3. Design of Primer and Probe From Gene Sequence

The entire DNA sequence of the candidate gene (*ZNF217* [Accession no. AF041259]) and the reference gene (*β globin* [Accession no. NG_000007.1]) were downloaded from the National Center for Biotechnology Information (http://www.ncbi.nlm.nih.gov/). Each gene sequence was copied into a Notepad file and then imported into the Primer Express program. Using Primer Express, optimum primers and probe were selected for each gene sequence based on guidelines recommended by PE Applied Biosystems (*see* **Subheadings 3.3.1.** and **3.3.2.** below).

### 3.3.1. Probe Design Guidelines

1. Keep the G-C content in the 20–80% range.
2. Avoid runs of identical nucleotide. This is especially true for guanine, where runs of four or more Gs should be avoided.
3. Do not put a G on the 5' end.
4. Select the strand that gives the probe a greater number of Cs than Gs.
5. Using Primer Express software, the melting point ($T_m$) should be 68–70°C.

### 3.3.2. Primer Design Guidelines

1. Choose the primers after the probe.

2. Design the primers as close as possible to the probe without overlapping the probe.
3. Keep the G-C content in 20–80% range.
4. Avoid runs of identical nucleotide. This is especially true for guanine, where runs of four or more Gs should be avoided.
5. Using the Primer Express software, the $T_m$ should be 58–60°C.
6. The five nucleotides at the 3' end should have no more than two G and/or C bases.

Following the guidelines in **Subheadings 3.3.1.** and **3.3.2.** the primer and probe sequences were made as follows (in all cases, the first sequence is the forward PCR primer, the second one is the Taqman probe, and the third sequence is the reverse PCR primer): (1) *ZNF217*, 5'-GAG GCG AGG AAG AAG GTG C, 6FAM5'-CCC ATC TGA GAT GCT CAA AGT TGC GA-3'TAMRA, 5'-CGG AAG CTG GCA GCA TTT T' and (2) *β globin*, 5'-CAA GAA AGT GCT CGG TGC CT, 6FAM5'-GTC CAG GTG AGC CAG GCC ATC ACT A-3'VIC, 5'-GCA AAG GTG CCC TTG AGG T.

In multiplex quantitative real-time PCR two different fluorogenic probes are used to allow each gene to be detected separately by the ABI PRISM as they are co-amplified. In this case the candidate gene probe was VIC-labeled while the reference gene was labeled with a FAM.

## 3.4. Quantitative Real-Time PCR

Before quantitative PCR can be performed it is necessary to optimize the primer and probe concentrations for use in the assay. Applied Biosystems' guidelines for primer (**Subheading 3.4.1.**) and probe (**Subheading 3.4.2.**) optimization are as follows.

### 3.4.1. Determining Optimum Primer Concentration

Using forward and reverse primer concentrations of 50 n$M$, 300 n$M$, and 900 n$M$, determine the minimum primer concentrations giving the maximum $\Delta R_n$ (**Fig. 2**). This is achieved by making several master-mix (MM) reactions where the forward and reverse primers concentrations of primer are varied. The fluorogenic probe designed to complement the forward and reverse primers should be included in the reaction (to allow detection of the PCR product) at a constant concentration (e.g., 50 n$M$) of all reactions. In all, 9 MM reactions are prepared, each containing different concentrations of forward and reverse primers as follows:

1. 50 n$M$ forward primer and 50 n$M$ reverse primer (for plate wells A1–A4).
2. 50 n$M$ forward primer and 300 n$M$ reverse primer (for plate wells A5–A8).
3. 50 n$M$ forward primer and 900 n$M$ reverse primer (for plate wells A9–A12).
4. 300 n$M$ forward primer and 50 n$M$ reverse primer (for plate wells B1–B4).
5. 300 n$M$ forward primer and 300 n$M$ reverse primer (for plate wells B5–B8).

6. 300 n*M* forward primer and 900 n*M* reverse primer (for plate wells B9–B12).
7. 900 n*M* forward primer and 50 n*M* reverse primer (for plate wells C1–C4).
8. 900 n*M* forward primer and 300 n*M* reverse primer (for plate wells C5–C8).
9. 900 n*M* forward primer and 900 n*M* reverse primer (for plate wells C9–C12).

Enough of each MM should be prepared for a quadruplicate reaction (*see* **Note 2**). Each well is a 50-µL reaction (i.e., $4 \times 50$ µL = 200 µL of each MM is required) (*see* **Note 3**).

### 3.4.2. Determining Optimum Probe Concentration

Following primer optimization, probe optimization should be initiated. To determine which probe concentration gives the maximum $\Delta R_n$, run four replicates at 50 n*M* intervals from 50–250 n*M* with the previously selected optimum primer concentrations as follows:

1. Optimum primers + 0.5 µL of probe (i.e., 50 n*M*) (for plate wells A1–A4).
2. Optimum primers + 1.0 µL of probe (i.e., 100 n*M*) (for plate wells A5–A8).
3. Optimum primers + 1.5 µL of probe (i.e., 150 n*M*) (for plate wells A9–A12).
4. Optimum primers + 2.0 µL of probe (i.e., 200 n*M*) (for plate wells B1–B4).
5. Optimum primers + 2.5 µL of probe (i.e., 250 n*M*) (for plate wells B5–B8).

Enough of each MM should be prepared for a quadruplicate reaction. Each well is a 50-µL reaction (i.e., $4 \times 50$ µL = 200 µL of each MM is required).

Optimum conditions in this assay were: 300 n*M* forward primer, 50 n*M* reverse primer, and 100 n*M* of probe for β *globin*, and 50 n*M* forward primer, 300 n*M* reverse primer, and 100 n*M* of probe for *ZNF217* (*see* **Notes 4–6**). Quantitative real-time PCR was performed using a 50-µL final reaction volume containing 2 µL volume of microdissected/proteinase K digested supernatant for each sample with 48 µL of Taqman Universal PCR master mix containing the optimized primer and probe concentrations for multiplex real-time PCR of β *globin* and *ZNF217*. The PCR conditions used in all reactions were: 2 min at 50°C, 10 min at 95°C with 40 two-step cycles of 95°C at 15 s and 60°C at 60 s.

## 3.5. Calculation of Gene Copy Number Using $C_t$ Values

The amount of information one gets out from the assay depends on whether there is another source of data regarding the cell samples.

**Subheading 3.5.1.** describes a situation in which gene amplification is determined relative to a sample know to be diploid for the candidate gene. This approach allows gene amplification to be detected and relative gene amplification between the different cases to be calculated but it cannot produce a precise copy number for the candidate gene in one sample. Specific gene copy num-

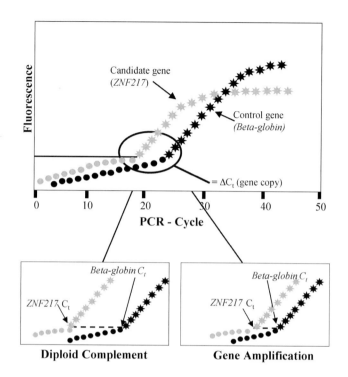

Fig. 3. Amplification plots observed in a multiplex real-time PCR. The position of the *ZNF217* amplification plot (and hence its $C_t$) will shift relative to the internal control (*β globin*) depending on the gene copy at this locus. The greater the shift to the right the greater the increase in gene copy and vice versa.

bers can be calculated only if a small proportion of the samples under investigation have been independently assigned a copy number by other methods, e.g., fluorescence *in situ* hybridization (FISH) analysis.

**Subheading 3.5.2.** describes an experiment in which FISH analysis of eight colorectal cell lines allowed the rapid and accurate assignment of a candidate gene (*ZNF217*) to 80 colon tumors.

## 3.5.1. Detection of Gene Amplification Relative to Known Diploid Samples

The difference between the $C_t$ of the control gene and the $C_t$ of the candidate gene varies depending on the gene copy number of the candidate gene (**Eq. 1**). By including known diploid samples in the assay it is possible to determine a $\Delta C_t$ value for which amplification of the candidate gene has occurred relative to the samples diploid for the candidate (**Fig. 3**) (*see* **Note 7**).

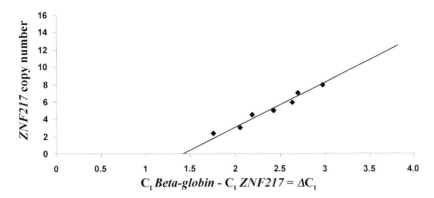

Fig. 4. A representative plot of $\Delta C_t$ determined by real-time PCR against *ZNF217* gene copy number calculated by FISH.

### 3.5.2. Calculation of Precise Gene Copy Using Cell Lines With Known Candidate Gene Amplification

Using cell lines for which the *ZNF217* gene copy number was known, as assessed by FISH (**Fig. 4**), it was possible to assign a gene copy number to the values of $C_t$ *ZNF217* minus $C_t$ *β globin* (i.e., $\Delta C_t$) generated for those same cell lines when assessed by quantitative real-time PCR. Real-time PCR of the *β globin* and *ZNF217* loci in colorectal cell lines for which the *ZNF217* gene copy had been independently established by FISH allowed the conversion of $\Delta C_t$ values to *ZNF217* gene copy number (*see* **Note 7**). Extrapolation of the line of "best fit" is required to calculate the *ZNF217* copy number in specimens with more than eight copies (the maximum amplification detected by FISH in the colorectal cell lines).

## 4. Notes

1. In this study, snap-frozen tissue was laser microdissected (**Subheading 3.1.**). A similar method can be used to laser microdissect paraffin-embedded material but it is recommended that hematoxylin stain be used in place of toluidine blue. The laser removes hematoxylin material more effectively than toluidine blue-stained wax sections. For snap-frozen tissue both stains are suitable.
2. Primer and probe concentration optimization should be performed in quadruplicate. However, when performing the assay on specimen DNA we have determined that performing the assay in triplicate is sufficient. This has three benefits: it reduces the amount of DNA required, it allows more samples to run in every 96-well plate, and it also reduces the cost by reducing the amount of reagents required per sample assessed.
3. The use of a digital pipet is recommended. Our laboratory has shown that a Finnipipette produces less variation between triplicates than, e.g., a manual

Gilson pipet. To determine if accurate pipetting is being performed the variation of the triplicate mean $C_t$ should be calculated for each well. Only sample $C_t$ values lying within 10% of the mean should be accepted.

4. One internal control used in multiplex PCR is better than two external controls in a singleplex assay. We have tested both methods and determined that any benefit incurred by the inclusion of two reference genes is removed when it is realized that it is not possible to add exactly the same amount of starting template to each well (*see* **Note 3**). If precisely the same volume of template is not added in wells containing the primers amplifying the candidate gene and the reference gene the $C_t$s generated from such a analysis are not comparable. By using a multiplex system it is ensured that the same amount of template is available to the reference gene as the candidate gene, because they both are amplified in the same well. In addition the multiplex method allows more samples to be assessed per 96-well plate and subsequently is more cost-effective.

5. It is recognized that for the vast majority of Taqman assays using a concentration of 900 n$M$ primers and a 250 n$M$ probe provides for a highly reproducible and sensitive assay when using cDNA or DNA as a substrate in a singleplex assay. This is not the case when using multiplex PCR and consideration must be given to the final primer concentration selected so that it encourages "primer starvation." Primer starvation is a term used to describe a deliberate reduction in the concentration of primers of the fastest amplifying gene in a multiplex reaction. In multiplex quantitative real-time PCR it is designed to stop the amplification of the fastest amplifying product after its $C_t$ has been attained, allowing optimum amplification of the second PCR product in a master-mix "rich" environment. Without primer starvation the fastest amplifying gene product will "steal" the master mix before the second gene has reached its $C_t$.

6. Investigators are encouraged to optimize primer and probe concentrations as described but maintain default PCR cycle conditions (2 min at 50°C, 10 min at 95°C, with 40 two-step cycles of [95°C at 15 s and 60°C at 60 s]).

7. A set of known standards should be included in every 96-well plate. Known standards are cases that have been assessed by other methods at the locus of interest. Ideally the known standards will have variable gene copy numbers (*see* **Subheading 3.5.2.**). If such specimens are not available, known diploid cases will suffice (*see* **Subheading 3.5.1.**). In either case these standards will allow any interplate variability to be detected.

## Acknowledgment

The author would like to thank Dr. M. C. E. McFadyen for helpful discussion and technical assistance with the described assay.

## References

1. Copur, S., Aiba, K., Drake, J. C., Allegra, C. J., and Chu, E. (1995) Thymidylate synthase gene amplification in human colon cancer cell lines resistant to 5-fluorouracil. *Biochem. Pharmacol.* **49,** 1419–1426.

2. Boonsong, A., Marsh, S., Rooney, P. H., Stevenson, D. A. J., Cassidy, J., and McLeod H. L. (2000) Characterization of topisomerase I locus in human colorectal cancer. *Cancer Genet. Cytogenet.* **121**, 56–60.

3. Curran, S., McKay, J. A., McLeod, H. L., Murray, G. I. (2000) Laser capture microscopy. *J. Clin. Pathol.: Mol. Pathol.* **53**, 64–68

4. Higuchi, R., Dollinger, G., Walsh, P. S., and Griffith, R. (1992) Simultaneous amplification and detection of specific DNA sequences. BioTechnol. **10**, 413–417.

5. Higuchi, R., Fockler, C., Dollinger, G., and Watson, R. (1993) Kinetic PCR analysis: Real-time monitoring of DNA amplification reactions. *BioTechnol.* **11**, 1026–1030.

6. Holland, P. M., Abramson, R. D., Watson, R., and Gelfand, D. H. (1991) Detection of specific polymerase chain reaction product by utilizng the 5' to 3' exonuclease activity of Thermus aquaticus DNA polymerase. *Proc. Natl. Acad. Sci. USA* **88**, 7276–7280.

7. Livak, K. J., Flood, S. J. A., Marmaro, J., Giusti, W., and Dertz, K. (1995) Oligonucleotides with fluorescent dyes at opposite ends provide a quenched probe system useful for detecting PCR product and nucleic acid hybridization. *PCR Methods and Applications* **4**, 357–362.

8. Glöckner, S., Lehmann, U., Wilke, N., Kleeberger, W., Länger, F., and Kreipe, H. (2000) Detection of gene amplification in intraductal and infiltrating breast cancer by laser-assisted microdissection and quantitative real-time PCR. *Pathobiol.* **68**, 173–179.

9. Lawrie, L. C., Curran, S., McLeod, H. L., Fothergill, J. E., and Murray, G. I. (2001) Applications of laser capture microdissection and proteomics in colon cancer. *J. Clin. Pathol.: Mol. Pathol.* **54**, 253–258.

# 4

## Comparative Genomic Hybridization Using DNA From Laser Capture Microdissected Tissue

### Grace Callagy, Lucy Jackson, and Carlos Caldas

### Summary

Comparative genomic hybridization (CGH) is a powerful screening technique that can identify regions of gain and loss within the whole genome in a single experiment. The combination of laser capture microdissection, whole-genome amplification, and CGH permits genomic screening with high specificity and sensitivity. This complement of techniques has enabled analysis of focal regions and subpopulations of cells within a tissue, which has previously been difficult, providing insight into disease progression and heterogeneity. This chapter outlines the techniques involved in producing labeled probes from DNA extracted from laser capture microdissected material and the methods for hybridization of these probes to metaphase chromosomes. This protocol can also be applied to the preparation of probes for CGH arrays.

**Key Words:** Laser capture microdissection (LCM); comparative genomic hybridization (CGH); degenerate oligonucleotide primed PCR (DOP-PCR); array CGH.

## 1. Introduction

Comparative genomic hybridization (CGH) is a powerful whole-genome screening technique that allows analysis of DNA copy-number losses (deletions) and gains across the whole genome in a single experiment. Metaphase-CGH, first described by Kallioniemi et al. *(1)*, is based on two-color fluorescence *in situ* suppression hybridization of equal amounts of test (e.g., tumor) and reference (e.g., normal) DNA to karyotypically normal metaphase chromosomes. This is done in the presence of $C_0t$-1 DNA, which suppresses repetitive elements, allowing them to be excluded from analysis. The smallest deletion that can be detected is approximately 10 Mb, although amplicon units (amplicon size × level of amplification) of approx 2 Mb can be resolved. An overview of the principle of CGH is given in **Fig. 1**.

From: *Methods in Molecular Biology, vol. 293: Laser Capture Microdissection: Methods and Protocols*
Edited by: G. I. Murray and S. Curran © Humana Press Inc., Totowa, NJ

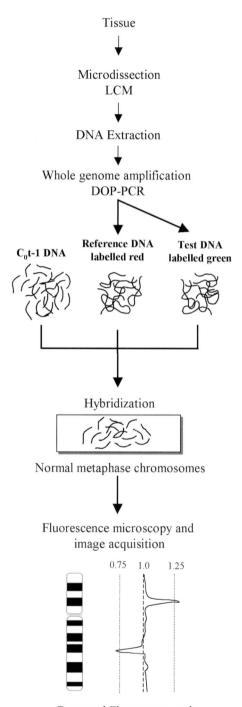

Tissue

Microdissection
LCM

DNA Extraction

Whole genome amplification
DOP-PCR

$C_0$t-1 DNA    **Reference DNA labelled red**    **Test DNA labelled green**

Hybridization

Normal metaphase chromosomes

Fluorescence microscopy and
image acquisition

0.75   1.0   1.25

Green:red Fluorescent ratio

An advantage of CGH is that only genomic DNA (gDNA) is required, allowing screening of formalin-fixed paraffin-embedded (FFPE) tissue archives. Approximately 1 μg of gDNA is required for each experiment, which until recently has necessitated manual dissection of whole tissue sections. As a result very small or focal lesions could not be studied without eliminating the effect of contaminating cells. The combination of laser capture microdissection (LCM), which allows precise dissection under direct visualization, and CGH has now facilitated genomic screening with higher specificity and sensitivity *(2–4)*. Microdissection of as few as five cells (~50 pg DNA) followed by whole-genome amplification (WGA) is sufficient for CGH *(3,5–10)*, but when very small quantities of DNA (<50 pg) are used the genome may be underrepresented, resulting in false CGH profiles *(11–13)*. Ultimately, the appropriate number of cells depends on the method of WGA that is used and also on the nature of the experiment. In our lab we use degenerate oligonucleotide primed polymerase chain reaction (DOP-PCR) as the preferential method of WGA.

This chapter describes the procedures involved in performing CGH using DNA probes from laser capture microdissected frozen or FFPE tissue. These include DNA extraction, WGA, DNA labeling, and hybridization. Although the protocols are intended for metaphase CGH, they are also applicable to array-based CGH.

## 2. Materials

1. Plain noncharged glass slides; absolute ethanol; xylene (flammable, harmful by inhalation and contact with skin); hematoxylin and eosin (H&E) staining reagents; desiccator; and Coplin jars.
2. PixCell II laser capture microdissection system and LCM caps with Capsure transfer film (Arcturus, Mountain View, CA); 0.5-mL Eppendorf tubes, with diameter appropriate for LCM caps, autoclaved/irradiated prior to use.
3. DNA lysis buffer (1X reaction buffer IV [Abgene, Surrey, UK]): 1.5 mM MgCl$_2$, 0.45% Nonidet P-40, 0.45% Tween-20; proteinase K, stored at –20°C.
4. Agarose gel electrophoresis equipment, 100 bp and 1 Kb DNA ladder.

---

Fig. 1. (*opposite page*) Schematic overview of the principle of CGH. Differentially labeled test and normal reference DNA are co-hybridised to normal male target metaphases in the presence of C$_0$t-1 DNA. Digital images for the hybridization pattern of test and reference DNA are acquired sequentially using fluorochrome-specific optical filters, and test (green)/reference (red) fluorescence ratio profiles along the axis of each chromosome are determined by computer-based image analysis packages. The method of image analysis shown assigns regions of loss and gain in the test genome on the basis of average test/reference fluorescence ratio profiles crossing predetermined threshold values.

5. DOP-PCR reagents: 6 MW primer 5'-CCG ACT CGA GNN NNN NAT GTG G-3' (MWG Biotech, Milton Keynes, UK); polyethylene glycol ether (PE W1) (Sigma-Aldrich, Dorset, UK); 5X Buffer D (Invitrogen, Paisley, UK); dNTP mix; PCR water; SuperTAQ DNA polymerase (HT Biotechnology, Cambridge, UK); MgCl$_2$; TAPS buffer (Sigma-Aldrich), all stored at –20°C.
6. PCR machine with ramping facility.
7. CGH nick translation kit (Vysis, Abbott Laboratories, Berkshire, UK), labeled dUTPs stored at –20°C; fluorescent labelled dUTPs, sensitive to light, stored in the dark at –20°C.
8. Deoxyribonuclease (DNAse) 1, 10X reaction buffer and stop solution (D5307, Sigma-Aldrich).
9. Sodium acetate.
10. Human C$_0$t-1 DNA stored at –20°C.
11. Hybrisol VI CGH hybridization buffer (S1370-30 Appligene Oncor) stored in the dark at 2–8°C.
12. Phase contrast microscope.
13. Normal metaphase CGH target slides (Vysis) stored at –20°C.
14. Deionized formamide (toxic and harmful to unborn children), stored in the dark at 4°C.
15. 20X standard saline citrate (SSC) pH 5.3, for denaturation solution.
16. Glass cover slips (22 × 22 mm).
17. Rubber cement (highly flammable).
18. HYBrite denaturing/hybridization system (Vysis).
19. Tween-20.
20. Molecular biology grade.
21. 20X SSC, pH 7.0, for all wash solutions made up fresh with distilled water.
22. Dried milk powder.
23. Fluorescent labeled antibodies, stored in the dark at 4°C.
24. VECTASHIELD® mounting medium with 4,6-diamidino-2-phenylindole (DAPI) (Vector Laboratories, Burlingame, CA), stored in the dark at 4°C.
25. Glass cover slips (22 × 60 mm).
26. Clear nail polish.
27. Fluorescent microscope; image capturing and CGH interpreter software.

## 3. Methods

The following sections outline the steps that are required for preparation of DNA from both frozen and FFPE tissue for CGH. The experiments are described in the following order: tissue preparation, microdissection, DNA extraction, whole-genome amplification, DNA labeling, hybridization, and analysis. The same process is applied to both the test and reference DNA to be used for each hybridization. We include a description of the two most commonly used DNA labeling protocols (DOP-PCR and nick translation) in **Subheading 3.6.**, although the former is the method of choice in our laboratory. This protocol is also applicable to array-based cases (*see* **Note 1**).

### 3.1.Tissue Preparation (see Note 2)

#### 3.1.1. Preparation of FFPE Tissue

1. Cut sections (3.5–4 μm thick) onto plain glass slides.
2. Dewax in xylene, three times for 3 min each, and rehydrate in 100%, 90%, 70% ethanol series and water for 2 min each.
3. Stain with H&E, toluidine blue, or methyl green using standard protocols.
4. Dehydrate in 70%, 90%, and 100% ethanol series and xylene, for 1–2 min each. Leave slides without a cover slip and store at room temperature in a desiccator until required.
5. Immediately prior to use, dehydrate in 100% ethanol, twice for 10 min each, and xylene, twice for 10 min each. Air-dry in a fume hood.

#### 3.1.2. Preparation of Frozen Tissue

1. Snap-freeze tissue and embed in OCT and cut sections at 6–10 μm on a cryostat.
2. Place slides immediately into 70% ethanol in a Coplin jar and leave for 2–4 min to fix. The most important step is the speed at which the slides are transferred to ethanol.
3. Stain slides as in **Subheading 3.1.1., step 3**.
4. Dehydrate in 100% ethanol twice for 10 min each and fresh xylene twice for 10 min.
5. Sections can be used immediately. Alternatively, the slides can be stored in a sealed desiccator at –80°C in batches appropriate for use (e.g., five slides). Slides should be gradually equilibrated to room temperature: at –20°C 24 h prior to use and at 4°C 4–5 h prior to use. Keep at room temperature in a desiccator while setting up the laser capture system.

### 3.2. Microdissection

We capture a minimum of 50 cells (~500 pg DNA) from test tissue. The reference and test DNA to be cohybridized must be of the same quality. When DNA extracted from FFPE tissue by LCM was cohybridized with DNA from peripheral blood that was diluted and amplified appropriately, we observed inconsistent and false CGH profiles. We use FFPE tissue as a reference for FFPE test tissue. Either LCM or manual dissection can be used to dissect the reference tissue. Capturing is performed according to the manufacturer's instructions for the system that is being used (PixCell II or PALM). We use the PixCell II system. The steps below outline points to be observed for successful capturing.

1. Wear gloves, avoid handling the cap, and use autoclaved/irradiated tubes. It is paramount that all attempts are made to avoid contamination.
2. Keep slides in a desiccator when not in use.
3. Avoid using excessive power (>55 W) for dissection, as it may damage the tissue.
4. Avoid folds in the section and repeated lifting and repositioning of the cap on the section, as the cap must make direct contact with the tissue for successful transfer.

5. Assess the adequacy and purity of each capture by visualizing the cap and residual tissue on the section directly by light microscopy.
6. The dissected material can be left on the cap in an Eppendorf tube at 4°C until required for extraction.
7. If reference DNA is to be obtained from manually dissected tissue, use a clean blade to scrape a 4-μm section from a plain glass slide prepared as in **Subheadings 3.1.1.** and **3.1.2.** Place the dissected material in a 1.5-mL Eppendorf tube.

## *3.3. DNA Extraction*

Enzymatic digestion is used to extract DNA as follows:

1. Add 0.2–0.25 mg/mL proteinase K to DNA lysis extraction buffer immediately prior to use and add 30–60 μL buffer to the microdissected tissue. The volume required depends on the yield of microdissection (*see* **Note 3**).
2. Invert tubes and incubate overnight at 55°C for optimal enzyme activity.
3. Centrifuge tubes at 13,400 ref. for 5 min, replace the LCM caps with a normal PCR tube lid.
4. To inactivate the proteinase K incubate the reaction at 90°C for 10 min, cool on ice for 2 min and centrifuge for 30 s. The extracted material can be stored indefinitely at –20°C.

## *3.4. DNA Quantitation*

1. The amount of DNA to be used for WGA (*see* **Subheading 3.5.**) and labeling reactions (*see* **Subheading 3.6.**) must be determined empirically.
2. When small amounts (pg) of DNA are used as a starting template indirect quantitation is used.
3. For primary (1°) DOP-PCR, 5-μL, 10-μL, and 15-μL aliquots of extracted DNA were amplified in parallel reactions and the product visualized on a 2% agarose 1X TBE gel.
4. Ten microliters of extracted DNA from FFPE material (equivalent to approx 10 ng DNA) was found to give optimum results for 1° DOP-PCR (*see* **Subheading 3.5.2.**).
5. The amount of template to be used for secondary (2°) DOP-PCR will need to be determined in a similar way and will depend on whether it is being used to label the DNA simultaneously (2.5 μL of 1° DOP-PCR product is found to be adequate) or if nick translation is being to label the DNA subsequent to amplification (12 μL of 1° DOP-PCR product is required). In either case, this will depend on the quality of the starting material (frozen or FFPE material).
6. We do not quantitate the amount of DNA directly; however, this can be done using the NanoDrop™ ND-1000 Spectrophotometer (NanoDrop Technologies, Rockland, DE).

## *3.5. Whole-Genome Amplification*

DOP-PCR is used to amplify test and reference DNA (*see* **Note 4**). In the first reaction (1° DOP-PCR), partially degenerate primers are used with initial low annealing temperatures to prime from multiple sites in the genome. This is followed by a high-stringency amplification step that ensures amplification from sites with the specific primer sequence. A second amplification step (2° DOP-PCR) is used to further amplify DNA and can also be used for labeling. Alternatively, DNA can be labeled by standard nick translation subsequent to 2° DOP-PCR.

### 3.5.1. Guidelines for All DOP-PCR Reactions

1. The importance of avoiding contamination cannot be emphasized enough. Wear gloves. Prepare all reactions in a PCR hood, ideally with pipets, tubes, water, and pipet tips dedicated to DOP-PCR.
2. Prepare the reaction mixture (number of samples + 0.2) in a PCR hood in a 1.5-mL Eppendorf tube. Add the enzyme last and vortex gently and centrifuge briefly before adding the enzyme.
3. Aliquot reaction mixture into PCR tubes in the PCR hood.
4. Add template outside PCR hood.
5. A negative (PCR water) and positive control (peripheral blood DNA) must be included for all reactions.

### 3.5.2. Primary DOP-PCR

1. Add the reaction components below in the order listed.

| Reagent | Volume (µL) | Final concentration |
|---|---|---|
| Water | 50–(26+X) | — |
| 5X Buffer D | 10 | 1X |
| PE W1 (20%) | 5 | 1% |
| dNTP mix (10 m$M$) | 5 | 1 m$M$ |
| 6 MW primer (20 m$M$) | 5 | 2 m$M$ |
| SuperTaq (5 U/mL) | 1 | 0.1 U/mL |
| Template DNA[a] | X | — |
| TOTAL | 50 | |

[a]10 µL of extracted DNA from FFPE tissue (*see* **Subheading 3.4.**)

2. The reaction cycles used are listed below.

**Pre-Amplification Step**

| Temp | Time | Ramp | Number of cycles |
|---|---|---|---|
| 94°C | 9 min | | |
| 94°C | 90 s | | |
| 30°C | 90 s | 0.2°C/s–30°C to 72°C | |
| 72°C | 180 s | | 8 |

*(continued)*

**Amplification step**

| Temp | Time | Number of cycles |
|------|------|------------------|
| 94°C | 90 s | |
| 62°C | 90 s | |
| 72°C | 90 s | 27–30 |

**Final extension**

| Temp | Time | |
|------|------|---|
| 72°C | 480 s | 1 |

3. Visualize the product by running 5-mL on a 2% agarose 1X TBE gel (**Fig. 2**). Highly degraded and small (<50 bp) product should not be used for CGH.

### 3.5.3. Secondary DOP-PCR

Secondary DOP-PCR amplifies the 1° DOP-PCR at high-stringency conditions. If it is to be used for simultaneous labelling of DNA then proceed directly to the labeling by the 2° DOP-PCR protocol in **Subheading 3.6.1.** (*see* **Note 5**). Use the following 2° DOP-PCR protocol if nick translation is to be used for DNA labeling.

1. Add the reaction components below in the order listed.

| Reagents | Volume (µL) | Final concentration |
|----------|-------------|---------------------|
| Water | 25–(9.5 + X) | — |
| 10X TAPS Buffer | 2.5 | 1X |
| dNTP mix (1.6 m$M$) | 2.5 | 0.16 m$M$ |
| 6 MW primer (20 m$M$) | 2.5 | 2 µ$M$ |
| MgCl$_2$ (25 m$M$) | 1.5 | 1.5 m$M$ |
| SuperTaq (5 U/mL) | 0.5 | 0.1 U/mL |
| Template DNA[a] | X | |
| TOTAL | 25 | |

[a]12 mL 1° DOP-PCR product from FFPE material is required to produce 0.8–1 µg DNA for nick translation; *see* **Subheading 3.4.**

2. Amplify using the identical high-stringency conditions above (amplification step and final extension in **Subheading 3.5.2., step 2**).
3. Visualize 2° DOP-PCR product by running 2 µL on a 2% agarose 1X TBE gel. Proceed to labeling by nick translation (*see* **Subheading 3.6.2.**).

---

Fig. 2. (*opposite page*) Examples of DOP-PCR and labelling. (**A**) Primary DOP-PCR amplified DNA from DNA extracted from laser captured tissue (lanes 3–12). Lane 10 is not suitable for 2° amplification and labeling. The positive (amplified peripheral blood DNA) and negative controls are in lanes 2 and 13, respectively. (**B**)

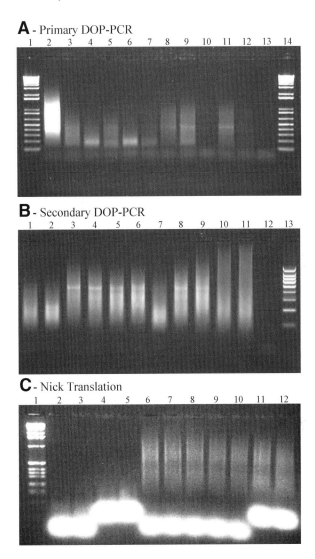

Fig. 2. (*continued*) Secondary DOP-PCR amplified 1° DOP-PCR product. During amplification DNA has been labeled by incorporation of dUTP conjugated to digoxigenin or biotin. Lanes 1–11 contain 2° DOP-PCR amplified product. Samples in lanes 2–9 are suitable for CGH; samples in lanes 10 and 11 need DNAse digestion to reduce the probe size to 200–2000 bp. Lane 12 contains the negative control and lane 13 a 100-bp ladder. (**C**) Secondary DOP-PCR amplified DNA labeled directly with fluorophores by nick translation. Labeled DNA appears as smears 100–1000 bp in lanes 2–12 (1 kb ladder in lane 1). DNA in lanes 6–12 is used for CGH. Labeled DNA in lanes 2–5 are unsuitable for CGH. Unincorporated fluorophores appear as intensely bright signals below the smears.

### 3.6. DNA Labeling

Adequate DNA labeling is one of the most critical steps in CGH, (*see* **Note 5**). For CGH, test DNA is always labeled green and reference red either directly, with green (SpectrumGreen-dUTP) and red (SpectrumRed-dUTP) fluorochromes, or indirectly, with digoxigenin-11-dUTP and biotin-16-dUTP. Indirect labeling is more cost-effective and gives a stronger signal but may give increased background fluorescence. The same labeling method should be used for both the test and reference DNA to be cohybridized.

#### 3.6.1. Labeling by DOP-PCR

DNA labeling occurs by substituting labeled dUTP for dTTP along the newly synthesized DNA strand.

1. Add the reaction components as follows:

| Reagent | Reaction volume (µL) | Final concentration |
|---|---|---|
| Water | $25-(8 + X + Y)$ | — |
| TAPS buffer (10X) | 2.5 | 1X |
| dNTP mix[a] | 2.5 | |
| 6 MW primer (20 µ$M$) | 2.5 | 2 µ$M$ |
| SuperTAQ (5 U/µL) | 0.5 | 0.05 U/µL |
| Template DNA[b] | X | |
| dUTP-label[c] | Y | |
| TOTAL | 25 | — |

[a]dA/C/GTP 2 m$M$; dTTP 1.6 m$M$. Final concentration: dA/C/GTP 0.2 m$M$; dTTP 0.16 m$M$

[b]Approximately 2.5 µL primary DOP-PCR product from FFPE material (*see* **Subheading 3.4.**).

[c]*See* **Note 6**

2. Amplify using the high stringency amplification step and final extension steps (*see* **Subheading 3.5.2., step 2**).
3. Proceed to checking the labeled DNA in **Subheading 3.7.**

#### 3.6.2. Labeling by Nick Translation

Nick translation labels DNA using two enzymes, DNA polymerase and DNAse 1 (*see* **Note 7**). Different nick translation kits are available and should be used according to the manufacturer's instructions. In our lab DNA is labeled using the CGH nick translation kit (Vysis) in the following steps:

1. Prepare the reaction mixture (number of samples + 0.2) on ice in a cooled 1.5-mL Eppendorf tube. Mix thoroughly with pipet tip before adding the enzyme. Tap and mix gently with pipet tip after adding the enzyme and centrifuge for a few seconds.

| Reagent | Volume (μL) | Final concentration |
|---------|-------------|---------------------|
| 10X Nick translation buffer | 5 | 1X |
| 0.1 m*M* dNTP mix | 10 | 0.02 m*M* |
| 0.1 m*M* dTTP | 5 | 0.01 m*M* |
| Nick translation enzyme | 10 | |

2. Aliquot 30 μL of reaction mixture into 0.2-μL PCR tubes (if using a thermocycler) or 1.5-mL Eppendorf tubes (if using a heat block). Add the following and vortex tube briefly:

| Reagent | Volume (μL) |
|---------|-------------|
| Nuclease-free water | 20–(X + Y) |
| Template DNA[a] | X |
| Labeled dUTP[b] | Y |

[a]Between 8–10 μL secondary DOP-PCR product, equivalent to 0.8–1 μg DNA (*see* **Subheading 3.4.**)
[b]*See* **Note 6**

3. Incubate, in the dark if using fluorochromes, for 2–4 h at 15°C.
4. Proceed to checking the labeled DNA (*see* **Subheading 3.7.**) and leave tube on ice.
5. Stop reaction at 80°C for 10 min, chill on ice, centrifuge.

### 3.7. Checking the Labeled DNA

Both the quality (probe size) and quantity of labeled test and reference DNA must be checked prior to each hybridization by gel electrophoresis as follows:

1. Run 1/10 volume of labeled DNA on a 2% agarose 1X TBE gel.
2. If probe size is too big (>2000 bp), digest as follows:
   a. For DOP-labeled DNA use DNAse treatment: to 10 μL DNA add: 1 μL DNAse 1 (diluted 1:70 with ddH$_2$O); 4 μL 10X DNAse 1 reaction buffer; and 25 μL ddH$_2$O. Incubate at 15°C for 10–45 min (depending on the initial size of DNA). Add 2 μL stop solution and vortex briefly to stop the reaction. Visualize by running 5 μL on a 2% agarose 1X TBE gel.
   b. For DNA labeled by nick translation: Add nick translation enzyme (e.g., 1 μL per sample) and incubate further (10 min) prior to stopping the reaction (*see* **Subheading 3.6.2.**, **step 5**).

### 3.8. Probe Precipitation

1. In a 1.5-mL Eppendorf tube, combine equal amounts of labeled test and reference DNA with human C$_0$t-1 DNA in excess (25–50:1, C$_0$t-1:probe DNA). For DOP-PCR labeled FFPE probes, 7.5 μL of each probe will suffice and 15 μg C$_0$t-1 DNA (i.e., 1 μg C$_0$t-1 DNA/μl 2° DOP-PCR DNA). If a probe has been digested

as described in **Subheading 3.7.**, **step 2a**, then 30–35 μL of the digested product is equivalent to 7.5 μL of nondigested DNA. If nick translation is used to label frozen DNA, then approx 10–15 μL of each probe is sufficient, depending on the probe quality, with $C_0t$-1 DNA at 50:1 μg.

2. Add 0.1 volume of 3 *M* sodium acetate (pH 5.5) and 2.5 volumes of ice-cold 100% ethanol. Vortex tube.
3. Leave at –20°C for at least 3 h or overnight for FFPE tissue.
4. Centrifuge at 13,400 ref for 30 min at 4°C to pellet the DNA.
5. Remove supernatant. Add ice-cold 70% ethanol to remove trace salt and centrifuge at 13,400 ref for 15 min at 4°C.
6. Remove the supernatant. Allow the pellet to dry at 37°C for 2–5 min or 15–20 min at room temperature.
7. Once dry, add 15 μL of hybridization buffer to each tube store at 4°C for at least 30 min or until required.

### 3.9. Preparation of Metaphase Slides and Probes for Hybridization

Four slides (eight hybridization areas) can be done at one time. Ideally, one control hybridization, normal vs normal (reference DNA labeled red with reference DNA labeled green) should be included with each batch.

1. Leave slides at room temperature for a few hours prior to use.
2. Check the number of metaphases by phase contrast microscopy (5 per 20X field required) and mark a suitable area for hybridization.
3. Take probes from 4°C. Denature at 72°C in a water bath for 10 min, mixing after 3–4 min to ensure the DNA is in solution. Cool on ice for 2 min, briefly centrifuge and incubate in a water bath at 37°C for 30 min to allow the $C_0t$-1 DNA to hybridize to repetitive elements within the probe DNA.
4. While the probe is denaturing (after 12–15 min), place the slides in prewarmed 50% formamide/50% 2X SSC at 72°C for 2–3 min.
5. Dehydrate slides in 70%, 90%, and 100% ice-cold ethanol series for 3 min each. Dry at 37°C for 1–2 min to evaporate the remaining ethanol.
6. Add 15 μL denatured probe mix to the marked area and immediately apply a glass cover slip (22 × 22 mm). Seal with rubber cement.
7. Incubate at 37°C for 2–3 d in a humidified dark box or on a HYBrite.

### 3.10. Washing of Slides and Detecting Hybridization

Following hybridization, the slides are washed in Coplin jars in a shaking water bath, at 42°C unless stated otherwise. Our wash protocol is outlined below but different wash protocols can be used (*see* **Note 8**). Solutions are prewarmed for 30 min prior to use.

1. Remove cement with forceps. Wash the slides in 50% deionized formamide/50% 2X SSC, three washes for 5 min each to remove excess probe and buffer. The cover slip falls off in the first wash.

2. Wash in 0.1X SSC, three washes for 5 min, to remove nonspecific DNA bound to the chromosomes.
3. Rinse in 4X SST (4X SSC/0.005% Tween-20), three washes for 5 min each. Slides hybridized with directly labeled probes are now ready for mounting (proceed to **step 8**). For indirectly labeled probe, the following steps (**4–7**) are required.
4. Incubate in 5% Marvel in 4X SST for 20 min at 37°C to block nonspecific binding of antibodies.
5. Rinse in 4X SST, for 5 min, three times, then drip-dry.
6. While the slides are still wet, apply 200 µL fluorescent-labeled antibodies, diluted appropriately in filtered 4X SST, to each slide (*see* **Note 9**). Cover with a plastic cover slip and incubate for 30 min at 37°C in a humidified chamber in the dark.
7. Remove cover slip and rinse in 4X SST, 5 min, three times.
8. Allow slides to almost dry, drop 20 µL DAPI with antifade onto slide, cover with 22 × 60 mm glass cover slip and seal with clear nail varnish.

The slides can be viewed under the fluorescence microscope or stored in the dark at 4°C.

### 3.11. Visualizing the Hybridization and Analysis

1. Images are captured using a cooled charge-coupled camera linked to a fluorescence Axioplan II microscope (Zeiss, Welwyn Garden City, UK) and VP SmartCapture software (Digital Scientific, Cambridge, UK).
2. The signals are captured independently using three single-band pass filters for tumor DNA (FITC/SpectrumGreen), reference DNA (Cy3/SpectumRed), and for chromosomes (DAPI).
3. The data from at least 10 images from each hybridization are analyzed by Quips™ CGH/Karyotyper and CGH/Interpreter software (Vysis).
4. For each experiment, a tumor:reference fluorescence ratio (FR) is calculated and compared to a control FR that is obtained from two independently labeled reference DNAs and serves as a reference standard for data analysis.
5. The resulting FR at each locus is equivalent to the ratio of the copy numbers of the corresponding DNA sequences in the test and reference genomes.
6. Therefore, increased green:red FR at a site indicates gain or amplification and reduced ratios represent deletions.
7. The FR change along each metaphase chromosome is quantified. FR values <0.80 and >1.20 are taken to indicate loss and gain, respectively. High-level gain is represented by a FR >1.5 *(1,8)*.

## 4. Notes

1. The protocols in **Subheadings 3.1.–3.5.** for tissue preparation, microdissection, and WGA have also been used to prepare DNA probes of frozen and FFPE material for array-based CGH (AmpliOnc 1, Vysis). For array-based CGH, we labeled DNA using the microarray nick translation kit (Vysis) but alternative labeling methods can be used, e.g., random priming PCR. The labeled product

should, however, be smaller than that required for metaphase CGH, i.e., 200–300 bp instead of 200–2000 bp to hybridize to the target clones on the array. The precipitation, hybridisation, and wash protocols will depend upon the type of array that is used.

2. Use a new blade for each case and do not use adhesive in the water bath. Bake slides briefly (e.g., 10 min) to adhere sections to the slide. If H&E is used, use the minimal concentration that is required to visualize the tissue. Excess stain may interfere with DNA extraction and interpretation of fluorescence *(12,14)*. The sections must be completely dehydrated immediately prior to use for efficient dissection: xylene steps can be increased to 30 min or slides can be dried at 30–37°C for 1 h. Dehydration is not a problem for FFPE tissue, but with frozen tissues a balance must be achieved between using increased drying times to dehydrate and the resulting increased adherence to the slide that will prevent capturing.

3. We have rarely required >60 µL extraction buffer when the cap has >50% surface covered (5000 pulses at 30 µm laser diameter). For manually dissected reference tissue, approx 200 µL of buffer is required for DNA extraction. Overnight incubation (14–18 h) is adequate. The caps should be viewed by light microscopy to assess the efficiency of extraction. If undigested material remains after overnight incubation, increase the concentration of proteinase K in the extraction buffer; however, proteinase K at 0.25 mg/mL is usually sufficient. For manually dissected tissue, if tissue is still visible in the extraction buffer after 14–18 h add another 5 µL of proteinase K (20 mg/mL), and incubate further while agitating gently and checking frequently for complete digestion of tissue.

4. DOP-PCR *(15)* requires minimal manipulation of sample DNA and reliably amplifies tiny amounts of DNA to produce probes that are the optimal size for CGH *(2,3,5,6,9,10,16–18)*. We modified the original DOP protocol for optimal results with FFPE tissue *(19)*. The optimal conditions may need to be altered depending on the tissue being used. We use 6 MW degenerate primers, but multiple specifically designed primers sets can be used to increase genome coverage to 96% *(20)*. The same amplification conditions must be applied to the test and reference DNA to be cohybridized. Hybridization of DOP-PCR amplified and labeled test DNA against reference DNA that has been labeled by nick translation but without amplification can show a false CGH profile with deviations at the heterochromatic regions and acrocentric p arms *(4,21)*.

5. The choice between 2° DOP-PCR and nick translation for DNA labeling depends on the quality and quantity of the DNA. The former is more suitable for archival material where the DNA fragments are already degraded, as DOP-PCR does not result in a smaller fragment size. If the DNA needs to be digested further an additional partial digestion with DNAse 1 can be done after labeling. Nick translation requires 0.8–1 µg of DNA and shortens fragment size to <500 bp because of the action of DNAse 1. This amount of DNA can be obtained using the DOP-PCR protocols described in **Subheadings 3.5.2. and 3.5.3.**; however, we use nick translation only for frozen material where the starting fragment length is 200–2000 bp.

6. Labeled haptens and fluorochromes are used in different concentrations in DOP-PCR and nick translation labeling reactions. The concentration that is used depends on the physiochemical properties of the dye, which affect its incorporation into DNA by DNA polymerase. Different labeling reactions should be performed with increasing concentrations of labeled nucleotide to determine the optimal concentration. Biotin- and digoxigenin-dUTP can be used at high concentrations but efficient labeling takes place when only 1/3 to 1/5 dTTP is replaced. Cy-3- and Cy-5-dUTP are required at much lower concentrations as they have an inhibitory effect at high concentration. We use commercially labeled dUTPs: SpectrumGreen-, SpectrumOrange-, digoxigenen-11-, and biotin-16-dUTPs (Vysis and RocheDiagnostics GmbH, Mannheim, Germany) at 0.01 m$M$ and at 0.04 m$M$ final concentration in our nick translation and DOP-PCR labeling protocols respectively.

7. Nick translation relies on a balance between DNAse and DNA polymerase I activity to label DNA. DNAse 1 cleaves double-stranded DNA shortening the DNA fragments. The polymerase 1 both removes nucleotides from the 5' end of the nick (5'-3' exonuclease activity) and adds nucleotides to the 3' end (5'-3' polymerase activity). The DNA is labeled when the dTTP in the nucleotide mix is partially replaced with labeled dUTP. The incubation time will vary for the different nick translation kits and will need to be varied for the tissue type (frozen or FFPE tissue). The optimal DNA fragment size for a metaphase CGH probe is 500–2000 bp *(1)* and will be seen for DNA extracted from frozen material; however, DNA from FFPE material is usually 200–1000 bp *(18)*. The size of the amplified fragment can be smaller if very small quantities of starting DNA are used as a starting template. Nevertheless, DNA fragments between 200 and 500 bp have yielded successful hybridisations. The single most important factor in assessing probe size is to ensure that the size is equivalent for the test and reference DNA to be cohybridized.

8. Different wash protocols can be used that vary in the concentration of the salt solutions, wash temperature, and wash time. The stringency of the washes can be altered depending on the signal intensity and level of background that is seen. In the protocol described in **Subheading 3.10.**, a salt concentration lower than 2X SSC should not be used with formamide in the first wash in **step 1**, as the resulting higher concentration of the formamide will result in loss of signal. If the loss of signal is too high then lower-stringency washes (0.1–2X SSC) can be used for the second wash in **step 2** but a temperature between 42–45°C is preferable. An alternative quick wash protocol can be used with 0.4X SCC, pH 7.0/0.3% nonidet P-40 at 74 ± 1°C for 2 min followed by 2X SSC, pH 7.0/0.1% nonidet P-40 at room temperature.

9. Fluorescent labeled antibodies must be in solution prior to application by centrifuging at 9300 ref for 10 min prior to use.
   If they have precipitated excess granular background may be apparent. We use anti-digoxigenin-FITC (200 µg/mL) and avidin-Cy3 (1 mg/mL) at dilutions of 1:200 and 1:400 respectively in filtered 4X SST.

## Acknowledgments

The authors would like to thank Dr. Suet-Feung Chin for her guidance, wisdom, and expertise and Dr. Mark Pett for his valued input into writing this manuscript. Cancer Research UK, Medical Research Council, the Breast Cancer Campaign, and the Sackler Foundation supported this work.

## References

1. Kallioniemi, A., Kallioniemi, O. P., Sudar, D., et al. (1992) Comparative genomic hybridization for molecular cytogenetic analysis of solid tumors. *Science* **258**, 818–821.
2. Aubele, M., Cummings, M., Walsch, A., e al. (2000) Heterogeneous chromosomal aberrations in intraductal breast lesions adjacent to invasive carcinoma. *Anal. Cell Pathol.* **20**, 17–24.
3. Aubele, M., Mattis, A., Zitzelsberger, H., et al. (1999) Intratumoral heterogeneity in breast carcinoma revealed by laser-microdissection and comparative genomic hybridization. *Cancer Genet. Cytogenet.* **110**, 94–102.
4. Zitzelsberger, H., Kulka, U., Lehmann, L., et al. (1998) Genetic heterogeneity in a prostatic carcinoma and associated prostatic intraepithelial neoplasia as demonstrated by combined use of laser-microdissection, degenerate oligonucleotide primed PCR and comparative genomic hybridization. *Virchows Arch.* **433**, 297–304.
5. James, L. and Varley, J. (1996) Preparation, labelling and detection of DNA from archival tissue sections suitable for comparative genomic hybridization. *Chromosome Res.* **4**, 163–164.
6. Speicher, M. R., du Manoir, S., Schrock, E., et al. (1993) Molecular cytogenetic analysis of formalin-fixed, paraffin-embedded solid tumors by comparative genomic hybridization after universal DNA-amplification. *Hum. Mol. Genet.* **2**, 1907–1914.
7. James, L. A., Mitchell, E. L., Menasce, L., and Varley, J. M. (1997) Comparative genomic hybridization of ductal carcinoma in situ of the breast: identification of regions of DNA amplification and deletion in common with invasive breast carcinoma. *Oncogene* **14**, 1059–1065.
8. James, L. A. (1999) Comparative genomic hybridization as a tool in tumour cytogenetics. *J. Pathol.* **187**, 385–95.
9. Aubele, M., Zitzelsberger, H., Schenck, U., Walch, A., Hofler, H., and Werner, M. (1998) Distinct cytogenetic alterations in squamous intraepithelial lesions of the cervix revealed by laser-assisted microdissection and comparative genomic hybridization. *Cancer* **84**, 375–379.
10. Huang, Q., Schantz, S. P., Rao, P. H., Mo, J., McCormick, S. A., and Chaganti, R. S. (2000) Improving degenerate oligonucleotide primed PCR-comparative genomic hybridization for analysis of DNA copy number changes in tumors. *Genes Chromosomes Cancer* **28**, 395–403.

11. Cheung, V. G. and Nelson, S. F. (1996) Whole genome amplification using a degenerate oligonucleotide primer allows hundreds of genotypes to be performed on less than one nanogram of genomic DNA. *Proc. Natl. Acad. Sci. USA* **93**, 14,676–14,679.

12. Hirose, Y., Aldape, K., Takahashi, M., Berger, M. S., and Feuerstein, B. G. (2001) Tissue microdissection and degenerate oligonucleotide primed-polymerase chain reaction (DOP-PCR) is an effective method to analyze genetic aberrations in invasive tumors. *J. Mol. Diagn.* **3**, 62–67.

13. Kuukasjarvi, T., Tanner, M., Pennanen, S., Karhu, R., Visakorpi, T., and Isola, J. (1997) Optimizing DOP-PCR for universal amplification of small DNA samples in comparative genomic hybridization. *Genes Chromosomes Cancer* **18**, 94–101.

14. Burton, M. P., Schneider, B. G., Brown, R., Escamilla-Ponce, N., and Gulley, M. L. (1998) Comparison of histologic stains for use in PCR analysis of microdissected, paraffin-embedded tissues. *Biotechniques* **24**, 86–92.

15. Telenius, H., Carter, N. P., Bebb, C. E., Nordenskjold, M., Ponder, B. A., and Tunnacliffe, A. (1992) Degenerate oligonucleotide-primed PCR: general amplification of target DNA by a single degenerate primer. *Genomics* **13**, 718–25.

16. Wiltshire, R. N., Duray, P., Bittner, M. L., et al. (1995) Direct visualization of the clonal progression of primary cutaneous melanoma: application of tissue microdissection and comparative genomic hybridization. *Cancer Res.* **55**, 3954–3957.

17. Griffin, D. K., Sanoudou, D., Adamski, E., et al. (1998) Chromosome specific comparative genome hybridization for determining the origin of intrachromosomal duplications. *J. Med. Genet.* **35**, 37–41.

18. Wells, D., Sherlock, J. K., Handyside, A. H., and Delhanty, J. D. (1999) Detailed chromosomal and molecular genetic analysis of single cells by whole genome amplification and comparative genomic hybridization. *Nucleic Acids Res.* **27**, 1214–1218.

19. Daigo, Y., Chin, S. F., Gorringe, K. L., et al. (2001) Degenerate oligonucleotide primed-polymerase chain reaction-based array comparative genomic hybridization for extensive amplicon profiling of breast cancers: a new approach for the molecular analysis of paraffin-embedded cancer tissue. *Am. J. Pathol.* **158**, 1623–1631.

20. Carter, N. P., Fiegler, H., and Piper, J. (2002) Comparative analysis of comparative genomic hybridization microarray technologies: Report of a workshop sponsored by the Wellcome Trust. *Cytometry* **49**, 43–48.

21. Voullaire, L., Wilton, L., Slater, H., and Williamson, R. (1999) Detection of aneuploidy in single cells using comparative genomic hybridization. *Prenat. Diagn.* **19**, 846–851.

# 5

## Detection of Ki-*ras* and *p53* Mutations by Laser Capture Microdissection/PCR/SSCP

### Deborah Dillon, Karl Zheng, Brina Negin, and José Costa

#### Summary

Efficient detection of somatic mutations is important for the development of clinical molecular diagnostic assays. However, the detection of somatic mutations in tissue is confounded by dilution of the tumor cell population by normal cells. Laser microdissection allows enrichment for tumor-associated genetic alterations to take place at the level of cell selection, eliminating the need to enrich for mutant alleles after amplification. In this chapter a method is described for somatic mutation analysis using cells acquired by laser capture microdissection.

**Key Words:** Mutation; neoplasm; *p53*; ki-*ras*; SSCP.

## 1. Introduction

The efficient detection of somatic mutations is important for the development of clinical molecular diagnostic assays, as well as for tissue-based translational research. Compared with the analysis of germline mutations, the detection of somatic mutations in tissue is confounded by dilution of the tumor cell population with non-neoplastic tissue elements. Currently available mutation detection assays typically require that a relatively large proportion of alleles be mutated or require time-consuming intermediate steps to enrich the polymerase chain reaction (PCR) product for the mutant allele prior to sequencing.

By allowing the separation of cells of interest from complex heterogeneous cell mixtures, laser capture microdissection (LCM) *(1,2)* provides an alternate strategy to post-PCR mutant-enrichment steps in somatic mutation detection assays. Laser microdissection allows enrichment for tumor-associated genetic

From: *Methods in Molecular Biology, vol. 293: Laser Capture Microdissection: Methods and Protocols*
Edited by: G. I. Murray and S. Curran © Humana Press Inc., Totowa, NJ

alterations to take place at the level of cell selection, eliminating the need to enrich for mutant alleles after amplification. Using LCM, clonal mutations can usually be detected and sequenced directly from original PCR products without band-cutting and reamplification for mutant enrichment *(3)*. This strategy is particularly effective for assaying genes with a limited number of expected mutation sites (such as *ras* genes), but can also be used for genes with a broader range of possible mutation sites. In the latter case, careful exclusion of adjacent non-neoplastic cells during microdissection is critical to optimize the percentage of mutated tumor cell alleles for detection.

Following PCR, amplification products can be either sequenced directly or further screened by single-strand conformational polymorphism (SSCP) analysis *(4,5)* to focus sequencing efforts on exons most likely to contain mutations. In SSCP analysis, mutations result in altered migration of the amplified mutant alleles relative to the amplified wild-type alleles in nondenaturing gels. Exons showing shifted bands are more likely to have mutations than exons showing the wild-type pattern of bands. SSCP also allows visual estimation of the relative contribution of mutant alleles to the total signal. When mutant alleles constitute at least 30% of the total signal, it is likely that interpretable sequence results can be obtained without further mutant enrichment.

## 2. Materials

### 2.1. Reagents

#### 2.1.1. Tissue Preparation and Laser Capture Reagents

1. Xylene, 100% ethanol, sterile gauze.
2. Hematoxylin, eosin, methyl green, double-distilled $H_2O$.
3. Proteinase K.
4. 1X PCR buffer: 50 m$M$ KCl, 10 m$M$ Tris-HCl, pH 8.3 (Perkin Elmer, Foster City, CA).

#### 2.1.2. PCR Reagents

1. Sterile double-distilled $H_2O$.
2. 10X PCR buffer: 500 m$M$ KCl, 100 m$M$ Tris-HCl, pH 9.0 (Perkin Elmer).
3. $MgCl_2$ (Perkin Elmer).
4. dNTPs.
5. Oligonucleotide primers (*see* **Table 1**).
6. AmpliTaq® Gold DNA Polymerase (Perkin Elmer).
7. Mineral oil.

#### 2.1.3. Gel Electrophoresis Reagents

1. Agarose (2:1 NuSieve:LE, FMC Bioproducts, Rockland, ME).
2. 100-bp ladder (Research Genetics, Huntsville, AL).

**Table 1**
**Primer Sequences, Annealing Temperatures, and Product Sizes**

| Gene/Exon | Primer sequence | Annealing temp | Product size |
|---|---|---|---|
| Ki-*ras* exon 1 | 5'-GAC TGA ATA TAA ACT TGT GG-3'<br>5'-AAT GGT CCT GCA CCA GTA AT-3' | 55°C | 152 bp |
| Ki-*ras* exon 2 | 5'-TGG CAA ATA CAC AAA GAA AG-3'<br>5'-GAC TGT GTT TCT CCC TTC T-3' | 55°C | 161 bp |
| *p53* exon 5 | 5'-CCG TGT TCC AGT TGC TTT AT-3'<br>5'-AGC CCT GTC GTC TCT CCA-3' | 58°C | 288 bp |
| *p53* exon 6 | 5'-GGG CTG GTT GCC CAG GGT-3'<br>5'-AGT TGC AAA CCA GAC CTC A-3' | 60°C | 181 bp |
| *p53* exon 7 | 5'-CCA CAG GTC TCC CCA AGG-3'<br>5'-TGG CAA GTG GCT CCT GAC-3' | 55°C | 183 bp |
| *p53* exon 8 | 5'-CCT ATC CTG AGT AGT GGT AA-3'<br>5'-TCC TCC ACC GCT TCT TGT-3' | 55°C | 182 bp |
| *p53* exon 9 | 5'-ACC TTT CCT TGC CTC TTT C-3'<br>5'-CGG CAT TTT GAG TGT TAG AC-3' | 55°C | 157 bp |

3. Ethidium bromide.
4. 40% MDE™ (mutation detection enhancing) nondenaturing acrylamide gel (FMC Bioproducts).
5. Ammonium persulfate (10% w/v).
6. TEMED.
7. Formamide dye solution: 95% formamide, 10 m$M$ NaOH, 0.05% xylene cyanol, 0.05% bromophenol blue.
8. 0.5X TBE gel running buffer: 45 m$M$ Tris-borate, 1 m$M$ NaEDTA, pH 8.3.
9. SYBR Green II (Molecular Probes, Eugene, OR).
10. TE: 5 m$M$ Tris-HCl, 0.5 m$M$ EDTA, pH 7.4.

## 2.2. Equipment

### 2.2.1. Equipment for Tissue Preparation and LCM

1. Plain (noncoated, noncharged) glass slides.
2. Microtome, forceps, picks, brushes, water bath.
3. Coplin jars for staining setup.
4. Laser capture microscope (Arcturus Engineering, Mountain View, CA).
5. Standard or high-sensitivity caps (Arcturus Engineering).

### 2.2.2. Equipment for PCR Amplification

1. Thermocycler.

### 2.2.3. Equipment for SSCP

1. Heat block maintained at 95°C.
2. 160 × 140 × 0.75 mm vertical gel apparatus (Hoefer Scientific SE600; Hoefer Scientific Instruments, San Francisco, CA), 18°C constant temperature.
3. Power supply.
4. Digital imager (Alpha-Innotech IS1000, San Leandro, CA) equipped with SG3 filter.

## 3. Methods

### 3.1. Tissue Preparation and Laser Microdissection

1. Using a microtome, cut several 5-μm sections of each formalin-fixed, paraffin-embedded tissue block and mount onto noncoated, noncharged plain glass slides, with thorough microtome decontamination prior to cutting each case (*see* **Note 1**). The clear part of the glass should not be touched with hands or fingers at any point in the procedure.
2. Stain the first section of each group of slides using a standard hematoxylin and eosin protocol, coverslip, and use as a reference for tissue morphology in later steps.
3. Stain the remaining sections with methyl green or eosin alone for subsequent microdissection (*see* **Note 2**).
4. Capture cell population of interest following manufacturer's protocols (*see* www.arctur.com for PixCell laser microscopes). 500–1000 30-μm captures are

generally adequate for obtaining 20–60 ng of DNA for mutation analysis (depending on the quality and type of fixation of the target tissue). This may be obtained on a single standard cap, generally requiring no more than 15 min of microdissection time, except in cases with difficult histology. Duplicate capture of sample onto two caps should be obtained for confirmation of results and to minimize potential contamination artifacts (*see* **Note 3**). **Figure 1** shows an example of the selective capture of invasive colonic adenocarcinoma cells in a formalin-fixed, wax-embedded tissue sample stained with methyl green.

## 3.2. DNA Extraction and PCR Amplification

### 3.2.1. Lysis and Digestion

1. Place 50 µL of DNA extraction buffer (5 mg/mL proteinase K in 50 m$M$ KCl, 10 m$M$ Tris-HCl, pH 8.3) in 0.5-mL microfuge tubes and cap each tube with a captured sample.
2. Incubate inverted sample at 37°C overnight.
3. Heat lysate at 95°C for 8 min. Lysate may be used directly into PCR without further purification, using 500–1000 cell equivalents into each 30-µL PCR reaction (*see* **Note 4**).
4. Store lysate or purified DNA at 4°C.

### 3.2.2. Amplification of Ki-ras and p53

1. Amplify exons individually using flanking primers (**Table 1**). Set up PCR reactions in a total volume of 30 µL, containing 0.5 µ$M$ of each primer, 250 µ$M$ of each dNTP, 50 m$M$ KCl, 10 m$M$ Tris-HCl (pH 8.3), 2.0 m$M$ MgCl$_2$, and 0.8 U of AmpliTaq® Gold DNA polymerase (*see* **Note 5**). For each specimen, two independent amplification reactions should be performed, containing 5–10 ng purified DNA (or 500–1000 cell equivalents, by calculation) from each of the laser captured samples. TE-only and wild-type controls (human placenta) should be included with each reaction (*see* **Note 6**).
2. Following a 12-min denaturation at 95°C, subject mixtures to 40 cycles (95°C/30 s, 55°C/45 s [Ki-*ras* exons 1 and 2, *p53* exons 7–9; *see* **Table 1** for other annealing temperatures], 72°C/45 s) with a final extension at 72°C for 10 min.
3. Amplification products (*see* **Table 1** for amplicon sizes) should be verified and quantity visually estimated on 3% agarose gels using a 100-bp ladder and staining with ethidium bromide.
4. Store amplification products at 4°C until further analysis by SSCP.

## 3.3. Single-Strand Conformational Polymorphism Analysis

1. Denature 30-ng aliquots of PCR product in 1–5X volume of formamide dye at 95°C for 5 min.
2. Chill samples immediately on ice.
3. Load samples into rinsed wells of 40% MDE™ nondenaturing acrylamide gels and perform electrophoresis in 0.5X TBE at 500 V for 3.0 h, 18°C constant temperature (Hoefer Scientific SE600 160 × 140 × 0.75 mm vertical gels).

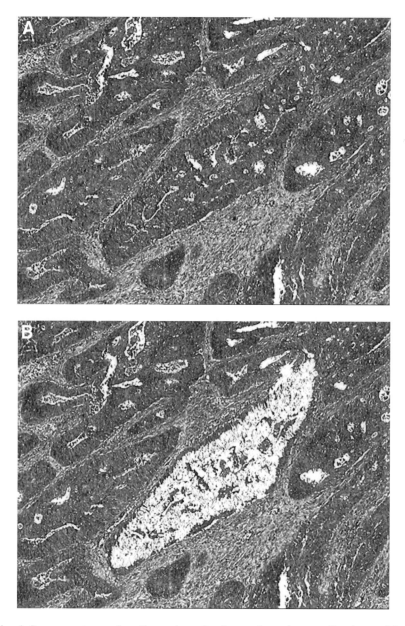

Fig. 1. Laser-capture microdissection of colorectal carcinoma. Sections of formalin-fixed, wax-embedded colorectal adenocarcinoma are stained with methyl green and positioned for laser capture (**A**). Tumor cells are selectively captured (**B**). Captured cells can be visualized on the cap (**C**). (Reprinted from **ref. 3** with permission from Elsevier.)

**C**

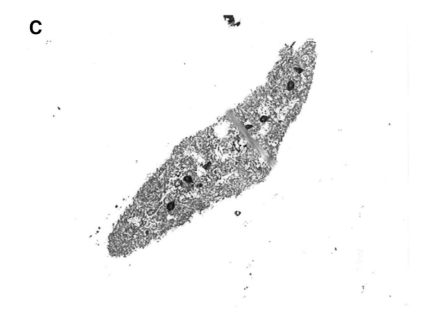

4. Stain gels for 20 min with gentle agitation and light protection using a 1:10,000 dilution of SYBR Green II in TE.

## 3.4. Image Collection and Interpretation

Using 254 nm ultraviolet transillumination, collect images through an SG3 filter on a digital imager. Mutated alleles show migration patterns different from the wild-type alleles (*see* **Note 7**). Relative mutant contribution can be estimated by densitometry of mutant to wild-type bands on the digital gel image. **Figure 2** shows representative gel images comparing the SSCP patterns from laser-capture microdissected tissues with hand-microdissected tissues and cut, reamplified aberrantly migrating bands. In the three examples shown, there is enrichment on SSCP in the laser-microdissected samples for the aberrantly migrating bands and in the sequences for the associated mutation. In Case A, there is a Ki-*ras* codon 13 mutation (GGC ♦ GAC; Gly ♦ Asp); in Case B, there is a Ki-*ras* codon 12 mutation (GGT ♦ AGT; Gly ♦ Ser); and in Case C, there is a *p53* exon 5 mutation at codon 179 (CAT ♦ GAT; His ♦ Asp).

In most cases in which laser microdissection can effectively separate the cell population of interest from potentially contaminating cell populations, mutations can be identified and verified by direct sequence analysis of the PCR product. However, SSCP may be used to focus sequencing efforts on exons

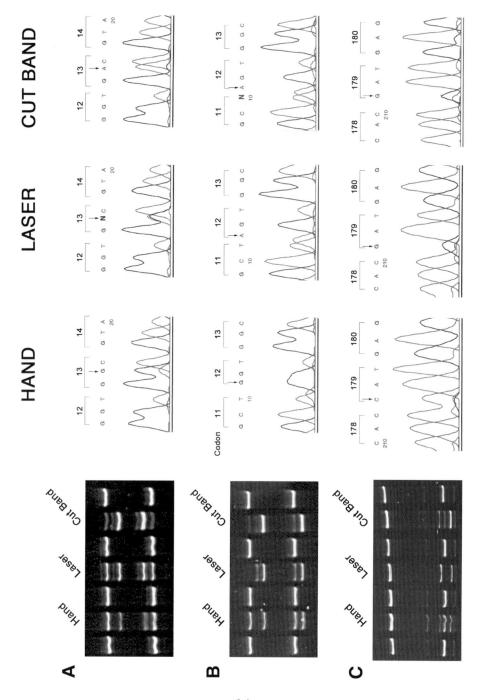

most likely to contain mutations, based on the presence of shifted bands. This is especially useful in cases in which microdissection is difficult due to very small numbers of the cells of interest and/or extensive infiltration of the cells of interest with another cell population (*see* **Note 8**).

## 4. Notes

1. To prevent cross-contamination, microtome surfaces must be thoroughly decontaminated prior to sectioning each new tissue block. Thoroughly clean the microtome stage and chuck of all excess material using xylene on sterile gauze, followed by 100% ethanol. Forceps, picks, and brushes should also be cleaned. A new knife blade or new section of the knife blade should be used for each block. The water bath should also be thoroughly cleaned with xylene, then ethanol, with a final rinse in distilled water. The water bath should then be filled with distilled water. Glue or other adhesives should not be used in the water bath. Frozen tissue blocks may also be used, in which case similar decontamination of the cryostat should be performed.

2. Eosin alone can be used satisfactorily to stain most tissues for microdissection. In our experience, there is a slight improvement in DNA yield with methyl green staining; however, its use is not justified except in cases in which the cell population of interest is very limited. Whichever stain is used, the staining process should end with at least 5 min in xylene.

3. It is essential to visually monitor the effectiveness of tissue lift during the capture stage, as inefficient capture will lead to failure of subsequent mutation detection procedures. High ambient humidity or failure to adequately dry the tissues during pre-LCM processing are the major causes for inability to capture selected tissue. Optimal humidity for LCM is below 30%. Tissues that are too dry, on the other hand, may result in additional nonspecific lift of surrounding tissues. Additional troubleshooting suggestions for poor tissue lift are available on the Arcturus web site (www.arctur.com).

4. The number of cell equivalents can be calculated for the total lysate sample by multiplying the average number of cell profiles in each 30-μm capture by the number of captures. We generally divide the number of visible cell profiles in

---

Fig. 2. SSCP gels and sequencing results comparing hand- and laser-microdissected samples for detection of mutations in Ki-*ras* and *p53*. Lanes 1, 3, 5, and 7 show the wild-type pattern (human placenta) for Ki-*ras* exon 1 (Cases **A** and **B**) and *p53* exon 5 (Case **C**). In each case, the intensity of the aberrantly migrating mutant bands is stronger in the laser-captured sample (lane 4) than in the hand-microdissected sample (lane 2). The reamplified cut band from the hand-microdissected sample (lane 6) shows similar shifted bands and is included for comparison. Sequences from the hand-microdissected, laser microdissected, and cut bands from the hand-microdissected are seen to the right of each gel with position of the altered bases noted (arrows). (Adapted and reprinted from **ref. 3** with permission from Elsevier.)

each capture by two to compensate for the effect of nuclear truncation in 5-μm sections. Alternatively, the DNA may be further purified using standard methods, or any one of a number of commercial kits. Purified DNA should be quantitated and 5–10 ng used in genotyping reactions. The use of limiting amounts of template facilitates the production of PCR artifacts and may result in the detection of artifactual mutations.

5. 1.5 μL Reload™ (Research Genetics, Huntsville, AL) may be added prior to amplification to facilitate visualization of reagents and products.

6. Standard precautions should be taken to avoid generation of PCR artifacts. All reagents, pipets, and working areas must be clean and free of contamination, with strict segregation of pre- and post-PCR areas. Pipet tips must contain an aerosol barrier to minimize cross-contamination. Use of two no-template controls are recommended: one to be closed prior to opening of any test samples for verification of reagent purity and one to be closed following addition of DNA to all sample tubes for procedural control. All reactions should be performed in duplicate.

7. Mutations result in conformational change of the renatured single-stranded PCR product, which then shows differential migration in nondenaturing gels (i.e., "shifted" bands). Sensitivity of this assay for the detection of mutations in Ki-*ras* has been established by analysis of serial dilutions of mutant cell line DNA. These studies demonstrate detection of less than 0.5% mutated alleles in a background of wild-type DNA *(6)*.

8. In the small number of cases in which shifted bands are strong and reproducible, yet no mutation is detected by sequencing, sequence analysis of reamplified bands cut from MDE gels may be performed. In this case, the shifted bands can be excised from MDE gels using a sterile blade, with subsequent elution of DNA by incubation for 5 min at 55°C in TE, and reamplification using the same primer sequences. The reamplified bands can then be gel-purified, bound to NA45 DEAE membrane, and eluted with 1 *M* NaCl in TE and reprecipitated in ethanol prior to sequencing. In our experience, it is rarely necessary to do this when the DNA is obtained by laser microdissection; however, it remains an option for confirmation purposes.

## Acknowledgments

We are grateful to Mr. Chuck Frederick, Mr. Amos Brooks, and Ms. Mary Helie for expert advice and technical assistance in optimizing the tissue preparation and staining protocols for laser capture microdissection.

## References

1. Emmert-Buck, M. R., Bonner, R. F., Smith, P. D., et al. (1996) Laser capture microdissection. *Science* **274**, 998–1001.
2. Bonner, R. F., Emmert-Buck, M. R., Cole, K., et al. (1997) Laser capture microdissection: molecular analysis of tissue. *Science* **278**, 1481–1483.

3. Dillon, D., Zheng, K., and Costa, J. (2001) Rapid, efficient genotyping of clinical tumor samples by laser-capture microdissection/PCR/SSCP. *Exp. Mol. Pathol.* **70,** 195–200.

4. Orita, M., Iwahana, H., Kanazawa, H., Hayashi, K., and Sekiya, T. (1989) Detection of polymorphisms of human DNA by gel electrophoresis as single strand conformation polymorphisms. *Proc. Natl. Acad. Sci. USA* **86,** 2766–2770.

5. Chaubert, P., Bautista, D., and Benhattar, J. (1993) An improved method for rapid screening of DNA mutations by nonradioactive single-strand conformation polymorphism procedure. *BioTechniques* **15,** 586.

6. Emanuel, J. R., Damico, C., Ahn, S., Bautista, D., and Costa, J. (1996) Highly sensitive nonradioactive single-strand conformational polymorphism detection of Ki-*ras* mutations. *Diag. Mol. Path.* **5,** 260–264.

# 6

## Whole-Genome Allelotyping
## Using Laser Microdissected Tissue

### Colleen M. Feltmate and Samuel C. Mok

#### Summary

Laser-based microdissection technologies have been recently developed and applied to procure homogenous populations of tumor cells from paraffin-embedded and frozen tissue sections. When combined with whole-genomic amplification techniques, sufficient amounts of DNA can be generated from a small number of tumor cells procured by laser-based microdissection. Amplified DNA can then be used to perform high-throughput genome-wide allelotyping using fluorescent-labeled microsatellite markers spanning the whole genome. Loss of heterozygosity can be assessed by Genescan and Genotyper software (ABI Prism).

**Key Words:** Allelotyping; loss of heterozygosity; ovarian cancer.

## 1. Introduction

Loss of heterozygosity (LOH) has been widely used in an effort to identify tumor-suppressor genes in various tissue types *(1–4)*. This technique utilizes markers specific for a particular chromosomal locus, which can differentiate both paternal and maternal alleles in normal tissue. The heterozygous pattern identified in normal non-neoplastic tissue is then compared with that of neoplastic or tumor DNA from the same patient. LOH is identified if one of the heterozygous alleles is missing.

Classical LOH studies used DNA extracted from whole-tissue samples *(3,5,6)*. Several studies have also used gross microdissection techniques in an effort to enrich specific cell populations of interest *(7,8)*. Contamination by stromal tissue, however, may still skew LOH patterns, decreasing sensitivity in the detection of loss of heterozygosity. To circumvent this problem, microdissection techniques, which can procure precisely defined homogenous tumor

From: *Methods in Molecular Biology, vol. 293: Laser Capture Microdissection: Methods and Protocols*
Edited by: G. I. Murray and S. Curran © Humana Press Inc., Totowa, NJ

cell population with low or even no contamination by nontumor cells, are required. Two laser-based microdissection systems have recently been developed. The PixCell II Laser Capture Microdissection (LCM) system (Arcturus Engineering, Mountain View, CA) uses direct contact with a transparent ethylene vinyl acetate (EVA) thermoplastic film to procure specific cells of interest *(9,10)*. The other (Leica, Germany) uses a frame foil slide and laser cutting technique. These laser-based microdissection techniques can be applied to both paraffin-embedded and frozen tissue sections. However, when only a small number of cells are available, the amount of DNA isolated may not be sufficient for large-scale allelotyping studies. Through whole-genome amplification methods *(11,12)* the entire genome can be amplified. High-throughput allelotyping can then be performed using fluorescent-labeled microsatellite markers spanning the whole genome *(1,2)*.

## 2. Materials

### 2.1. Tissue Processing

1. OCT compound.
2. Formalin.
3. Ethanol.
4. Xylene.
5. Paraffin wax.
6. Cryomold.
7. Liquid nitrogen.
8. Cryostat and microtome.
9. Uncoated glass slides.
10. Polyethylenenaphthalate (PEN) framed foil slides (Leica, Germany).
11. 1% Methyl green.

### 2.2. Microdissection and DNA Extraction

1. PixCell II LCM System (Arcturus Engineering).
2. LMD-Laser Microdissecting System (Leica, Germany).
3. 0.5-mL Eppendorf tube.
4. Digestion buffer: 1X Expand High Fidelity Buffer (Roche), 4 mg/mL proteinase K, and 1% Tween-20 at 55°C.

### 2.3. Whole-Genome Amplification

1. Reaction mixture: 0.05 mg/mL gelatin, 40 m$M$ 15-mer random primers (Operon Technologies, Alameda, CA), 0.2 m$M$ dNTP, 2.5 m$M$ $MgCl_2$, 1X Expand High Fidelity Buffer, and 3.5 U of Expand High Fidelity Polymerase (Roche).
2. Thermocycler.

## 2.4. Allelotyping (Applied Biosystems, Foster City, CA)

1. ABI Prism Linkage Mapping Set.
2. True Allele Buffer (10X GeneAmp PCR buffer II, GeneAmp dNTP mix [2.5 m*M*], AmpliTaq Gold DNA Polymerase [5 U/mL], 25 m*M* MgCl$_2$).
3. ABI310-Genetic Analyzer (Applied Biosystems).
4. 1X Buffer.
5. POP-4 Polymer.
6. Capillary (47 cm × 50 μm).
7. Rox standard.
8. Deionized formamide.
9. Genescan and Genotyper software (Applied Biosystems).

## 3. Methods

The methods described below outline (1) tissue preparation, (2) micro-dissecting procedures, (3) extraction of DNA, (4) whole-genome amplification procedures, and (5) the allelotyping process.

### 3.1. Tissue Preparation

Tissue processing is initiated within 1 h of surgical removal from patients. Samples can be snap-frozen in liquid nitrogen or fixed in formalin. The two types of processing technologies are described below.

#### 3.1.1. OCT Preparation

Tissue is cut into 0.5-cm$^3$ sections and placed centrally in a plastic cryomold. OCT compound is then used to fill the mold. The entire cryomold is placed on the surface of liquid nitrogen until OCT is solidified. Tissue blocks can then be stored in a –80°C freezer.

#### 3.1.2. Formalin Fixation

Tissue is placed in formalin overnight on a rocking platform. Tissue is dehydrated through an ascending series of ethanol, cleared in xylene, and embedded in paraffin.

#### 3.1.3. Tissue Sectioning and Staining

1. Frozen sections (7 μm) are cut using a cryostat and mounted on either a plain slide or a PENfoil slide.
2. They are immediately fixed in 70% ethanol for 30 s, stained with 1% methyl green (*see* **Note 1**) for 5 s, and washed in distilled water.
3. Sections on plain slides are dehydrated through an ascending series of ethanol: 70%, 30 s; 95% × 2, 30 s; 100% × 2, 5 min.

4. Sections are then cleared in xylene for 5 min and allowed to air-dry.
5. Sections on PENfoil slides are fixed in a similar manner. The dehydration process utilizes 70% and 95% ethanol for 30 s each.
6. Paraffin sections (7 μm) are cut using a microtome.
7. Sections on plain slides are dewaxed by routine methods in xylene and ethanol.
8. Staining and dehydration are performed similarly as above.
9. Sections on PENfoil slides are dewaxed in xylol for 2 min.
10. A descending series of ethanol is used: 100% × 2, 30 s; 95% ×2, 30 s; 70%, 10 s; dH$_2$O, 10 s. Stain as above, wash in dH$_2$O, and air-dry sections.

## 3.2. Microdissection and DNA Extraction

### 3.2.1. PixCell II LCM System

1. Sections utilizing plain glass slides are required for the PixCell II LCM system.
2. This microscope focuses the laser beam to discrete spot sizes (7, 15, or 30 μm), delivering precise pulsed doses that are controlled by the operator, to the targeted film. Targeted cells are transferred to the cap surface (**Fig. 1**).
3. The operator must control both the movement of the stage and the laser firing.
4. A special film adhesive is applied to the slide prior to microdissection to remove poorly bound tissue fragments.
5. The cap should be reviewed periodically during dissection, using a blank slide, to ensure there is no contamination by unprocured tissue fragments.
6. Approximately 5000 cells are procured for the digestion process (*see* **Note 2**).
7. After completing the dissection, the cap can be placed directly into an Eppendorf tube for DNA extraction.

### 3.2.2. Leica LMD System

1. This laser microdissecting system requires the use of special PENfoil slides. Once the specific magnification is determined, the cells of interest are circumscribed on the screen image and then cut automatically.
2. The laser beam moves over the specimen to make the cut.
3. The specimen remains stationary so that it can be clearly observed during the cutting process.
4. The material cut out by the laser falls into the cap of an Eppendorf tube below (**Fig. 2**).
5. There is an inspection mode to visually confirm that the dissected cells have been deposited in the tube.

### 3.2.3. DNA Extraction

1. Fifty μL of digestion buffer are added either directly to the Eppendorf tube and inverted for the PixCell cap dissection or directly to the cap containing the dissected material from the Leica dissection process.
2. The digestion takes place at 55°C over 72 h in a humidified incubator.
3. After 24–36 h, an additional microliter of 20 mg/mL proteinase K solution is added to ensure complete digestion.

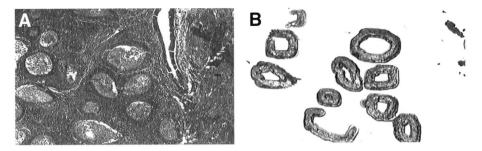

Fig. 1. Laser capture microdissection of mucinous epithelial ovarian tumor (**A**). Tumor cells procured on cap are shown in (**B**).

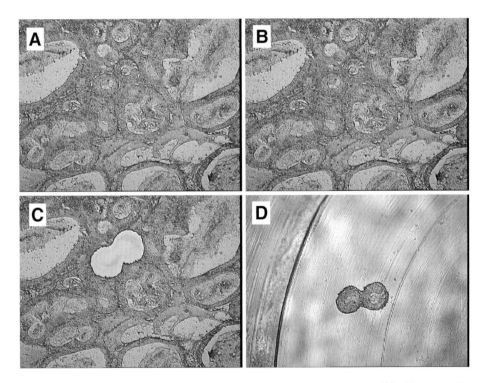

Fig. 2. Photomicrograph of mucinous epithelial ovarian tumor (**A**). Tumor cells selected are outlined by red line (**B**). After laser cut (**C**), cut section in Eppendorf cap (**D**).

4. The digest is then spun briefly, heated to 95°C for 10 min to inactivate proteinase K, spun briefly again before transferring to a new tube, and stored at 4°C (*see* **Notes 3** and **4**).

## *3.3. Whole-Genome Amplification*

1. Whole-genome amplification is performed by using a modified primer-extension preamplification (I-PEP) process *(12)*.
2. Ten µL of digested microdissected DNA is added to 50 µL of the I-PEP reaction mixture.
3. Fifty primer extension cycles are carried out in a thermocycler after an initial 94°C, 3 min denaturation step.
4. Each cycle consists of a 1-min at 94°C, 2 min at 37°C, a ramping step of 0.1°C/s up to 55°C, a 4-min primer extension step at 55°C, and 30 s at 68°C.
5. Following amplification the I-PEP DNA can be quantified using a fluorometer. I-PEP product can be diluted to 50 ng/µL and stored at –20°C.

## *3.4. Allelotyping*

1. Microsatellite markers are obtained from ABI Prism Linkage Mapping Sets. These sets may vary in average interval between loci, e.g., 5 or 10 cM, and are optimized for high-throughput allelotyping.
2. Microsatellite markers consist of fluorescent primer pairs end-labeled with fluorochromes FAM, VIC, or NED that amplify dinucleotide repeat fragments.
3. Cytogenetic location of the markers can be determined by using the following websites: UCSC's Genome Browser (http://genome.ucsc.edu), NCBI's Map Viewer (http://www.ncbi.nlm.nic.gov/genome/guide) or Ensemble (http://www.ensembl.org) (*see* **Notes 5–7**).

### *3.4.1. PCR Reactions*

1. PCR reactions are carried out in a 10-µL volume utilizing 1 µL of I-PEP DNA, 6.6 µL of True Allele Premix, 2.54 µL dH$_2$O, and 0.66 µL of primer pair.
2. Thermocycler settings for amplification are as follows: 95°C for 12 min; 10 cycles of 94°C for 15 s, 55°C for 15 s, 72°C for 30 s; 30 cycles of 89°C for 15 s, 55°C for 15 s, 72°C for 30 s.
3. PCR product is then stored in a –20°C freezer until ready for the pooling process.

### *3.4.2. Pooling PCR Products*

1. PCR products for each sample can then be pooled by panel such that the fluorescent primer products are pooled together by volume in a ratio of 1:1:2 (HEX:VIC:NED).
2. A stock solution of formamide and Rox standard is then made using 12 µL of formamide and 0.55 µL Rox for each sample (this solution can be stored for 2 wk in a –20°C freezer).
3. Two µL of pooled fluorescently labeled PCR product is then added to 12.5 µL of the formamide-Rox stock solution.
4. The mixture is then heated to 95°C for 5 min, cooled on ice, and placed in the ABI310-Genetic Analyzer for data acquisition.

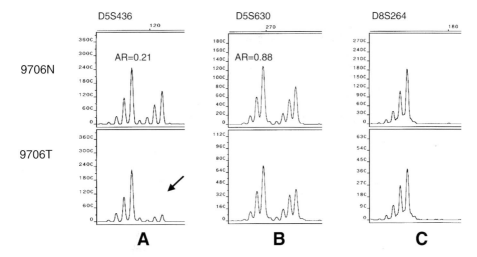

Fig. 3. Representative example electropherogram traces for normal (9706N) and tumour (9706T) genotypes. Allelic ratios (AR) were calculated as described and are shown at the top. **(A)** Arrow, LOH for marker D5S436. **(B)** marker D5S630 shows heterozygous without loss and **(C)** marker D6S264 noninformative NI or homozygous.

### 3.4.3. Data Acquisition and Analysis

1. Data are initially processed using Genescan 2.1 software.
2. Result files can then be imported into Genotyper (version 2.5) and the data tabulated according to allele size, allele peak height, and allele area.
3. An allelic ratio is calculated using allelic peak volume in a formula: $(N1/N2)/(T1/T2)$, where $N$ is normal allele and $T$ is tumor allele.
4. For each particular marker locus, LOH is assessed in the corresponding tumor sample if it is informative (heterozygous) in the normal DNA sample. LOH is present when the decrease in one allele is greater than 50% (normal/tumor allelic ratios <0.5 or >2.0).
5. Individual results are classified into homozygous, heterozygous with no loss, and heterozygous with loss. **Figure 3** demonstrates the electropherogram traces for various allelic patterns.
6. Software package developed by the University of Texas Southwestern Medical Center supports a clustering algorithm as well as basic calculations for percent LOH and fractional allelic loss ( *1,13*).

## 4. Notes

1. The choice of staining depends on the tissue type under study. It is important to note that hematoxylin may affect the DNA template such that subsequent PCR yield may be compromised (*14*).

2. In order to eliminate loosely bound contaminate tissue from the slide, a Prep Strip (Arcturus) can be first applied to the slide. If contamination is noted on the cap itself, a CapSure Pad (Arcturus) can be used to remove the unprocured tissue fragment(s).
3. DNA quality may vary in archival formalin fixed tissues. Both quality and quantity of DNA may require increased amount of dissected tissue prior to DNA digestion in order to provide adequate DNA template for subsequent reactions.
4. Digestion of microdissected tissues can be accomplished over different time courses. We prefer a 72-h protocol with the addition of 1 mL of proteinase K (20 mg/mL) after 24 h.
5. Both PCR reaction and pooling for allelotyping are labor-intensive. Robotics can increase both accuracy and speed in these processes.
6. ABI provides pooling panels for their fluorescent markers. The majority of times these can be followed, however, we do not recommend pooling more than eight markers especially when DNA isolated from formalin-fixed tissue is utilized. When making decisions about pooling, consideration should be given to the dye color, fragment size, and specific loci of interest.
7. Although ABI recommends that each capillary can be used for 100 reactions, we are able to obtain reliable electropherograms through 300 reactions.

## Acknowledgments

This study was supported by Ovarian Cancer Research Program grant no. DAMD17-99-1-9563 from the Department of Defense, Early Detection Research Network Grant CA86381 from the National Institute of Health, Department of Health and Human Services, Gillette Center for Women's Cancer, Adler Foundation, Inc., the Ovarian Cancer Research Fund, Inc., the Morse Family Fund, and the Natalie Pihl Fund, and DF/HCC Ovarian Cancer SPORE program.

## References

1. Girard, L., Zochbauer-Muller, S., Virmani, A. K., Gazdar, A. F., and Minna, J. D. (2000) Genome-wide allelotyping of lung cancer identifies new regions of allelic loss, differences between small cell lung cancer and non-small cell lung cancer, and loci clustering. *Cancer Res.* **60,** 4894–4906.
2. Shen, C. Y., Yu, J. C., Lo, Y. L., et al. (2000) Genome-wide search for loss of heterozygosity using laser capture microdissected tissue of breast carcinoma: an implication for mutator phenotype and breast cancer pathogenesis. *Cancer Res.* **60,** 3884–3892.
3 Lu, K. H., Weitzel, J. N., Kodali, S., Welch, W. R., Berkowitz, R. S., and Mok, S. C. (1997) A novel 4-cM minimally deleted region on chromosome 11p15.1 associated with high grade nonmucinous epithelial ovarian carcinomas. *Cancer Res.* **57,** 387–390.

4. Schorge, J. O., Muto, M. G., Welch, W. R., et al. (1998) Molecular evidence for multifocal papillary serous carcinoma of the peritoneum in patients with germline BRCA1 mutations. *J. Natl. Cancer Inst.* **90,** 841–845.
5. Vogelstein, B., Fearon, E. R., Kern, S. E., et al. (1989) Allelotype of colorectal carcinomas. *Science* **244,** 207–211.
6. Huang, L. W., Garrett, A. P., Schorge, J. O., et al. (2000) Distinct allelic loss patterns in papillary serous carcinoma of the peritoneum. *Am. J. Clin. Pathol.* **114,** 93–99.
7. Radford, D. M., Fair, K., Thompson, A. M., et al. (1993) Allelic loss on a chromosome 17 in ductal carcinoma in situ of the breast. *Cancer Res.* **53,** 2947–2949.
8. Fearon, E. R., Hamilton, S. R., and Vogelstein, B. (1987) Clonal analysis of human colorectal tumors. *Science* **238,** 193–197.
9. Bonner, R. F., Emmert-Buck, M., Cole, K., et al. (1997) Laser capture microdissection: molecular analysis of tissue. *Science* **278,** 1481, 1483.
10. Emmert-Buck, M. R., Bonner, R. F., Smith, P. D., et al. (1996) Laser capture microdissection. *Science* **274,** 998–1001.
11. Zhang, L., Cui, X., Schmitt, K., Hubert, R., Navidi, W., and Arnheim, N. (1992) Whole genome amplification from a single cell: implications for genetic analysis. *Proc. Natl. Acad. Sci. USA* **89,** 5847–5851.
12. Dietmaier, W., Hartmann, A., Wallinger, S., et al. (1999) Multiple mutation analyses in single tumor cells with improved whole genome amplification. *Am. J. Pathol.* **154,** 83–95.
13. Eisen, M. B., Spellman, P. T., Brown, P. O., and Botstein, D. (1998) Cluster analysis and display of genome-wide expression patterns. *Proc. Natl. Acad. Sci. USA* **95,** 14,863–14,868.
14. Burton, M. P., Schneider, B. G., Brown, R., Escamilla-Ponce, N., and Gulley, M. L. (1998) Comparison of histologic stains for use in PCR analysis of microdissected, paraffin-embedded tissues. *Biotechniques* **24,** 86–92.

# 7

## Microdissection for Detecting Genetic Aberrations in Early and Advanced Human Urinary Bladder Cancer

**Arndt Hartmann, Robert Stoehr, Peter J. Wild, Wolfgang Dietmaier, and Ruth Knuechel**

### Summary

Laser microdissection is an essential method for the investigation of the multistep carcinogenic process in the urinary bladder. Reliable detection of tumor-specific alterations which can be compromised by the presence of normal cells, requires microdissection of pure tumor cell populations (>80%) to detect loss of heterozygosity (LOH) by either fluorescence *in situ* hybridization (FISH) or sequence analysis. Multiple molecular methods need to be performed in the course of studying often-small lesions. This chapter describes in detail the use of laser microdissection, whole-genome amplification by improved primer extension preamplification (I-PEP)-polymerase chain reaction, and subsequent LOH, FISH, and sequencing analysis in the investigation of urothelial tumors and their precursor lesions. The combination of the described methods allows a wide spectrum of molecular investigations of tumor cells and helps to understand the fundamental alterations involved in urothelial carcinogenesis.

**Key Words:** Bladder cancer; whole-genome amplification; oncogene; tumor suppressor gene.

## 1. Introduction

Bladder cancer has two hallmarks that make it an interesting model for cancer research. Transitional cell carcinomas are mostly multifocal, recurring tumor entities, and are endoscopically accessible. Multifocal tumors, which are mostly derived from the urothelium (>95%), are found at the same stage or different stages of tumor progression *(1)*. Recent methods of fluorescence-guided tumor diagnosis allow better detection, especially of precancerous and flat, e.g., nonpapillary, urothelial lesions *(2,3)*. Because the hypothesis for molecular carcinogenesis of human bladder cancer was mainly derived from

From: *Methods in Molecular Biology, vol. 293: Laser Capture Microdissection: Methods and Protocols*
Edited by: G. I. Murray and S. Curran © Humana Press Inc., Totowa, NJ

manifest or even advanced tumor stages, the investigation of genetic aberrations in early tumor stages was considered to be interesting *(4)*.

The assessment of genetic changes in advanced invasive tumors is important and is described in several chapters of this book. Besides the marked inflammatory reaction in early stages of bladder cancer (pT1), the high stromal component in muscle-invasive tumors (pT2) has to be eliminated as well. In addition, precancerous lesions of the bladder typically show an abrupt transition from normal to dysplastic urothelium; even a pagetoid growth pattern may occur, i.e. intraepithelial low quantity areas of highly abnormal cells (**Fig. 1A** = CIS abrupt; **B** = CIS pagetoid). Therefore, laser microdissection provides new options for gaining pure cell populations in tumors with low amounts of tumor cells.

## 2. Materials

### 2.1. Laser Microdissection

1. PALM Robot-Microbeam (PALM, Wolfratshausen, Germany).
2. Stereo microscope DRC (Zeiss, Germany).
3. Microlance3 sterile needles (Braun, Germany).
4. Polyethylen membrane (1.35 µm; PALM).
5. Glass object slides (Menzel, Germany).
6. Poly-L-lysine (0.1%).
7. TESA® house and garden universal cover strip (Beiersdorf, Germany).
8. Xylene.
9. Ethanol (70%, 96%, 100%).
10. $H_2O$ (Millipore, Bedford, MA).
11. Methylene blue solution (1%).

### 2.2. DNA Isolation and Whole-Genome Amplification (WGA)

1. SpeedVac SC110 (Savant Instruments Inc., NY).
2. QIAamp DNA Mini Kit (Qiagen, Germany).
3. Proteinase K (Merck, Darmstadt, Germany).
4. Tween-20 (Merck).
5. 1X Expand HiFi buffer No. 3 (Roche, Mannheim, Germany).
6. Expand™ High Fidelity PCR System (Roche).
7. 15-mer random primer (NNNNNNNNNNNNNNN, synthesized by Proligo, France).
8. Thermocycler PTC100 (MJ Research, MA).

### 2.3. Fluorescence In Situ Hybridization

1. Carnoy's solution (3 parts methanol, 1 part acetic acid).
2. Maxi Prep DNA Isolation Kit (e.g. NucleoBond® PC 500 Kit, MN GmbH, Dueren, Germany).

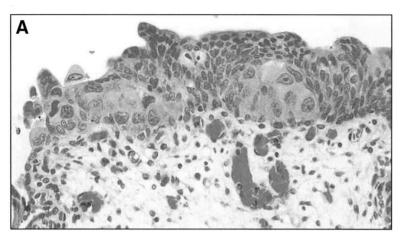

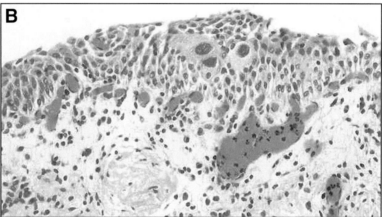

Fig. 1. Carcinoma *in situ* of the bladder. (**A**) Carcinoma *in situ* with distinct areas of atypical cells within normal urothelium. (**B**) Extreme variant with single atypical cells in normal urothelium, called pagetoid CIS. Photomicrographs (hematoxylin and eosin staining) are taken from the WHO classification on urothelial tumors *(5)*.

3. Standard nick translation kit (e.g., Roche).
4. Xylene.
5. Carlsberg solution: 0.1% proteinase K solution (20 mg/mL), 1.4% NaCl (5 M), in 10 mL Tris-HCl base (0.1 M), pH 7.2.
6. Ethanol (70%, 85%, 96%, 100%).
7. PBS-buffer: 8 g NaCl, 1.15 g $Na_2HPO_4$, 0.2 g KCl, 0.24 g $KH_2PO_4$, make up to 1 L with $H_2O$.
8. 20X standard saline citrate (SSC): 175.32 g NaCl, 88.24 g sodium citrate (dihydrate), make up to 1 L with $H_2O$, pH 5.3.

9. 2X SSC, pH 7.0–7.1.
10. 2X SSC; 0.3% NP40, pH 7.0–7.5.
11. C100T: citric acid with 0.5% Tween-20.
12. Proteinase K (0.8 µg/mL).
13. Methylene blue solution (1%).
14. RNase solution: 1 m$M$ Na-EDTA, 0.3% NP40 (10%), 1% RNase (10 mg/mL) in 10 mL of 10 m$M$ Tris-HCl, pH 7.5.
15. Zytospin centrifuge.

## 2.4. Loss of Heterozygosity Analysis

1. Oligonucleotide primers specific for the appropriate microsatellite markers.
2. PCR reagents including Taq polymerase, dNTPs, MgCl$_2$.
3. 10X TBE: 0.9 $M$ Tris base, 0.9 $M$ boric acid, 4% 0.5 $M$ EDTA, pH 8.0.
4. Gel mix for 6.7% polyacrylamide gel: 7.5 $M$ urea, 18% acrylamide (37%), 10.8% 10X TBE.
5. $N,N,N',N'$-Tetramethylethylenediamine (TEMED).
6. Ammonium persulfate (APS, 10%).
7. Polyacrylamide gel electrophoresis (PAGE) equipment.
8. Loading buffer: 10 mL formamide, 0.01 g xylene xyanol FF, 0.01 g bromophenol blue, 200 µL EDTA (0.5 $M$, pH 8.0).
9. Silver staining reagents: 10% ethanol, 1% nitric acid, 0.012 $M$ silver nitrate, 0.28 $M$ sodium carbonate including 0.1% formaldehyde, 10% acetic acid.
10. Whatman filter paper.
11. Gel dryer.

## 2.5. Gene Sequencing

1. Oligonucleotide primers specific for the individual gene of interest.
2. PCR reagents including Taq polymerase, dNTPs, MgCl$_2$.
3. Agarose gel equipment.
4. PEG Mix: 26% PEG8000, 0.6 $M$ sodium acetate, pH 5.5, 6.6 m$M$ magnesium chloride.
5. 100% ethanol, 70% ethanol, 3 $M$ sodium acetate, pH 4.6.
6. PRISM Ready Dye Terminator Cycle Sequencing Kit (Applied Biosystems GmbH, Weiterstadt, Germany).
7. Applied Biosystems 373 DNA sequencer.

## 3. Methods

The methods described below outline (1) laser microdissection of relevant tumor areas, (2) DNA isolation and whole-genome amplification, (3) fluorescence *in situ* hybridization (FISH), (4) LOH analysis, and (5) gene sequencing. Emphasis is put on laser microdissection of fresh-frozen material; FISH analysis has been described in frozen and wax-embedded material, respectively.

### 3.1. Laser Microdissection

With many precancerous lesions and early-stage tumors presenting as intraluminal formations, conventional manual microdissection under a stereomicroscope using sterile needles is the method of choice for the majority of cases. However, laser microdissection is appropriate if accurate segregation of low cell amounts is required, for example: (1) small areas of dysplasia or carcinoma *in situ*, (2) denuding carcinoma *in situ*, (3) intraepithelial migration of tumor cells, (4) invasive tumors with surrounding fibrosis and inflammation, and (5) small high-grade foci (G3) in well-differentiated carcinomas, e.g., at the invasive interface.

#### 3.1.1. Fresh-Frozen Material

1. Genomic DNA should be prepared from 5-μm frozen sections.
2. Pure tumor cell populations (>80% of tumor cells) are obtained using a PALM Robot-MicroBeam laser microdissection device.
3. For laser microdissection, the specimens are mounted on a $2 \times 3$ cm polyethylene membrane (1.35 μm), which is taped onto a supporting object slide (TESA house and garden universal cover strip) (*see* **Notes 1** and **2**).
4. Adhesion of tissues to the polyethylene membrane can be preserved by preparation of the membrane with 80 μL of poly-L-lysine before tissue sections are applied (*see* **Notes 3–5**).
5. After staining with 1% methylene blue, the selected tumor region is cleared of nontumorous cells (e.g., inflammatory cells) and microdissected precisely following its irregular shape.
6. With one single laser shot the entire microdissected membrane tissue area is then ejected and catapulted toward the collector, i.e., the lid of the reaction tube (*see* **Note 6**).
7. Catapulted specimens in the lid are morphologically well preserved, and can be documented with a digital camera.

#### 3.1.2. Wax-Embedded Tissue

1. Before microdissection, 5-μm histologic sections are dewaxed (1 h at 65°C) followed by incubation in xylene at room temperature for $2 \times 15$ min.
2. The sections are then rehydrated in ethanol (100%, 96%, 70%; at least $2 \times 5$ min each).
3. After incubation for 5 min in $H_2O$, specimens are stained with 1% methylene blue and used for microdissection.

### 3.2. DNA Isolation and Whole-Genome Amplification

The investigation of small premalignant lesions containing few aberrant cells often requires a preamplification of the DNA using whole-genome amplifica-

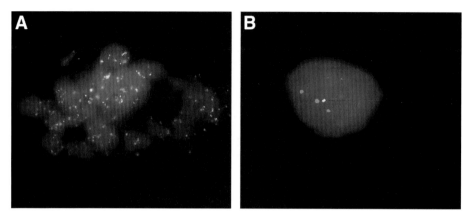

Fig. 2. FISH analysis of bladder tumor specimens. (**A**) Microdissected cells of a pT1G3 bladder tumor. Red signals: centromeric probe for chromosome 8; green signals: locus specific probe (BAC) for chromosome 8p11–12; blue staining: DAPI counterstaining of the nucleus. Tumor cells show a high-level amplification of chromosome 8. (**B**) Bladder tumor with aneuploidy of chromosome 8 and deletion of chromosomal region 8p11–12. Three red signals, but only two green signals, can be seen.

tion (WGA). The combination of microdissection and WGA allows multiple analysis of small cancer precursor lesions, providing additional information about the first chromosomal events in carcinogenesis. Details of the method used, i.e., improved primer extension preamplification polymerase chain reaction (I-PEP-PCR) including data on minimal cell numbers required are described in detail in Chapter 8 of this book.

### 3.3. FISH

The detection of chromosomal losses is of great importance in the molecular analysis of human malignancies. Regions frequently affected by deletions might contain important tumor suppressor genes (TSGs), which may play an essential role in carcinogenesis. Comparative genomic hybridization (CGH) has low resolution and gives only an overview of the chromosomal changes occurring in tumor cells. Multicolor FISH (*see* **Fig. 2**) allows the investigation of specific deletions using BAC clones and (for chromosome enumeration) commercially available centomere probes. The reliable detection of deletions requires microdissection from thick (>20 μm) sections to avoid artefacts because of sectioning of nuclei and loss of chromosomal material. The method described here is for formalin-fixed and wax-embedded tissue, but also works in tissue microdissected from frozen sections or in touch preparations of fresh tumor tissue. For these cases pretreatment can be omitted from the protocol, and digestion is performed by using C100T incubation for approx 90 min. BAC clones can be selected from several commercially available BAC libraries and labeled by nick translation. Several pretreatment kits are available from suppliers of FISH probes.

### 3.3.1. Probe Preparation

1. Centromeric probes already labeled with fluorescent dyes (e.g., FITC, Texas Red) are commercially available from several suppliers (e.g., Vysis, Abbott Laboratories, Berkshire, UK) and are ready to use following the manufacturer's instructions.
2. BAC clones can be used for the investigation of specific chromosomal regions.
3. After isolation from bacterial culture, digoxigenin-11-dUTP labeling of the BAC using standard nick translation is necessary.
4. The BAC can then be visualized using fluorescent immunostaining.

### 3.3.2. Pretreatment of Tissue Sections

1. Incubate the slides for 25 min at 72°C.
2. Afterward, remove the wax from the tissue.
3. Incubate the slides $2 \times 10$ min in fresh xylene at room temperature (RT) and, for dehydration, a few minutes in 100% ethanol.
4. After drying at RT, slides are put in freshly prepared 2X SSC for 2 min at 73°C.
5. After washing for 5 min in PBS buffer at RT, incubate slide in proteinase K solution for 15 min at 39°C.
6. Wash the slides in for 5 min in PBS buffer at RT followed by a dehydration of the tissue in graded ethanol (70%, 85%, 100%, each for 1 min at RT).
7. Slides are dried at RT.

### 3.3.3. Denaturing of Probe and Tissue DNA

1. All steps described here should be performed in the dark.
2. Denature the probe(s) for 5 min at 73°C in a water bath.
3. Apply the probe to the tissue section and cover the tissue with a cover slip.
4. Seal the cover slip with a vulcanized rubber and heat the slide for 9 min at 96°C.
5. Incubate the slide overnight at 37°C in a humid chamber.

### 3.3.4. Washing the Slides

1. Incubate the slides for 30 s in 2X SSC/0.3% NP40 at RT to detach cover slip.
2. Carefully remove cover slip.
3. Wash slides in 2X SSC/0.3% NP40 for 2 min at 73°C followed by an incubation for 1 min in deionized water at RT.
4. Carefully dry the slides and add 10 µL DAPI in Vectashield mounting medium (Vector Laboratories, Burlingame, CA).
5. Store the slide in the dark at 2–8°C (*see* **Note 7**).

### 3.3.5. Microdissected Formalin-Fixed, Wax-Embedded Tissue

1. After removal of wax, incubate slides in 80% ethanol, methylene blue, and water, each for approx 15 s at RT.
2. Carefully microdissect the cells of interest and immediately transfer the tissue into 50-µL Carlsberg solution (on ice).
3. Incubate the cells for 30 min at 37°C to digest the cytoplasm of the cells.
4. Add 50 µL RNase to the cells and incubate another 15 min at RT.
5. Apply the cell suspension to a standard cytospin centrifuge.
6. Air-dry the cells for 5 min.
7. Incubate the cells for 10 min in Carnoy's solution at RT for fixation.
8. Air-dry the cells and store the slides at –20°C/–80°C until use.

### *3.4. LOH Analysis*

LOH analysis using microsatellite markers (highly polymorphic repetitive DNA sequences) located within specific chromosomal segments is a different method that allows a detailed deletion analysis of a chromosomal region of interest. The investigation of a large number of microsatellite markers of a specific chromosomal region allows the narrowing of a minimally deleted region facilitating the detection of a putative TSG.

The detection of chromosomal losses in malignant tumors gives new insights into the correlation between chromosomal alterations and tumor progression and prognosis. The investigation of premalignant tissue offers the chance to detect the deletions occurring early in the carcinogenic process. The following description provides a short protocol for the rapid establishment of a microsatellite analysis.

### 3.4.1. Selection of Microsatellite Markers

1. The first prerequisite for the LOH analysis is the selection of a panel of microsatellite markers localized within the chromosomal region of interest (*see* **Note 8**).
2. Most available microsatellite markers with amplification primers and information about size of amplicon and literature data can be found at the Genome Database (www.gdb.org).

### 3.4.2. Thermogradient PCR

1. It is advisable to determine optimal PCR conditions before starting LOH analysis.
2. The establishment of PCR amplification conditions can be done using a gradient-thermocycler (*see* **Notes 9–11**).
3. A standard PCR protocol (25 µL reaction volume containing 0.2 mmol/L dNTP, 0.3 µmol/L of each primer (forward and reverse primer), 1.5 mmol/L MgCl$_2$, 0.5 U DNA Taq polymerase, 50–100 ng template DNA) should be used.
4. The following cycle conditions are satisfactory in most instances: 3 min preheating at 95°C, 35 cycles of 1 min at 95°C for denaturation—1 min at specific tem-

perature range (gradient) for primer annealing—1 min at 72°C for primer elongation, 10 min at 72°C for final elongation, cooling to 10°C for termination of PCR
reaction.
5. PCR conditions (annealing temperature, MgCl$_2$ concentration) resulting in a specific PCR product should be used for further analysis.

### 3.4.3. Microsatellite Amplification

1. PCR amplification of the microsatellite markers chosen for the LOH analysis
   should be performed according to optimized PCR conditions (*see* **Subheading 3.4.2.**).
2. If a gradient-thermocycler is not available, PCR protocol and cycle conditions
   shown above should be used.
3. Optimal annealing temperature should be assessed by several test PCR reactions
   with varying temperatures and, if necessary, varying MgCl$_2$ concentrations
   (0.25 *M*–2.5 *M*).

### 3.4.4. Polyacrylamide Gel Electrophoresis

1. For separation of PCR products standard polyacrymide gel electrophoresis
   (PAGE) equipment (e.g., Sequi-Gen Sequencing Cell, Bio-Rad, Hercules, CA)
   can be used.
2. For a standard 32.5 × 38 cm denaturing gel (6.7% PAA, 1 mm thickness) the
   following protocol is recommended: add to 80 mL Gel-Mix 1 mL APS and
   100 µL TEMED to start polymerization of the gel; complete polymerization process takes 60 min.
3. Before sample loading mix equal amounts of PCR reaction and loading buffer
   and denature the sample for 5 min at 95°C.
4. After denaturation place the sample on ice immediately.
5. Electrophoresis should be started after preheating to 40–50°C for about 5 min
   (2500 V).
6. PAGE should be performed at a temperature of 55°C for best results.

### 3.4.5. Silver Staining

1. After PAGE, silver staining of the gel is performed to detect PCR products *(6)*.
2. The following procedure is recommended:
   a. 5 min 10% ethanol.
   b. 3 min 1% nitric acid.
   c. Rinse with water three times.
   d. 0.012 *M* silver nitrate for at least 20 min (up to 1 h) in complete darkness.
   e. Rinse with water several times.
   f. Incubate gel with 0.28 *M* sodium carbonate with 0.1% formaldehyde several
      times until PCR products become visible (*see* **Fig. 1**).
   g. Stop reaction with 10% acetic acid for 2 min.
   h. Rinse with water several times.
3. After staining, transfer gel to Whatman filter paper and dry gel.

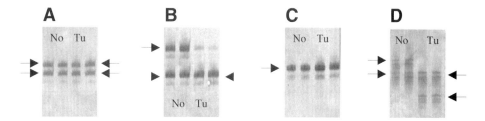

Fig. 3. Microsatellite analysis. No, normal urothelium; Tu, papillary bladder tumor. (A–C) microsatellite marker D8S591 on chromosome 8p; (D) microsatellite marker D8S255 on chromosome 8p. (A) Informative patient sample. Two alleles can be distinguished (see arrows), no deletion in the tumor. (B) Informative patient sample, loss of one allele in the tumor → chromosomal deletion (LOH). (C) Patient sample is not informative, the two alleles cannot be distinguished, a normal to tumor comparison is not possible. (D) Informative patient, band-shift in the tumor → microsatellite instability (MSI).

### 3.4.6. Evaluation of LOH Analysis

1. Direct comparison between normal tissue and premalignant or tumor tissue reveals chromosomal alterations (*see* **Fig. 3**).
2. Informative cases are scored as allelic loss (LOH) when intensity of the signal for the tumor allele is decreased to at least 50% relative to the matched normal allele (**Fig. 1B**).
3. To prevent errors because of preferential amplification of one allele during the PCR reaction, all LOH analyses should be run in duplicate. Following independent PCR reactions for the DNA aliquots detected LOH should be verified in a second, independently performed PCR reaction.

## 3.5. Gene Sequencing

Sequencing analyses have revealed a large number of mutations in tumor-related genes. Activating mutations in oncogenes (e.g., FGFR3) or knockout of tumor suppressor genes (e.g., *p53*) by single-point mutations, deletions, insertions, or chromosomal rearrangement are well studied in most human malignancies. Both aberrant constitutive activation of oncogenes and loss of function of TSGs caused by gene mutations are important genetic alterations in carcinogenesis. Sequence analysis is therefore an essential part of tumor characterization. The method described below is a short protocol for automated genomic sequencing (using DNA from snap-frozen or formalin-fixed, wax-embedded tissue). The described protocol is also adapted for DNA after WGA (*see* **Notes 12–14**).

## 3.5.1. PCR Amplification and PEG Precipitation

1. For amplification of the target region for sequence analysis, PCR conditions described above are recommended (*see* **Subheading 3.4.2.**).
2. Before the PCR product can be used for sequencing, PEG (polyethylenglycol) precipitation is necessary for purification.
   a. Add equal amount of PEG Mix to PCR reaction; mix well.
   b. Incubate at room temperature for 20 min.
   c. Centrifuge for 30 min at 15,000*g*.
   d. Remove supernatant carefully (do not disturb pellet).
   e. Add 100 µL 100% ethanol without mixing.
   f. Centrifuge for 10 min at 15,000*g*.
   g. Remove supernatant.
   h. Dry pellet (SpeedVac, 5 min).
   i. Resuspend DNA pellet in 20 µL water and mix well.
3. After PEG precipitation a DNA aliquot (3 µL) should be used for agarose gel electrophoresis to control precipitation and to assess DNA concentration.

## 3.5.2. Sequencing Reaction

1. For sequence analysis, a PRISM Ready Dye Terminator Cycle Sequencing Kit (Applied Biosystems GmbH, Weiterstadt, Germany) is advisable using the manufacturer's protocol.
2. Before sequencing reactions are analyzed (Applied Biosystems 373 or 377 sequencer), sequencing products must be purified using an ethanol precipitation.
   a. Add 55 µL 100% ethanol and 2 µL sodium acetate to sequencing reaction (20 µL volume); mix well.
   b. Incubate at room temperature for 20 min in complete darkness.
   c. Centrifuge for 30 min at 15,000*g*.
   d. Remove supernatant.
   e. Dry pellet for 5 min at room temperature (cup upside down).
   f. Add 500 µL 70% ethanol without mixing.
   g. Centrifuge for 10 min at 15,000*g*.
   h. Remove supernatant.

## 3.6. Comparison of Methods Used

The most important prerequisite to achieving reliable results from the application of the methods described above is the analysis of a pure cell population without contamination. The use of laser microdissection ensures a purity of >90% of the microdissected cells and is essential for the investigation of small preneoplastic lesions, flat malignancies (e.g., CIS), or invasive tumors. An advantage of the methods described is the applicability to both snap-frozen and formalin-fixed, wax-embedded tissue, allowing the investigation of archival material. In addition, the methods complement one another. FISH analysis is ideal for detecting chromosomal deletions, losses of whole chromosomes, and amplifications of specific chromosomal regions. Gene-specific probes enable the investigation of tumor tissue for high-level amplification of known oncogenes or deletions of tumor suppressor genes. The use of several gene-specific and centromer-specific probes labeled with different colors (multicolor FISH) allows the analysis of interrelated members of a whole signalling pathway. However, FISH needs a standardized protocol and a very experienced observer. Weak hybridization, tissue damage, destroyed or overlapping nuclei, or splitting of signals could complicate scoring of the signals.

LOH analysis is one of the most frequently used methods for detection of chromosomal alterations in cancer research. Direct comparative analysis of polymorphic microsatellite markers from normal and tumor tissue shows chromosomal deletions and microsatellite instability (MSI), genomic changes found in most malignancies. Using LOH analysis, amplification of specific genes can not be determined. This restricts the field of application to detection of allelic losses or MSI. Analysis of LOH is an essential tool for definition of a minimal deleted chromosomal region in a tumor and therefore is excellent preparatory work for FISH analysis. Both methods can also be used for verification of each other. To achieve reliable results from microsatellite analysis, minimal contamination of the tumor cells with stromal or inflammatory cells, for example, is acceptable. Using laser-assisted microdissection, it should be very easy to achieve a clean cell population for analysis.

Another crucial point is the absolutely necessary verification of the results from LOH analysis. Poor DNA quality and a low DNA quantity could cause a preferential PCR amplification of only one allele, mimicking an allelic loss. This leads to a misinterpretation of the result and the detection of a "false-positive" LOH. In our experience at least 10–20 cells from frozen sections and at least 100 cells from wax-embedded tissue should be used for PCR analysis to avoid preferential amplification. Our own experience showed a frequency of this PCR failure in about 5–10% of all LOH analysis. This is a particular problem when trying to define a minimally deleted region on a chromosome. To avoid this source of error each identified deletion or MSI should be verified by a second independently performed PCR reaction.

Gene sequencing is an ideal complement to the methods described above. Mutation analysis of a (putative) TSG (e.g., *p53*) that is located within a chromosomal region showing allelic deletion in LOH or FISH analysis could reveal specific mutations within the coding region or splice sites of this gene, causing a complete knockout. Sequence analysis of known oncogenes which showed no significant alterations by FISH analysis (e.g., no high-level amplification) might discover activating mutations.

Genomic sequencing can be used for both snap-frozen and formalin-fixed, wax-embedded tissue. It is a more sensitive method than LOH analysis. Again it is very important to verify all found mutations with a second, independent generated PCR amplicon. To confirm the results, sequencing of both DNA strands is highly advisable. Often a PCR failure in one strand mimics a mutation that is not visible in the antisense strand. Also the quality and quantity of the DNA are important factors. Using formalin-fixed tissue, often only short PCR amplicons can be generated (max. approx <500 bp) caused by degradation of the DNA during the fixation. The problem of low DNA quantity can be solved using WGA before sequencing analysis. It has been shown *(7)* that there is only a minimal risk of incorrect amplification by PCR during WGA. However, it is advisable to split the DNA in half before WGA to provide the possibility of repeating the whole process (WGA and gene sequencing) for verification.

Taken together, all three described methods are ideal for the analysis of both snap-frozen and formalin-fixed, wax-embedded tissue. The combined application allows the rapid detection of the most important alterations in tumors and provides the basis for more intensive investigations.

## 4. Notes

1. The polyethylene membrane should be cut in pieces of $2 \times 3$ cm with a surgical blade between two sheets of paper on a metal surface. After putting the object slide into 100% ethanol, the polyethylene membrane adheres to the object slide easily by capillary effects. The membrane should be dry and as flat as possible before fixation with a cover strip ($0.5 \times 2$ cm) at each ending.
2. For wax-embedded tissue, fixation of the polyethylene membrane with the cover strip (TESA house and garden universal) is indispensable before removal of the wax with xylene.
3. Poly-L-lysine must always be handled under ventilated conditions.
4. The slides prepared with a polyethylene membrane and poly-L-lysine should be completely dry before tissue sections are applied.
5. To avoid unintentional fixation of the polyethylene membrane, poly-L-lysine should not contaminate the space between the membrane and the object slide.
6. At least 200 cells are prepared from each specimen to prevent preferential monoallelic amplification.
7. The scoring of FISH signals can be hampered by a strong and diffuse background staining. In this case agitation of the slides during the washing steps decreases the background.
8. The microsatellite markers used for LOH analysis should show a high rate of heterozygosity. This minimizes the problem of too many noninformative samples during deletion analysis.
9. For the initial optimization of PCR conditions use DNA from cultured cells or blood to avoid wasting tumor DNA.

10. If only weak PCR signals are visible after PAGE, PCR cycle number can be increased up to 50 rounds to achieve a larger amount of PCR amplicons.
11. PCR amplification from formalin-fixed, wax-embedded tissue is sometimes unsuccessful. Initial PCR preheating for 5–10 min at 95°C often improves results.
12. Evaluation of the sequencing results should be done very carefully. Often only the graphs indicate the presence of a mutation that is not visible in the shown sequence (≥ high degree of contamination with cells from normal tissue).
13. A nested PCR can be used to obtain reliable sequencing results if only a small amount of DNA is available for analysis.
14. The results from the analysis described above are ideal preparatory work for the application of additional techniques. Mutations detected by genomic sequencing can be used for allele-specific PCR analysis *(8)*. Immunohistochemical staining of tissue specimens allows direct analysis of an oncogene or tumor suppressor gene on protein level in close correlation to mutational status.

## References

1. Epstein, J. I., Amin, M. B., Reuter, V. R., and Mostofi, F. K. (1998) The World Health Organization/International Society of Urological Pathology consensus classification of urothelial (transitional cell) neoplasms of the urinary bladder. Bladder Consensus Conference Committee. *Am. J. Surg. Pathol.* **22,** 1435–1448.
2. Filbeck, T., Pichlmeier, U., Knuechel, R., Wieland, W. F., and Roessler, W. (2002) Clinically relevant improvement of recurrence-free survival with 5-aminolevulinic acid induced fluorescence diagnosis in patients with superficial bladder tumors. *J. Urol.* **168,** 67–71.
3. Zaak, D., Kriegmair, M., Stepp, H., et al. (2001) Endoscopic detection of transitional cell carcinoma with 5-aminolevulinic acid: results of 1012 fluorescence endoscopies. *Urology* **57,** 690–694.
4. Veltman, J. A., van Weert, I, Aubele, M., et al. (2001) Specific steps in aneuploidization correlate with loss of heterozygosity of 9p21, 17p13 and 18q21 in the progression of pre-malignant laryngeal lesions. *Int. J. Cancer* **91,** 193–199.
5. Mostofi F. K, Davis C. J., and Sesterhenn, I. A. (1999) Histological typing of urinary bladder tumors. WHO, International Histological Classification of Tumors. Springer, New York.
6. Schlegel, J., Stumm, G., Scherthan, H., et al. (1995) Comparative genomic in situ hybridization of colon carcinomas with replication error. *Cancer Res.* **55,** 6002–6005.
7. Dietmaier, W., Hartmann, A., Wallinger, S., et al. (1999) Multiple mutation analyses in single tumor cells with improved whole genome amplification. *Am. J. Pathol.* **154,** 83–95.
8. Stoehr, R., Knuechel, R., Boecker, J., et al. (2002) Histologic-genetic mapping by allele-specific PCR reveals intraurothelial spread of p53 mutant tumor clones. *Lab Invest.* **82,** 1553–1561.

# 8

# Laser Microdissection for Microsatellite Analysis in Colon and Breast Cancer

Peter J. Wild, Robert Stoehr, Ruth Knuechel, Arndt Hartmann, and Wolfgang Dietmaier

## Summary

Microsatellite analysis is a frequently used method for detection of chromosomal deletions by loss of heterozygosity studies and for detection of microsatellite instability. For reliable microsatellite analyses, a tumor cell content of at least 80% is required. Therefore, laser microdissection is an important prerequisite for those studies, allowing the contamination-free isolation of morphologically defined pure tumor cell populations. The combination of exact microdissection and subsequent whole-genome amplification by improved primer extension preamplification polymerase chain reaction (I-PEP-PCR) facilitates the analysis of multiple microsatellite loci in small tumor samples. This is especially important for the investigation of malignant tumors with low tumor cellularity. This chapter describes in detail the use of whole genome amplification by I-PEP PCR and microsatellite analysis in laser microdissected specimens of colon and breast cancer.

**Key Words:** Laser microdissection; whole-genome amplification; microsatellite instability; breast cancer; colon cancer.

## 1. Introduction

Molecular analysis in tumor pathology should be performed in precisely determined areas with homogenous tumor cells. Analysis of tumor-specific genetic alterations can be compromised by the presence of normal cells. Thus, contamination by stromal and inflammatory cells should be minimal. For reliable microsatellite analyses and detection of chromosomal deletions by loss of heterozygosity (LOH) studies, a tumor cell content of at least 80% is required. In order to obtain such high-pure tumor DNA from small tissue samples or carcinomas with a prominent desmoplastic stroma, e.g., breast cancer or colon

From: *Methods in Molecular Biology, vol. 293: Laser Capture Microdissection: Methods and Protocols*
Edited by: G. I. Murray and S. Curran © Humana Press Inc., Totowa, NJ

cancer, accurate tissue microdissection is important in molecular pathology. The spectrum of techniques ranges from manual microdissection or the use of a micromanipulator to laser-assisted microdissection systems, allowing the contamination-free isolation of morphologically defined pure cell populations from histologically heterogenous tumor areas under microscopic control.

Manifestations of a mutator phenotype in cancer are at least two types of genetic instability, microsatellite instability (MSI) and chromosomal instability (CIN). In colorectal cancer, a set of five microsatellite markers was defined to determine microsatellite instability *(1,2)*. Tumors showing instabilities in at least two of the five examined loci ( 40% unstable loci) are defined as MSI-high (MSI-H). Tumors with only one unstable marker (20% unstable loci) are declared as MSI-low (MSI-L). In colorectal cancer, MSI is independently predictive of a relatively favorable patient outcome, reduced likelihood of metastases, and better survival following adjuvant chemotherapy with fluorouracil-based regimens.

In contrast to the rigorous diagnostic guidelines established in the hereditary nonpolyposis colorectal cancer (HNPCC) model, there are no well-defined criteria for the assesment of MSI in breast and other noncolorectal cancers. This phenomenon has been investigated in numerous breast cancer studies, but large disparities in the percentage of microsatellite unstable tumors are reported, ranging from 0% to 100%. Heterogeneous patient cohorts, suboptimal choices of MSI markers, and ambiguous diagnostic criteria for MSI likely contribute to such disparities. There is growing evidence that MSI in breast cancer patients may represent a type of genetic instability different from that seen in colorectal cancer. MSI in breast cancer has typically been found in tri- and tetra-nucleotide-repeats, whereas diagnosis of HNPCC-associated replication errors is most easily determined by examining a panel of simple mono- and dinucleotide repeat sequences. MSI-positive breast cancers seem to show normal expression of the DNA mismatch repair proteins hMSH2 and hMLH1, and lack of MSI in breast cancer occurring in HNPCC families has led to the exclusion of breast cancer as an integral tumor of the HNPCC syndrome. A similar pattern of MSI, termed EMAST (elevated MSI at selected tetranucleotide markers) has been detected in non-small-cell lung cancer, bladder cancer, and skin cancer. This type of MSI was associated with p53 mutations and displayed a phenotype inconsistent with defects in the MMR pathway *(3)*.

## 2. Materials

### 2.1. DNA Isolation and I-PEP Whole-Genome Amplification

1. QIAamp DNA Mini Kit (Qiagen, Germany).
2. Proteinase K (Merck, Darmstadt, Germany).
3. Tween-20 (Merck).

4. 1X Expand HiFi buffer No. 3 (Roche, Mannheim, Germany).
5. Expand™ High Fidelity PCR System (Roche).
6. 15-mer random primer (NNNNNNNNNNNNNNN, synthesized by Metabion, Germany).
7. Thermocycler PTC100 (MJ Research, Boston, MA).
8. SpeedVac SC110 (Savant, Farmingdale, NY).

## 2.2. Microsatellite PCR

1. ABI310 capillary electrophoresis system (Applied Biosystems, Weiterstadt, Germany).
2. Thermocycler PTC100 (MJ Research).
3. Taq-DNA-Polymerase (Roche).
4. Oligonucleotide microsatellite primers (synthesized by Applied Biosystems; **Table 1**).
5. 10X PCR buffer containing $MgCl_2$ (Roche).
6. dNTP mix (10 m$M$; Roche).

## 3. Methods

The methods described below outline (1) isolation of genomic DNA with subsequent whole-genome amplification and (2) fluorescence-based detection of MSI by capillary electrophoresis using ABI sequencing equipment. Laser microdissection with laser pressure catapulting or with manual tissue transfer (**Fig. 1**) and polyacrylamide gel electrophoresis (PAGE) with silver nitrate staining (**Table 1**) are described in detail in Chapter 7.

## 3.1. DNA Isolation and Whole-Genome Amplification

The investigation of laser microdissected lesions resulting in relatively low tumor cell numbers requires preamplification of the DNA using whole-genome amplification (WGA) when multiple microsatellite markers are to be analyzed.

### 3.1.1. Isolation of Approximately 50–1000 Cells

Microdissected tissue samples are lysed by adding a mixture of 9 µL 1X Expand HiFi buffer No. 3 containing 1.5% Tween-20 and 1 µL Proteinase K (20 µg). After vortexing, a 4-h incubation at 50°C followed by a final 15-min incubation at 94°C is performed. This lysate can be directly used for I-PEP PCR.

### 3.1.2. Isolation of More Than 1000 Cells

If more than 1000 cells are prepared, a purification step using a DNA purification kit (e.g., QIAamp DNA Mini Kit) is recommended according to the manufacturer's specifications. Samples are transferred into 1.5-mL tubes, provided with the QIAamp DNA Mini Kit. To increase DNA yield, elution with 2 × 100 µL of 70°C preheated water should be performed (*see* **Note 1**).

## Table 1
## Characteristics of Oligonucleotide Microsatellite Primers

| Repeat-type | Marker | Primer sequence (5' to 3') | Annealing temperature | PAGE[a] MSI | PAGE[a] LOH |
|---|---|---|---|---|---|
| Mononucleotide | BAT-25 | TCG CCT CCA AGA ATG TAA GT<br>TCT GCA TTT TAA CTA TGG CTC | 58°C |  | NA[b] |
| | BAT-26 | TGA CTA CTT TTG ACT TCA GCC<br>AAC CAT TCA ACA TTT TTA ACC C | 58°C | 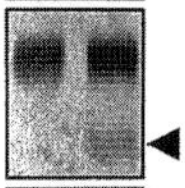 | NA |
| Dinucleotide | D5S346 (APC) | ACT CAC TCT AGT GAT AAA TCG<br>AGC AGA TAA GAC AGT ATT ACT AGT T | 55°C | 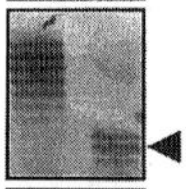 | 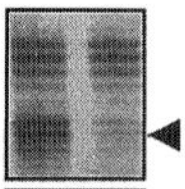 |
| | D17S250 (Mfd15) | GGA AGA ATC AAA TAG ACA AT<br>GCT GGC CAT ATA TAT ATT TAA ACC | 52°C | 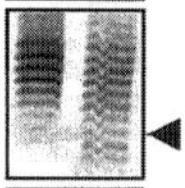 | 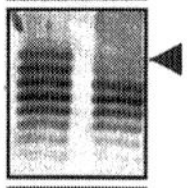 |
| | D2S123 | AAA CAG GAT GCC TGC CTT TA<br>GGA CTT TCC ACC TAT GGG AC | 60°C |  | 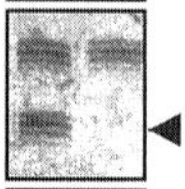 |
| Tetranucleotide | Mycl1 | TGG CGA GAC TCC ATC AAA G<br>CTT TTT AAG CTG CAA CAA TTT C | 53°C | 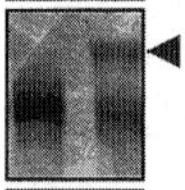 | 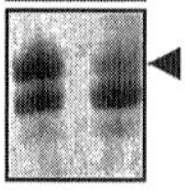 |
| | D18S51 | GAG CCA TGT TCA TGC CAC TG<br>CAA ACC CGA CTA CCA GCA AC | 58°C | 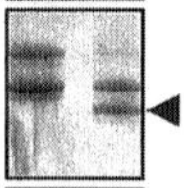 | NA |
| | D1S549 | CAA AGA GGA CAT GTG TTT GTG<br>TAC CAG CAA TGG GTA GTA TGG | 58°C | 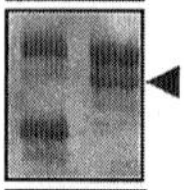 | 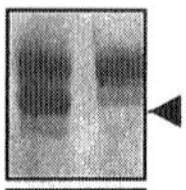 |
| | D2S443 | GAG AGG GCA AGA CTT GGA AG<br>ATG GAA GAG CGT TCT AAA ACA | 58°C | 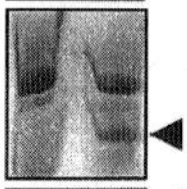 | 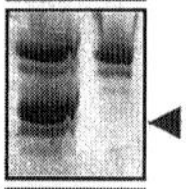 |
| | D21S1436 | AGG AAA GAG AAA GAA AGG AAG G<br>TAT ATG ATG AAA GTA TAT TGG GGG | 58°C | 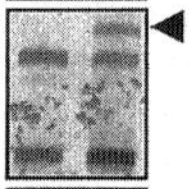 | 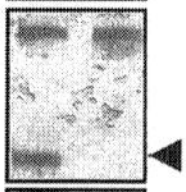 |
| Pentanucleotide | TP53.Alu | TCG AGG AGG TTG CAG TAA GCG GA<br>AAC AGC TCC TTT AAT GGC AG | 60°C | 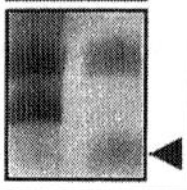 | 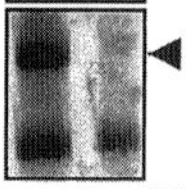 |

[a]PAGE, polyacrylamide gel electrophoresis; MSI, examples for microsatellite instability (left column); LOH, examples for loss of heterozygosity (right column); normal tissue left, tumor right.

[b]NA, not available; t, relevant allele.

---

Fig. 1. (*opposite page*) Laser microdissection of an intraductal comedocarcinoma of the breast with laser pressure catapulting (left column) and an adenocarcinoma of the colon at the invasive interface with manual tissue transfer (right column).

**Laser microdissection &
laser pressure catapulting
(LPC)**
specimen mounted on a
polyethylene membrane

**Laser microdissection &
manual tissue transfer**

specimen mounted on
a regular glass slide

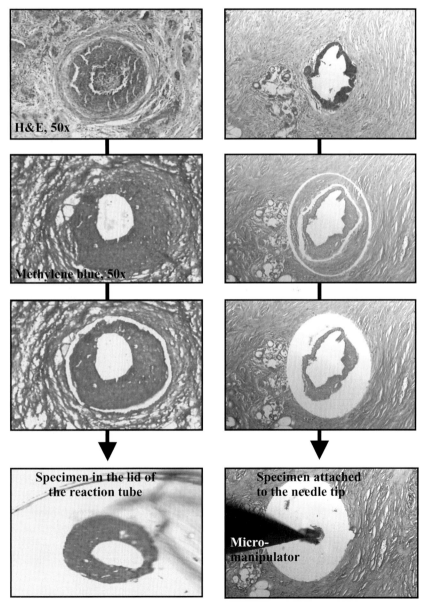

Each elution step includes a 5-min incubation of the QIAamp spin column with the preheated water at 70°C before centrifugation. A 10 µL aliquot can then be used for the whole-genome preamplification PCR.

### 3.1.3. Improved Primer-Extension-Preamplification (I-PEP) PCR

1. WGA by I-PEP *(4)* provides template DNA from laser microdissected material for many PCR amplifications (e.g., microsatellite analyses, DNA sequencing). However, successful I-PEP depends on high quality of starting DNA. In case of ineffective direct PCR amplification without preamplification, negative results after I-PEP are also to be expected.

2. For the initial evaluation of DNA quality, a standard control PCR (e.g., with D2S123 marker) and DNA from non-laser microdissected tissue should be performed prior to I-PEP (*see* **Notes 2** and **3**).

3. I-PEP PCR can be set up by combining the following components:
   a. 3.4 µL 280 µ*M* I-PEP primer (16 µ*M* final).
   b. 6 µL 25 m*M* MgCl$_2$ (2.5 m*M* final).
   c. 3 µL 1 mg/mL gelatine (0.05 mg/mL final).
   d. 6 µL 10X PCR-buffer No. 3 (Expand HiFi buffer; Roche).
   e. 0.6 µL 10 m*M* dNTP-Mix (0.1 m*M* final).
   f. 29.6 µL H$_2$O.
   g. 1.4 µL 3.6 U/µL Expand HiFi-polymerase. (0.06 U/µL final).

4. 50 µL I-PEP reaction mix is added to 10 µL lysed cells and used for WGA performed by the following PCR scheme:

   | | |
   |---|---|
   | Initial denaturation step: | 2 min at 94°C |
   | 50 amplification cycles: | 1 min at 94°C |
   | | 2 min at 28°C |
   | | ramp of 0.1°C/s to 55°C |
   | | 4 min at 55°C |
   | | 30 s at 68°C |
   | Final elongation step: | 8 min at 68°C |

5. Alternatively, a faster amplification avoiding the 4-min incubation step at 55°C can be achieved by a protocol with only 20 cyclic ramps from 28 to 55°C followed by 30 cycles with an annealing temperature of 60°C. This temperature, which is relatively high for 12-mer random primers, can be used because the primers are elongated during the preceeding 20 ramp cycles. The complete PCR scheme is as follows:

   | | |
   |---|---|
   | Initial denaturation step: | 4 min at 94°C |
   | 20 amplification cycles: | 30 s at 94°C |
   | (ramps) | 1 min at 28°C |
   | | ramp of 0.1°C/s to 55°C |
   | | 45 s at 55°C |
   | 30 amplification cycles: | 30 s at 94°C |

|                    |                  |
|--------------------|------------------|
|                    | 45 s at 60°C     |
|                    | 1 min at 72°C    |
| Final elongation step: | 8 min at 68°C |

## *3.2. Detection of Microsatellite Instability After I-PEP*

1. For most accurate results, fluorescence-based microsatellite analysis using capillary electrophoresis is recommended (*see* **Notes 4** and **5**).
2. Only one primer of each primer pair has to be labeled by a fluorescence dye (e.g., BAT25 Forward-FAM, BAT26 Forward-NED, D5S346 Reverse-HEX, D17S250 Forward-NED, D2S123 Forward-FAM).
3. A fluorescence-based multiplex microsatellite PCR kit is commercially available (HNPCC-Microsatellite Instability Test, Roche) with which all five first-choice reference markers can be tested simultaneously.
4. However, separately performed PCR amplifications after I-PEP are recommended due to higher amplification efficiency in cases where few cells (<500 cells) are microdissected.
5. Preamplified DNA can be used directly for subsequent PCR analyses such as microsatellite analysis or DNA sequencing. The most favorable technique for MSI detection is the combined use of fluorescence-labeled amplification primers and capillary electrophoresis with ABI sequencing equipment (e.g., ABI310, Applied Biosystems).
6. 1/30 aliquots (i.e., 2 µL) of preamplified DNA should be taken for each microsatellite amplification. However, in some cases the amount of template DNA should be adapted to the amount of microdissected tissue.
7. Routinely, the following PCR protocol is recommended:

PCR Master Mix for 1 reaction

|                     | µL   | Final conc. | Stock solution |
|---------------------|------|-------------|----------------|
| $H_2O$              | 20.2 |             |                |
| DMSO                | 1.5  | 5%          | 100%           |
| $MgCl_2$            | 1.8  | 1.5 µ$M$    | 25 µ$M$        |
| 10X Rx Buffer       | 3.0  | 1X          | 10X            |
| DNTPs               | 0.6  | 0.2 m$M$    | 10 m$M$        |
| Upstream primer[a]  | 0.4  | 0.3 µ$M$    | 25 µ$M$        |
| Downstream primer[a]| 0.4  | 0.5 µ$M$    | 25 µ$M$        |
| Taq-polymerase      | 0.1  | 0.02 U/µL   | 5 U/µL         |
| Total               | 28.0 |             |                |
| Add 28 µL Master Mix to 2 µL template DNA | | | |

[a]Loci to be analyzed for MSI detection in colorectal and breast cancer (*see* **Table 1**).

8. Microsatellite amplification is achieved by using the following cycle program:
   Initial denaturation step:   2 min at 94°C

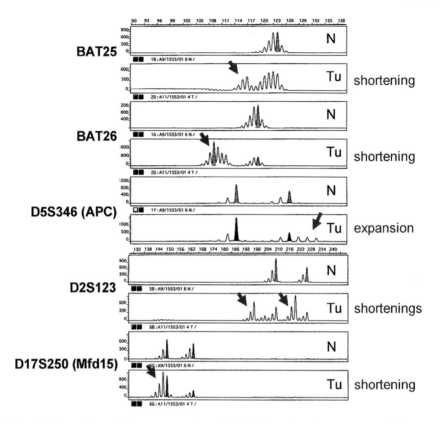

Fig. 2. Examples of fluorescence-based MSI detection by automated fragment analysis (GeneScan2.1) using ABI310 sequencing equipment. Arrows indicate shortened or expanded microsatellite repeats.

| 50 amplification cycles: | 45 s at 94°C |
|---|---|
| | 45 s at 55–60°C (*see* **Table 1**) |
| | 45 s at 72°C |
| Final elongation step: | 8 min at 72°C |

9. MSI is defined by the presence of novel peaks after PCR amplification of tumor DNA that is not present in PCR products of the matching normal DNA (for examples *see* **Table 1** and **Fig. 2**).

10. Characteristics of microsatellite primers are given in **Table 1**.

## 4. Notes

1. After the elution of DNA with $2 \times 100$ μL water, PCR template concentration can further be increased by reducing the elution volume to 50 μL (SpeedVac SC110, Savant), when required.

**Table 2**
**Examples of False-Positive Results in Microsatellite Analysis**

| Problem | Primer [a] | Repetetive PCR & PAGE [b] | Comment |
|---|---|---|---|
| Unspecific bands | Mycl1 | | The unspecific band of the first round pretending MSI could not be reproduced. |
| Preferential amplification | D2S123 | | Preferential amplification of the smaller allel of normal tissue has feigned MSI of the tumour; this could not be confirmed. |

[a]Examples of false-positive MSI results (left column) with subsequent PCR repetition (right column), revealing a normal heterozygous allelic status.
[b]Polyacrylamide gel electrophoresis.

2. For testing the quality of DNA from starting tissue material a conventional microsatellite amplification using D2S123 marker should be performed with non-laser microdissected material.

3. If PCR performance is poor, the amount of DNA for I-PEP and microsatellite PCR should be increased.

4. Pre-amplified DNA should be tested for reliable biallelic microsatellite amplification with laser microdissected control normal cells. Five independently prepared samples should be analyzed in parallel and no divergent microsatellite patterns should be visible.

5. At least 200 cells should be prepared from each specimen to prevent preferential monoallelic amplification. To further prevent false-positive results due to monoallelic amplification of unspecific bands, all positive results should be validated. Examples for false-positive MSI results are given in **Table 2**.

## References

1. Dietmaier, W., Wallinger, S., Bocker, T., Kullmann, F., Fishel, R., and Rüschoff, J. (1997) Diagnostic microsatellite instability: Definition and correlation with mismatch repair protein expression. *Cancer Res.* **57,** 4749–4756.

2. Boland, C. R., Thibodeau, S. N., Hamilton, S. R., et al. (1998) A National Cancer Institute Workshop on Microsatellite Instability for cancer detection and familial predisposition: Development of international criteria for the determination of microsatellite instability in colorectal cancer. *Cancer Res.* **58,** 5248–5257.

3. Ahrendt, S. A., Decker, P. A., Doffek, K., et al. (2000) Microsatellite instability at selected tetranucleotide repeats is associated with p53 mutations in non-small cell lung cancer. *Cancer Res.* **60,** 2488–2491.

4. Dietmaier, W., Hartmann, A., Wallinger, S., et al. (1999) Multiple mutation analyses in single tumor cells with improved whole genome amplification. *Am. J. Pathol.* **154,** 883–895.

# 9

## Assessment of RET/PTC Oncogene Activation in Thyroid Nodules Utilizing Laser Microdissection Followed by Nested RT-PCR

### Giovanni Tallini and Guilherme Brandao

#### Summary

Single palpable nodules of the thyroid gland are common in clinical practice; the majority of such lesions are benign. However, noninvasive thyroid nodules that exhibit borderline morphological signs of papillary cancer represent a diagnostic challenge. Rearrangements of the RET oncogene have been proposed as a marker for papillary thyroid cancer. In this chapter, methods for the analysis of the RET oncogene in laser microdissected papillary thyroid cancer tissue are described.

**Key Words:** Oncogene; RNA extraction; thyroid neoplasm.

## 1. Introduction

Solitary palpable nodules of the thyroid gland are frequently encountered in clinical practice with an incidence in the United States of 2–4% among adults. Females are more commonly affected than males. The majority of such lesions are benign processes, either follicular adenomas or non-neoplastic conditions such as nodular hyperplasia (colloid nodules), simple cysts, or thyroiditis, including patients with Hashimoto's thyroiditis *(1–4)*. Noninvasive thyroid nodules that exhibit borderline morphological signs of papillary cancer are difficult to diagnose and represent a challenge to the pathologist that will establish the final diagnosis, eventually leading to an accurate medical intervention *(5)*. In these cases, the morphological signs of papillary carcinoma are incomplete, in terms of "quality" when the cytologic alterations are not convincing enough and/or "quantity" when they are not uniformly present throughout the nodule. Regardless, they are superimposed on thyroid nodules that have otherwise benign histologic features and either a well-defined tumor capsule—as in the

From: *Methods in Molecular Biology, vol. 293: Laser Capture Microdissection: Methods and Protocols*
Edited by: G. I. Murray and S. Curran © Humana Press Inc., Totowa, NJ

case of follicular adenomas—or a poorly defined or incomplete capsule, as in the case of hyperplastic nodules. Rearranged versions of the RET proto-oncogene called RET/PTC (for papillary thyroid carcinoma) *(6)* are a marker for papillary thyroid cancer *(7)*. RET/PTC results from the fusion of the RET tyrosine-kinase (TK) domain with the 5' terminal region of heterologous genes, which leads to the formation of RET/PTC chimeric oncogenes. To date, at least 16 such chimeric mRNAs involving 11 different genes have been reported, of which RET/PTC1 (resulting from the fusion of RET with the H4 gene) and RET/PTC3 (resulting from the fusion of RET with the RFG gene) are by far the most common *(6,8)*. Reverse transcriptase-polymerase chain reaction (RT-PCR) for RET/PTC1 and RET/PTC3 can be performed successfully on RNA extracted after laser capture microdissection (LCM) of thyroid tissue. Evaluation of thyroid nodules exhibiting borderline histological morphology for papillary thyroid carcinoma and analysis of RNA obtained after LCM demonstrates that these lesions often harbor rearranged RET alleles and that RET/PTC activation is restricted to those areas where the classical cytological papillary carcinoma features (nuclear clearing, overlapping, and contour irregularities in the form of indentations, grooves, and pseudo-inclusions) are better developed *(9)*. It may be clinically useful to perform LCM for RET/PTC transcript analysis in thyroid nodules with incomplete papillary thyroid carcinoma cytologic alterations. Cases with detectable RET/PTC oncogene activation should be considered as papillary carcinoma when the aberrant transcripts are detected after LCM of areas with *well-developed* cytologic changes, no matter how focal these changes may be within the thyroid nodule. In fact, these cases fulfill both morphologic (presence of the diagnostic alterations) and molecular (aberrant RET/PTC transcripts) criteria for such a diagnosis. The diagnosis of cancer, however, should be limited to these areas and it is not correct to consider the entire nodule as papillary cancer, an observation that is clinically relevant since tumor size influences thyroid carcinoma staging *(10)*. Cases in which the PTC cytologic alterations are only *poorly developed* and focal are more difficult to understand. One may argue that if the molecular markers of papillary carcinoma are there, these lesions should be considered malignant as well. Since cells with RET rearrangement represent a clone (or a subclone) within the nodule *(9)*, its "size" (i.e., the relative proportion of the nodule with the aberrant RET/PTC transcripts) should probably also be taken into account. On the other hand, the finding of tumor-specific molecular alterations *per se*, without full support of clinicopathologic data, does not necessarily imply malignancy, and the term "well differentiated tumor of uncertain malignant potential" appears justified for these lesions, also considering that we know little about their clinical behavior *(11)*. Limited follow-up information indicates that an extremely good prognosis has to be expected, but it is possible that these noninvasive thyroid nodules with focal RET/PTC transcriptional activation and poorly developed cytologic changes represent a new class of papillary carcinoma precursor.

## 2. Materials
### 2.1. LCM and RNA Extraction

1. Serial 5-μm sections of paraffin-embedded tissue.
2. Plain, nonadhesive glass slides.
3. Methyl green.
4. PixCell I system (Arcturus Engineering, Mountain View, CA).
5. Thermoplastic film-coated cap.
6. Eppendorf tube.
7. Proteinase K (Sigma Chemical, St. Louis, MO).
8. Guanidinium thiocyanate.
9. β-Mercaptoethanol.
10. 0.5% Sarkosyl.
11. Tris-HCl, pH 7.5.
12. 2 $M$ Sodium acetate, pH 4.0.
13. Water-saturated phenol.
14. Chloroform-isoamyl alcohol.
15. Glycogen solution (10 μg/μL).
16. Isopropanol.
17. Ethanol.
18. DNase.
19. Random hexamers.
20. Murine leukemia virus (MuLV) RT.
21. $MgCl_2$.
22. dNTP.
23. RNase inhibitor.
24. PCR buffer II (Perkin-Elmer, Foster City, CA).
25. Primers specific for the human aldolase gene (for mRNA control).
26. AmpliTaq DNA polymerase.
27. Perkin-Elmer 9700 thermal cycler.
28. 3% Agarose gel.
29. Probe covering the tyrosine-kinase domain of RET.
30. Previously characterized papillary carcinoma samples (for positive control).
31. Undifferentiated thyroid carcinoma cell line ARO that lacks RET/PTC rearrangement.

## 3. Methods
### 3.1. LCM and RNA Extraction

Cases with discrete foci of PTC-type cytologic alterations can be processed for RNA extraction and nested RT-PCR after LCM (**Fig. 1**) following the gen-

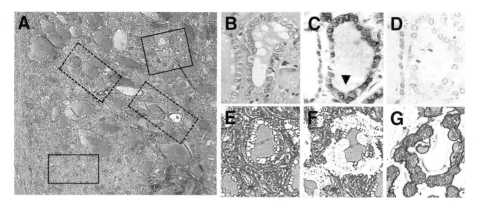

Fig. 1. Histologic appearance, immunohistochemical staining with RET(TK) antibodies, and laser capture microdissection of tissue from one noninvasive thyroid nodule with poorly developed papillary thyroid carcinoma-type nuclear changes (*9*). In the low-magnification image (**A**) the rectangles with solid lines indicate the areas of the tumor with papillary carcinoma features while the rectangles with dotted lines indicate the areas lacking them. The areas with and those without papillary carcinoma features were targeted for laser capture microdissection and separately processed for RNA extraction. A higher magnification of the foci with papillary carcinoma features is shown in (**B**). Cells with cytologic alterations of papillary carcinoma were immunohistochemically positive with RET(TK) antibodies (**C**); the corresponding negative control is shown in (**D**). The nuclear features of papillary carcinoma are poorly developed with minor degrees of nuclear clearing as seen in **B**, **C**, **D**, and occasional grooves (**C**) (arrowhead). The cells with nuclear alterations were positioned for laser capture microdissection (**E**), selectively captured (**F**), and visualized on the thermoplastic coated caps (**G**) before being processed for RNA extraction. Laser capture microdissection was similarly performed in areas of the nodule lacking the cytologic alterations of papillary carcinoma. RET/PTC3 transcripts were detected after nested RT-PCR of RNA extracted from areas with cytologic alterations but not after nested RT-PCR of RNA extracted from areas lacking cytologic changes. (Reproduced from **ref. 9** with permission.)

eral procedures outlined at the National Institute of Health LCM web site (http://dir.hichd.nih.gov/lcm/lcm.htm).

1. Serial 5 μm sections are mounted on plain nonadhesive glass slides, dewaxed, and stained with methyl green.
2. The microtome and the water bath should be decontaminated before cutting in each case.
3. The number of serial sections will depend on the amount of material available (usually four to eight sections will suffice).
4. Areas from the same thyroid nodule lacking papillary carcinoma-type nuclear changes can also be targeted for LCM and used as negative controls.

5. Approximately 1000 30-μm shots are used to transfer on the thermoplastic film-coated cap cells obtained from each thyroid lesion targeted for microdissection.
6. RNA is extracted according to established protocols (*see* **Notes 1** and **2**).
7. Briefly, each cap is placed in an Eppendorf tube containing 200 μL of 6 mg/mL proteinase K, 1 *M* guanidinium thiocyanate, 25 m*M* β-mercaptoethanol, 0.5% Sarkosyl, and 20 m*M* Tris-HCl, pH 7.5.
8. The Eppendorf tube should be inverted multiple times to fully digest the tissue off the cap.
9. Twenty μL (0.1 × vol) of 2 *M* sodium acetate, pH 4.0, and 220 μL (1 × vol) of water-saturated phenol are added to the RNA extraction solution followed by chloroform-isoamyl alcohol (0.3 × vol).
10. After vigorous vortexing and cooling on wet ice, the samples are centrifuged to separate the aqueous and organic phases.
11. The aqueous phase is transferred to a new tube containing 1 μL of glycogen solution (10 μg/μL) used as a carrier and to facilitate pellet visualization.
12. After adding an equal volume of cold isopropanol the RNA is precipitated at –20°C overnight, centrifuged, washed with ethanol, treated with DNase, and reextracted.
13. The pellets should be stored at –80°C.

### 3.2. Nested RT-PCR for RET/PTC1 and RET/PTC3 Rearrangements

1. Three μ*M* of resuspended RNA from each case to be analyzed should be reverse-transcribed with 2.5 μ*M* of random hexamers in a 20-μL reaction mix containing 2.5 U/μL murine leukemia virus (MuLV) RT, 5 m*M* MgCl$_2$, 1 m*M* each dNTP, and 1 U/μL RNase inhibitor in 1X PCR buffer II.
2. The thermoprofile for cDNA generation used is 25°C for 10 min, 42°C for 60 min, 99°C for 5 min, and 5°C for 5 min.
3. RT-PCR with primers specific for the human aldolase gene should be used for mRNA control.
4. The aldolase + primer is 5'-CGC AGA AGG GGT CCT GGT GA-3' (nucleotides 18 to 37 of exon 1), the aldolase – primer is 5'- CAG CTC CTT CTT CTG CTC CG-3' (nucleotides 175 to 194 of exon 2) *(12)*.
5. The expected 176-bp product for aldolase should be obtained to verify RNA integrity.
6. RET/PTC1 and RET/PTC3 transcripts are investigated using nested RT-PCR. The primer sequence and location are shown in **Fig. 2**.
7. For PCR, 3 μL of the cDNA template are used for the first round of amplification with the external primer sets (**Fig. 2**) in a 30-μL reaction volume with 0.1 μ*M* for each primer, 200 μ*M* each dNTP, and 0.8 U AmpliTaq DNA polymerase in Buffer II containing 2.0 m*M* MgCl$_2$.
8. After a 12-min hot start at 94°C, nine cycles of touchdown amplification are performed (progressively lowering the annealing temperature from 61°C to 55°C), followed by 40 cycles of amplification (94°C for 30 s, 55°C for 45 s, and 72°C for 45 s) with a Perkin-Elmer 9700 thermal cycler.

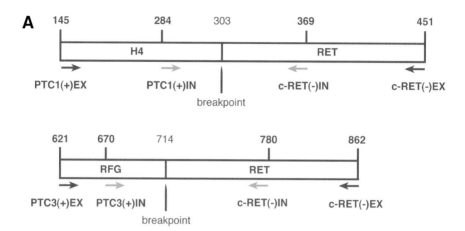

Fig. 2. **(A)** Schematic of the nested primer location relative to RET/PTC1 and RET/PTC3. Primer positions are according to the RET/PTC1 (**5**, GB no. M31213) and RET/PTC3 (**7**, GB no. X77548) standard sequences with the numeration beginning from the start codon. **(B)** Primer sequence and expected size of the amplicons for nested RT-PCR. EX: external primer set. IN, internal primer set. (Reproduced from **ref. 9** with permission.)

9. For the second round of amplification, 2 μL of first-round PCR product are used with the internal primer sets (**Fig. 2**) and the same reaction conditions described for the first amplification round.
10. The nested RT-PCR products for RET/PTC1 and RET/PTC3 are analyzed on a 3% agarose gel and hybridized with a probe covering the tyrosine-kinase domain of RET (*13*).

11. RNA extracted from previously characterized papillary carcinoma samples can be used as positive controls.

12. Amplification in the absence of RT, or in the presence of RNA extracted from the undifferentiated thyroid carcinoma cell line ARO that lacks RET/PTC rearrangement, can be used as a negative control *(13,14)*.

## 4. Notes

1. When using this technique, especially in wax-embedded 10% neutral-buffered formalin-fixed archive material, the integrity of the mRNA can be compromised to a point that no suitable RNA will be available for the assay. In our experience, amplification of a housekeeping gene (aldolase) is successful in approximately two thirds of LCM microdissected tissue from formalin-fixed, paraffin-embedded routinely processed samples of thyroid tissue. Because RNA from such specimens is significantly degraded compared with snap-frozen material, target amplicons involving small stretches of RNA sequence are more successful than larger ones *(15,16)*.

2. Although a number of methods for extracting RNA are available, proteinase K digestion of microdissected tissue is an essential step *(16)*. Histological dyes bind to cellular components such as DNA and RNA, and can also interfere with nuclei acid retrieval. Cellular RNA appears to be particularly more sensitive and dependent on the histological stain employed *(15)*. Methyl green does not bind to RNA *(17)* and is a valid alternative to hematoxylin or toluidine blue for RNA expression analysis of thyroid tissue.

3. Precipitating fixatives such as ethanol improve preservation of nucleic acids *(16)* and thyroid samples are no exception to this general rule; however, the overwhelming majority of archival material in surgical pathology practice is formalin-fixed. Published protocols are available to optimize nucleic acid recovery from LCM microdissected tissue *(15)*.

It is important to keep in mind that although fixed and frozen specimens render different amount and quality of nucleic acid, pre-LCM tissue preparation and microdissection itself are influenced by the type of starting material *(16)*. A recent work designed to detect RT-PCR products from paraffin-embedded tissues, using different fixatives, points out that the method employed to detect the RT-PCR product obtained from wax-embedded tissues may also influence the result. In fact, real-time RT-PCR analysis shows a higher success rate in RNA recovery compared with ethidium bromide stain of conventional agarose gels, although the size of the amplicon may also be a factor *(18)*. In summary, as synthesized by Goldsworthy et al., "shortening the duration of fixation, increasing the size sample, amplifying products less than 100 bp, using a nested primer method, and increasing the number of amplification cycles all can be used to achieve positive results" *(18)* and these general concepts fully apply to the recovery of RNA from thyroid tissue samples.

## References

1. Vander J. B., Gaston E. A., and Dawber T. R. (1968) The significance of non-toxic nodules: final report of a 15-year study of the incidence of thyroid malignancy. *Ann. Intern. Med.* **69,** 537–540.
2. Mazzaferri E. L. (1992) Thyroid cancer in thyroid nodules: finding a needle in the haystack. *Am. J. Med.* **93,** 359–362.
3. Rosai, J. (1996) Thyroid gland, in *Ackerman's Surgical Pathology,* Mosby, St. Louis, MO, pp. 493–567.
4. Andreoli, T. E., Carpenter, C. C. J., Bennett, J. C., and Plum, F. (1997) The thyroid gland, in *Cecil Essentials of Medicine,* W. B. Saunders, Philadelphia, pp. 487–496.
5. Rosai, J., Carcangiu M. L., and Delellis, R. A. (1992) Tumors of the thyroid gland, in *Atlas of Tumor Pathology, Third Series, Fascicle 5,* Armed Forces Institute of Pathology, Washington, DC.
6. Grieco, M., Santoro, M., Berlingieri, M. T., et al. (1990) PTC is a novel rearranged form of the ret proto-oncogene and is frequently detected in vivo in human thyroid papillary carcinomas. *Cell* **60,** 557–563.
7. Santoro, M., Sabino, N., Ishizaka, Y., et al. (1993) Involvement of RET oncogene in human tumors: specificity of RET activation to thyroid tumors. *Br. J. Cancer* **68,** 460–464.
8. Santoro, M., Dathan, N. A., Berlingeri, M. T., et al. (1994) Molecular characteristics or RET/PTC3: a novel rearranged version of the RET protooncogene in a human thyroid papillary carcinoma. *Oncogene* **9,** 509–516.
9. Fusco, A., Chiapetta, G., Hui, P., et al. (2002) Assessment of RET/PTC oncogene activation and clonality in thyroid nodules with incomplete morphological evidence of papillary carcinoma—A search for the early precursor of papillary cancer. *Am. J. Path.* **160,** 2157–2167.
10. Greene, F.L., Page, D.L., Fleming, I.D., et al. (2002) *AJCC Cancer Staging Manual,* 6th edition. Springer-Verlag, New York, pp. 77–87.
11. Williams, E. D. (2000) Guest editorial: two proposals regarding the terminology of thyroid tumors. *Internat. J. Surg. Pathol.* **8,** 181–183.
12. Izzo, P., Costanzo, P., Lupo, A., et al. (1987) A new human species of aldolase A mRNA from fibroblasts. *Eur. J. Biochem.* **164,** 9–13.
13. Viglietto, G., Chiappetta, G., Fukunaga, et al. (1995) RET/PTC oncogene activation is an early event in thyroid carcinogenesis. *Oncogene* **11,** 1207–1210.
14. Tallini, G., Santoro, M., Helie, M., et al. (1998) RET/PTC oncogene activation defines a subset of papillary thyroid carcinomas lacking evidence of progression to poorly differentiated or undifferentiated tumor phenotypes. *Clin. Cancer Res.* **4,** 287–294.
15. Huang, L. E., Luzzi, V., Ehrig, T., Holtschlag, V., and Watson, M. A. (2002) Optimized tissue processing and staining for laser capture microdissection and nucleic acid retrieval. *Methods Enzymol.* **356,** 49–62.
16. Fend, F., Specht, K., Kremer, M., and Quintanilla-Martinez, L. (2002) Laser capture microdissection in pathology. *Methods Enzymol.* **356,** 196–206.

17. Sheehan, D. C. and Hrapchak, B. B. (1980) *Theory and Practice of Histotechnology*, Battle Press, Columbus, OH, 2nd edition 1980, p. 151.
18. Goldsworthy, S. M., Stockton, P. S., Trempus, C. S., Foley, J. F., and Maronpot, R. R. (1999) Effects of fixation on RNA extraction and amplification from laser capture microdissected tissue. *Mol. Carcinog.* **25,** 86–91.

# 10

# Combined Laser-Assisted Microdissection and Short Tandem Repeat Analysis for Detection of *In Situ* Microchimerism After Solid Organ Transplantation

## Ulrich Lehmann, Anne Versmold, and Hans Kreipe

## Summary

Following the transplantation of a solid organ leukocytes of donor origin migrate out of the organ, contributing to a chimeric blood cell population ("peripheral microchimerism"). At the same time, leukocytes and pluripotent precursor cells of the recipient migrate into the organ, creating an "*in situ* microchimerism." A method is described for the identification of cells with the recipient's genotype in the transplanted organ by combining laser-assisted microdissection and short tandem repeat analysis. The microdissection allows the contamination-free isolation of morphologically and immunohistochemically characterized cells or groups of cells from histological tissue sections. The subsequent analysis of highly polymorphic short tandem repeats enables unequivocal genotyping in nearly all donor-recipient instances. Employing this new methodological approach, we could identify in individual transplanted organs differentiated parenchymal cells of recipient's origin, which most probably are derived from circulating precursor cells from the bone marrow.

**Key Words:** Organ transplantation; STR analysis; *in situ* microchimerism; microdissection.

## 1. Introduction

The transplantation of a solid organ inevitably creates a chimeric organism ("macrochimerism"), because donor and recipient have a different genotype (with the exception of identical twins). In addition to the transplanted organ, leukocytes are also transferred from the donor to the recipient. This creates a chimeric leukocyte population in the blood circulation of the recipient, a phenomenon called "peripheral microchimerism" *(1,2)*. At the same time circulating cells from the recipient migrate into the new organ, thereby creating a so-called "*in situ* microchimerism."

From: *Methods in Molecular Biology, vol. 293: Laser Capture Microdissection: Methods and Protocols*
Edited by: G. I. Murray and S. Curran © Humana Press Inc., Totowa, NJ

It has been 40 yr since the hypothesis was proposed that cells from the recipient may stably populate the transplanted organ *(3,4)*, thereby modulating the antigenicity of the foreign tissue in the recipient's body. However, to date experimental evidence has been lacking. In particular the question of whether parenchymal cells (e.g., hepatocytes or bronchial epithelium) of recipient origin can be found in the transplanted organ was denied until quite recently *(*e.g., **ref. 5***)*.

In animal models the discrimination between donor and recipient cells and the identification of parenchymal cells of recipient's origin can be achieved by selecting animals with differing genetic markers for transplantation experiments or by utilizing genetically engineered animals with easily detectable markers, such as the green fluorescent protein (GFP) *(6–8)*. In humans this is obviously not possible. Earlier studies relied on the immunohistochemical analysis of blood group antigens; however, under many circumstances this methodology cannot discriminate unequivocally between donor and recipient, because the blood group antigens are usually very similar *(9)*.

More recently, the detection of the Y-chromosome by fluorescence or chromogenic *in situ* hybridization (FISH or CISH) *(10)* in case of cross-gender transplantations has been successfully used by several groups *(11–13)*. The limitation of this approach is the requirement for a sex mismatch. The preferred situation for Y-chromosome hybridization is the transplantation of a female organ into a male recipient because in this situation the cells of interest (the recipient's cells in the donor organ) are marked positively by the presence of a Y-chromosome signal. Because the majority of organ donors are younger males, less than one quarter of the archived biopsies can be analyzed by this method.

In order to develop a methodology that is completely independent of gender mismatch and that has the potential to analyze all available biopsies (especially formalin-fixed and paraffin-embedded biopsies), we combined the isolation of the cells of interest from histological sections using laser-assisted microdissection and the subsequent genotyping utilizing highly polymorphic genetic markers (so-called short tandem repeats or STRs *[14]*) for the identification of recipient's cells in the transplanted organ. Short tandem repeats are nucleotide repeats (di-, tri-, and tetranucleotide repeats), which are dispersed throughout the human genome. The lengths of these repeats show a quite high allelic variability and the combination of several STRs can be utilized for the unequivocal identification of any individual. In the situation of a transplanted organ only two individuals have to be differentiated by their genotype. Therefore, the analysis of only one highly polymorphic marker, such as the tetranucleotide repeat at the SE33 locus, is sufficient.

Immunohistochemical labeling prior to laser microdissection enables the clear distinction of circulating leukocytes from the recipient in the transplanted organ (CD45 or CD68 positive) from parenchymal cells of recipient's origin (e.g., cytokeratin positive).

Using this new methodology we sought to find out when and to what extent liver cell chimerism after transplantation occurs, how it is correlated with rejection, and whether it influences the long-term fate of the graft *(15)*.

## 2. Materials

Manufacturers or distributors are specified only if reagents or laboratory equipment might be important for the outcome, or if a source might be difficult to identify. All chemicals were purchased in analytical grade quality from Merck, Roth, or Sigma and kept strictly separate from the postamplification area in our institute.

1. Glass slides.
2. Cyanacrylate glue (UHU, Buehl, Germany).
3. Soft brush, for smoothing the foil on the glass slides before fixing it with glue.
4. Poly-L-lysine (0.1% aqueous solution, Sigma, Taufkirchen, Germany), stored at 4°C.
5. Poly-propylene foil, 1.2 µm (PALM, Bernried, Germany).
6. "Xylol-Ersatz," a xylene substitute, which is less toxic and smells less unpleasant (Vogel, Karlsruhe, Germany).
7. Ethanol (100%, 96%, 70%).
8. Glass cuvets.
9. ABC Vectastain-Kit (Vector Laboratories, Burlingame, CA).
10. Methylene blue (Loeffler's Methylene blue, Merck, Darmstadt, Germany).
11. Methyl green (Merck). Staining solutions are stored at room temperature in the dark.
12. Antibodies: LCA (1:100, Dako, Hamburg, Germany), CD68 (1:100, Dako), Cytokeratin 8/18 (1:100, Dako).
13. Liquid wax (MJ Research, Boston, MA).
14. Proteinase K-buffer: 50 m$M$ Tris-HCl, pH 8.1, 1 m$M$ EDTA, 0.5% Tween-20.
15. Proteinase K, stock solution: 20 mg/mL in water, aliquots stored at –20°C (Merck).
16. TE buffer: 10 m$M$ Tris-HCl, pH 8.1, 1 m$M$ EDTA
17. 0.5-mL Tubes with transparent lid and lowered inner lid, for collecting dissected and catapulted cells (PALM).
18. Taq-Polymerase: Hot Start Taq (Qiagen, Hilden, Germany).
19. Polymerase chain reaction (PCR)-buffer: supplied with the enzyme.
20. Nucleotides for PCR: 10 m$M$ dNTP-Mix (MBI, Fermentas, St. Leon-Roth, Germany).
21. PCR tubes: 0.2-mL PCR-tubes, colorless (no. 710900, Biozym, Hessisch Oldendorf, Germany).

22. PCR primer: 5'-6FAM-AGAGAGAGAAAGGAAGGAAGG 5'-CTACC GCTATAGTAACTTGC One primer is labelled at the 5' end with 6-FAM (in our case the forward primer, but the reverse primer can also be labeled) (*see* **Note 1**).

23. Formamide: deionized, minimum 99.5% (GC), for molecular biology (F-9037, Sigma).

24. Septa for 0.5-mL sample tubes (no. 401956, Applied Biosystems, Darmstadt, Germany).

25. GeneScan350 ROX Size Standard (no. 401735, Applied Biosystems).

26. Matrix for capillary: Performance Optimized Polymer (POP-4, Part no. 402838, Applied Biosystems).

27. Tips with aerosol protection, DNAse-, RNAse-free (Sarstedt, Nümbrecht, Germany).

28. HPLC-water (no. 4218, JT Baker, Deeventer, Holland).

29. 3 *M* Sodium acetate, pH 7.0, containing 100 µg/mL dextran T500 (Sigma).

30. Hypochlorite solution, for use diluted 1:4 with water (Roth).

31. PCR bench with UV lamp, for decontamination of racks and irradiation of polypropylene foil.

32. Laser microdissection system (PALM).

33. Sequencer ABI310 (Applied Biosystems).

34. Refrigerated table-top centrifuge for 0.2- to 2.0-mL tubes (max. 14,000$g$).

35. Vortex.

36. 40°C Incubator.

37. Thermoshaker with heated lid (CLF, Emersacker, Germany).

## 3. Methods

The protocols described below for STR analysis after laser-assisted micro-dissection concentrate on the following steps:

1. Organization of the laboratory.
2. Preparation of foil-coated glass slides for microdissection.
3. Cutting and staining histological tissue sections.
4. Laser microdissection and specimen recovery.
5. Isolation of DNA from microdissected cells.
6. STR-PCR.
7. Analyzing fluorescent-labeled PCR products.

### 3.1. Organization of the Laboratory

In order to prevent any cross-contamination of samples that could lead to false-positive results (indicating chimerism of cell populations that are actually not chimeric), strict guidelines for the laser-assisted microdissection and the setup of the PCR mixtures, as well as for the physical separation of the analysis of reaction products (postamplification) from all stages of sample preparation (preamplification), have to be implemented.

For these reasons, strictly enforced protocols concerning the cleaning of instruments and the handling of samples before and after amplification must be followed by all personnel involved. We perform all preamplification steps, including the laser-assisted microdissection, in a separate laboratory consisting of two rooms: one for setting up the PCR master mix (a "template-free" room) and the other for preparation of tissue sections, microdissection, nucleic acid extraction, and adding DNA to the PCR mixes. Plastic labware and the benches are cleaned regularly using a 3% hypochlorite solution. The PCR products are analyzed in a separate laboratory. Under no circumstances should amplified samples or equipment from this working area be brought back to the pre-PCR area.

## 3.2. Preparation of Foil-Coated Glass Slides

1. The tissue section is mounted on a glass slide coated with a very thin foil. This foil serves as a carrier for the tissue and is glued to the glass slide only at the four corners. Therefore, after microdissection the dissected piece of tissue can be easily removed together with the supporting foil, which is completely inert and does not interfere with any subsequent analysis.
2. Because manufactured slides coated with polypropylene foil are very expensive, we buy the foil from PALM and prepare the coated slides ourselves.
3. For coating with poly-L-lysine a drop of the solution (0.1% in sterile water) is spread over the membrane with a sterile pipet tip, carefully avoiding any damage of the very thin foil. (Any leakage underneath the membrane might result in problems with the laser pressure catapulting technology, because the poly-L-lysine will glue the foil onto the glass slide.)
4. The foil, pretreated in this way, is cut into appropriately sized pieces and mounted onto the glass slides.
5. It is very important to use a very sharp scalpel blade to avoid damaging the foil.
6. Any fold in the foil is removed with a clean soft brush and it is fixed at the four corners with cyanacrylate glue from UHU (Buehl, Germany) (*see* **Note 2**).
7. After curing overnight, the foil-coated glass slides are irradiated for 45 min with short-wavelength UV light to destroy all traces of DNA or RNA (*see* **Note 3**).

## 3.3. Specimen Preparation and Staining

1. Formalin-fixed paraffin-embedded biopsies are cut using a conventional microtome and sections are mounted on foil-covered slides.
2. After cutting a biopsy the cryotome blade and the sample holder are cleaned meticulously before the next biopsy is cut to avoid any cross-contamination.
3. To improve adhesion, the slides with sections are incubated for 15 min at 55°C.
4. Afterward, the sections are dewaxed and rehydrated using "Xylol-Ersatz" and ethanol (2X xylene substitute for 10 min, 2X 100% ethanol for 5 min, 2X 96% ethanol for 5 min, 1X 70% ethanol for 5 min, sterile water for 5 min).

### 3.3.1. Conventional Staining

For staining with methylene blue, the hydrated sections are covered with staining solution for 30 s, rinsed twice with sterile water, dehydrated with a drop of absolute ethanol, and allowed to dry at room temperature.

### 3.3.2. Immunohistochemical Staining

1. For labeling cells with antibodies before microdissection we use the Vectastain kit from Vector Laboratories (*see* **Note 4**). As outlined above, all reagents brought into contact with samples before amplification (pre-PCR) have to be strictly separated from all other reagents used in the laboratory (post-PCR). The great advantage of ready-to-use kits is that the components are free of any potentially contaminating PCR products and completely separated from all reagents normally used in the laboratory. This justifies the higher costs.
2. Antibodies directed against CK8/18 and LCA and CD68 are used in a 1:100 dilution.
3. Nonspecific binding to the sections is prevented by preincubation for 20 min with horse serum provided with the kit.
4. For incubation with antibodies and the detection reaction, the manufacturer's instructions are followed.

## 3.4. Microdissection Using PALM Laser Microdissection Microscope

1. The width of the laser cut can be altered by adjusting the laser energy and/or the focus of the laser.
2. In our laboratory, the optimal focus for using the 40× long-distance objective from Zeiss (40x/0.60 Korr /0–2) is around 980 (arbitrary units); for using the 10× objective from Zeiss (10×/0.50 /0.17) the optimal value is 680.
3. Depending on the thickness of the section and the tissue type, the energy for cutting is between 920 and 1000 (arbitrary units), for catapulting greater than 1020.
4. The actual numbers for the energy and focus setting may vary slightly for different instruments (*see* **Note 5**).
5. For recovery of dissected cells using the laser pressure catapulting technology of the PALM system, the 0.5-mL tubes distributed by PALM itself are most suitable because of the lowered inner lid, which shortens the distance the catapulted cells have to travel. Also the specimens are readily visible in the transparent lid (*see* **Fig. 1**).
6. For catapulting a dissected cell or a group of cells, the laser is focused slightly below the section and the laser energy is increased.
7. A single short laser pulse is sufficient for catapulting the specimen into the lid of a reaction tube placed directly above the section.
8. The lid of the reaction tube is conveniently positioned in the holder of the micromanipulator (*see* **Note 6**).
9. Despite having the very sophisticated (and also often necessary) laser pressure catapulting technology we recover dissected pieces of tissue with a sterile needle

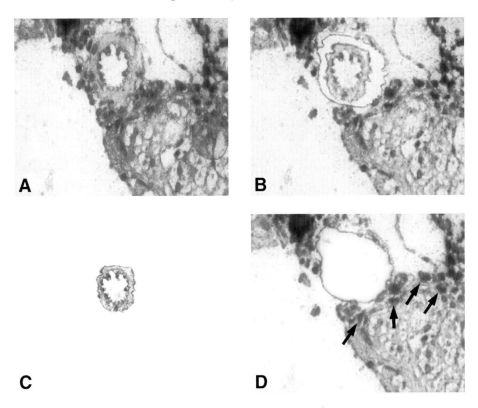

Fig. 1. Laser-assisted microdissection and isolation of endothelial cells from a liver vessel. Inflammatory cells are labelled immunohistochemically (CD68 and LCA, *see* arrows). (**A**) Section before microdissection. (**B**) Endothelial lining dissected. (**C**) Isolated endothelial cells in the lid of a reaction tube. (**D**) Section after removal of endothelial cells (fast red and hematoxylin counterstain; original magnification: ×100). The reduced optical quality is because the tissue section is dried and not cover-slipped.

if they are large enough. This can be done easily by hand without any technical support. After some practice this turns out to be straightforward for each person working with the laser microscope.

## 3.5. Isolation of DNA

1. Small numbers of cells are lysed in the lid of the reaction tube by adding 10–30 µL TE-buffer containing 40 µg proteinase K.
2. The closed tubes are incubated in a small incubator in an inverted position at 45°C overnight.
3. The next day samples are centrifuged and heated for 8 min at 95°C for inactivation of proteinase K and efficient denaturation of the DNA.
4. This lysate is used directly for subsequent PCR analysis.

5. If thousands of cells are isolated the samples are lysed in a larger volume of proteinase K buffer (100–300 µL) containing 500 µg/mL proteinase K.
6. The samples are incubated in a vigorously shaking thermoshaker at 56°C overnight.
7. The next day the lysate is transferred to a new tube (*see* **Note 7**) and the DNA is precipitated by adding sodium acetate (pH 7.0) containing dextran T500 as a carrier (100 µg/mL) and ethanol.
8. This precipitation step almost completely removes contaminating dyes and cell debris.
9. After centrifugation and washing of the pellet with 70% ethanol the DNA is air-dried and dissolved in 30–50 µL of sterile water.

### 3.6. Short Tandem Repeat-PCR

1. The amplification reaction is performed in a final volume of 25 µL containing 200 n$M$ of each primer, 0.5 U Hot Start Taq Polymerase, 1.5 m$M$ MgCl$_2$, 250 n$M$ of dNTP, and up to 10 µL DNA lysate.
2. The forward primer is labeled with 6-FAM at the 5' end.
3. The reaction mixture is preheated at 95°C for 10 min, followed by 35 cycles at 95°C for 30 s, 56°C for 30 s, and 72°C for 1 min, with a final elongation step at 72°C for 10 min (*see* **Note 8**).
4. As described above (**Subheading 3.1.**), the PCR is set up in a completely separate room under strict guidelines to prevent cross-contamination.

### 3.7. Analysis of Fluorescence-Labeled PCR Products

1. For the analysis of PCR products, 1 µL of the reaction mix is mixed with 0.3 µL size standard (GeneScan350) and 12 µL formamide.
2. This mixture is heated for 2 min at 90°C and chilled immediately on ice (*see* **Note 9**).
3. The samples are placed in the sample holder and electrophoresis is started and analyzed as described in detail by the manufacturer (*see* **Fig. 2**).
4. We use routinely an injection time of 5 s.

## 4. Notes

1. This new primer pair for the SE33 locus was designed to reduce the length of the PCR products by 85 bp compared to the original SE33 primers (*16,17*) to obtain fragments ranging from 140 to 236 bp in length.
2. We tested several glues and tapes, following tips and hints from colleagues, and PALM. Cyanacrylate glue ("two-component glue") turned out to be the best, almost completely resisting xylene and 100% ethanol. All other glues or tapes tested dissolved in one of these solvents. But even this glue starts dissolving after prolonged incubations of longer than 45 min.
3. It is very important to wait until the glue is completely cured, before UV irradiation starts, because otherwise the glue will never set and will dissolve rapidly in xylene or the xylene substitute.

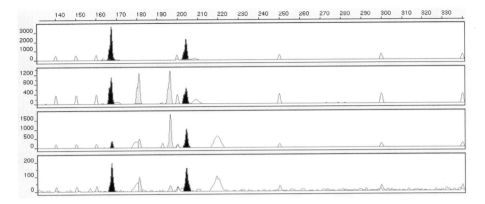

Fig. 2. Examples of PCR product analysis employing the ABI310 capillary sequencer from Applied Biosystems. (**A**) Recipient's genotype (explanted organ). (**B**) Genotyping of a whole section. Chimerism is clearly discernible: All four alleles from the donor and the recipient are visible. (**C**) Laser microdissected cholangiocytes displaying *in situ* microchimerism. (**D**) Laser microdissected leukocyte infiltrate displaying primarily the genotype of the recipient.

4. Only if one routinely performs immunohistochemical staining for laser microdissection on a large scale is the "in-house" set-up of detection reagents under PCR contamination-free conditions cost-effective. We didn't perform a comprehensive comparison of commercially available staining kits; therefore other kits might work as well.

5. The width of the laser cut depends very much on the thickness of the section and the type of tissue structure that has to be cut (e.g., adipose tissue is very easily cut; connective tissue is often quite resistant). Therefore, the energy sufficient for a fine cut through fat tissue will not be sufficient for cutting connective tissue structures and the energy adjusted to the latter tissue type will create a quite broad, irregular cut through fat tissue. However, adjustment of the appropriate laser focus and energy must be learned by trial and error; the values given in the text are only a guide.

6. Dissected cells can also be catapulted into the lid of a reaction tube without changing the focus of the laser, but this will create a "bullet hole" in the specimen. This is no problem if larger structures are dissected and the laser "bullet hole" can be placed in an irrelevant part of the specimen (e.g., the lumen of a vessel dissecting the vessel lining endothelial cells). When dissecting single cells or very small cell clusters it is essential to change the focus of the laser. This adjustment of the laser focus for catapulting has to be learned by trial and error; the correct adjustment is a delicate balance between the size of the specimen and the laser focus and the laser energy. A tiny drop of liquid wax from MJ Research is distributed in the lid of the reaction tube. This wax film ensures that the catapulted specimens will adhere firmly to the lid.

7. The transfer of the lysate of larger groups of cells to a new tube before precipitation is necessary in order to separate the pieces of supporting membranes from the cell lysate. These pieces are isolated together with the dissected cells. They are not lysed and interfere physically with precipitation of nucleic acids by preventing the formation of a compact pellet at the bottom of the tube during centrifugation.

8. Following this PCR protocol we can achieve detection sensitivity of 2–4%, i.e., one to two cells of recipient's origin are reproducibly detected in a background of 50 donor cells.

9. For the analysis of the PCR products utilizing the capillary sequencer from Applied Biosystems we use ordinary colorless 0.5-mL tubes from which we cut off the lid. But we recommend the use of specialized (and quite expensive) septa from Applied Biosystems (Part no. 401956) to close these tubes before placing them in the sample holder.

## Acknowledgments

The authors would like to thank Britta Hasemeier for expert assistance in preparing the illustrations, Wolfram Kleeberger for providing the original microscopic pictures for **Fig. 1**, and Thomas Rothämel for initial advice concerning STR-PCR analysis.

The work of the authors is supported by grants from the Deutsche Forschungsgemeinschaft (SBF265, Project C11) and the Deutsche Krebshilfe (10-1842-Le I).

## References

1. Starzl, T. E. and Demetris, A. J. (1998) Transplantation tolerance, microchimerism, and the two-way paradigm. *Theor. Med. Bioeth.* **19,** 441–455.
2. Wekerle, T. and Sykes, M. (1999) Mixed chimerism as an approach for the induction of transplantation tolerance. *Transplantation* **68,** 459–467.
3. Medawar, P. B. (1965) Transplantation of tissues and organs: introduction. *Br. Med. Bull.* **21,** 97–99.
4. Williams, G. M. and Alvarez, C. A. (1969) Host repopulation of the endothelium in allografts of kidneys and aorta. *Surg. Forum.* **20,** 293–294.
5. Bittmann, I., Baretton, G. B., and Schneeberger, H. (1998) Chronic transplant reaction of the kidney. A interphase cytogenetic and immunohistologic characterization of the involved cells in relation to donor and recipient origin. *Pathologe* **19,** 129–133.
6. Kopen, G. C., Prockop, D. J., and Phinney, D. G. (1999) Stromal cells migrate throughout forebrain and cerebellum, and they differentiate into astrocytes after injection into neonatal mouse brains. *Proc. Natl. Acad. Sci. USA* **96,** 10,711–10,716.
7. Lagasse, E., Connors, H., Al-Dhalimy, M., et al. (2000) Purified hematopoietic stem cells can differentiate into hepatocytes in vivo. *Nat. Med.* **6,** 1229–1234.

8. Krause, D. S., Theise, N. D., Collector, M. I., et al. (2001) Multi-organ, multi-lineage engraftment by a single bone marrow-derived stem cell. *Cell* **10**, 369–377.

9. O'Connell, J. B., Renlund, D. G., Bristow, M. R., and Hammond, E. H. (1991) Detection of allograft endothelial cells of recipient origin following ABO-compatible, nonidentical cardiac transplantation. *Transplantation* **51**, 438–442.

10. Tanner, M., Gancberg, D., Di Leo, A., et al. (2000) Chromogenic in situ hybridization: a practical alternative for fluorescence in situ hybridization to detect HER-2/neu oncogene amplification in archival breast cancer samples. *Am. J. Pathol.* **157**, 1467–1472.

11. Alison, M. R., Poulsom, R., Jeffery, R., et al. (2000) Hepatocytes from non-hepatic adult stem cells. *Nature* **406**, 257.

12. Theise, N. D., Nimmakayalu, M., Gardner, R., et al. (2000) Liver from bone marrow in humans. *Hepatology* **32**, 11–16.

13. Lagaaij, E. L., Cramer-Knijnenburg, G. F., van Kemenade, F. J., van Es, L. A., Bruijn, J. A., and van Krieken, J. H. (2001) Endothelial cell chimerism after renal transplantation and vascular rejection. *Lancet* **357**, 33–37.

14. Clayton, T. M., Whitaker, J. P., Sparkes, R., and Gill, P. (1998) Analysis and interpretation of mixed forensic stains using DNA STR profiling. *Forensic Sci. Int.* **91**, 55–70.

15. Kleeberger, W., Rothamel, T., Glockner, S., Flemming, P., Lehmann, U., and Kreipe, H. (2002) High frequency of epithelial chimerism in liver transplants demonstrated by microdissection and STR-analysis, *Hepatology* **35**, 110–116.

16. Polymeropoulos, M. H., Rath, D. S., Xiao, H., and Merril, C. R. (1992) Tetranucleotide repeat polymorphism at the human beta-actin related pseudogene H-beta-Ac-psi-2 (ACTBP2). *Nucleic Acids Res.* **20**, 1432.

17. Moller, A. and Brinkmann, B. (1994) Locus ACTBP2 (SE33). Sequencing data reveal considerable polymorphism. *Int. J. Legal Med.* **106**, 262–267.

# III

# RNA AND GENE EXPRESSION STUDIES USING MICRODISSECTED CELLS

# 11

## Laser-Assisted Microdissection of Membrane-Mounted Tissue Sections

### Lise Mette Gjerdrum and Stephen Hamilton-Dutoit

#### Summary

Biological tissues (in particular those affected by disease) are inherently complex mixtures of different cell types and matrices. This heterogeneity can complicate the interpretation of molecular biological studies performed on whole-tissue extracts if the precise cellular origin of the molecules being tested is not known. Laser-assisted microdissection (LAM) has emerged as a leading histological technique for obtaining samples enriched for specific target cell populations or tissue components for subsequent molecular (especially polymerase chain reaction-based) analysis. This method allows the identification and study of target-specific molecular alterations in heterogeneous specimens, and enables more accurate detection and quantification of target molecules. In this chapter, we focus on tissue microdissection performed with an ultraviolet laser system and describe protocols for the basic procedure and for handling of the samples.

**Key Words:** Laser-assisted microdissection; MOMeNT; frozen tissue; formalin-fixed; paraffin-embedded tissue.

## 1. Introduction

Advances in molecular biology have provided new tools for the analysis of the genetic processes that govern disease. A crucial factor for the reliability of the data obtained using tests based on tissue extracts is the cellular homogeneity of the study samples. This is particularly true for cancer tissues, which are inherently complex. In addition to the main tumor clone and subclones, they typically include a variable mixture of dead or dying cells, stroma, blood vessels, inflammatory cells, and other non-neoplastic tissue components. Many molecular biological assays (including detection of loss of heterozygosity, comparative genomic hybridization, high-throughput cDNA arrays, and proteomics) require relatively homogeneous test material to unmask tumor cell-specific genetic alterations (1–4). In cancers, the presence in the test material

From: *Methods in Molecular Biology, vol. 293: Laser Capture Microdissection: Methods and Protocols*
Edited by: G. I. Murray and S. Curran © Humana Press Inc., Totowa, NJ

of as a few as 20% nontumor cells may confound the correct result *(5)*. Therefore, accurate and meaningful molecular analysis of tumors requires precise
documentation of the cellular origin of the test DNA, RNA, or proteins in tissue extracts.

Laser-assisted microdissection (LAM) is a general term for methods that
use a laser to selectively obtain homogeneous cell specimens, both from cytological preparations and from frozen or formalin-fixed, paraffin-embedded
(FFPE) tissue sections.

These microdissected samples are suitable for a range of downstream
molecular analyses of DNA, RNA, or proteins *(1,6–9)* (*see* **Note 1**). In laser
microbeam microdissection (LMM), a pulsed ultraviolet (UV) laser is used to
photoablate unwanted tissue and to cut out the area of interest (*see* **Note 2**).
Mounting the section on a thin polyethylene membrane facilitates the microdissection and makes it possible to remove both large tissue fragments and
individual target cells intact, while reducing the risk of contamination *(1)*. The
transfer of dissected material is accomplished in a variety of ways (*see* **Note 2**;
**Fig. 1**). In this protocol, we use an UV laser system supplied by Molecular
Machines & Industries (MMI, Glattbrugg, Switzerland) for microdissection.
General protocols for microdissection of membrane-mounted native tissue
(MOMeNT) using both FFPE and frozen tissues are described, together with
instructions for improving tissue morphology during microdissection and notes
on the storage of sections (*see* **Note 3**). Further, we provide protocols for
sample collection with a needle and using the single-step collection device
supplied by MMI (*see* **Fig. 1**). Protocols for RNA and DNA extraction from
frozen and paraffin-embedded tissues are given, suitable for conventional polymerase chain rection (PCR)-based analysis, as well as for real-time quantitative PCR.

## 2. Materials

### 2.1. LMM Using Disposable Needles for Sample Collection

1. Polyethylene membranes (MMI).
2. Fixogum rubber cement (MMI).
3. 31 G × 5/16 8-mm Microfine Insulin pen needles (Becton Dickinson, Broendby,
   Denmark).
4. 0.2-mL MicroAmp tube (PE Biosystems, Foster City, CA) or 1.5-mL
   Eppendorf tube.
5. Cytotec (Schuco International, London, UK).

### 2.2. LMM Using Single-Step Collection

1. Membrane-mounted metal slides (MMI).
2. Tubes with adhesive lid, with or without diffuser (MMI).
3. Single-step collection robot-stage and cap-holder (MMI).

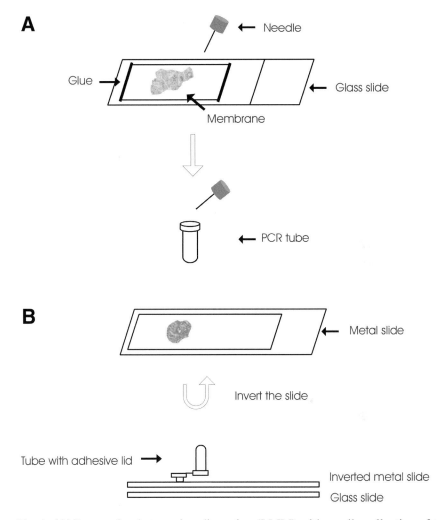

Fig. 1. **(A)** Laser microbeam microdissection (LMM) with needle collection of target. **(B)** LMM with single-step target collection.

## 2.3. Disposables Shared by Both Methods

1. Laser microdissection system: SLμCUT (MMI) (*see* **Note 2** for suitable alternative systems).
2. Noncharged Superfrost glass slides (Menzel-Glaser, Braunschweig, Germany).
3. Xylene; store at room temperature (RT).
4. Graded alcohols (99%, 96%, 70%); store at RT.
5. Absolute acetone store at 4°C.
6. Histological stain (Mayer's hematoxylin, methyl green, or toluidine blue).

## 2.4. DNA Extraction

1. DNA extraction buffer containing 50 m$M$ Tris-HCl, pH. 8.1, 1 m$M$ EDTA, pH.8.0, and 0.5% Tween-20; store at RT.
2. Proteinase K, 5 mg/mL, stock solution at –20°C.
3. Glycogen 20 mg/mL (Invitrogen A/S, Taastrup, Denmark) at –20°C for single-cell collection.

## 2.5. RNA Extraction

1. RNA digestion buffer containing 20 m$M$ Tris-HCl, pH 7.5, 20 m$M$ EDTA, pH.8.0, and 1% SDS at RT.
2. Proteinase K, 20 mg/mL, stock solution at –20°C.
3. TRIzol (Life Technologies, Gaithersburg, MD) at 4°C.
4. Chloroform at RT.
5. Isopropanol at RT.
6. 3 $M$ Sodium acetate, pH 5.2, at –20°C.
7. Glycogen 20 mg/mL (Invitrogen) at –20°C.
8. Absolute and 70% ethanol at RT.
9. Sterile H$_2$O.

## 3. Methods

The methods described include: (1) LMM performed with disposable needles, including preparation of the membrane-mounted glass slides and instructions for improving the quality of morphology; (2) single-step collection; (3) preparation of sections from FFPE tissues and preparation of frozen sections; (4) laser microdissection using both sample collection methods; (5) DNA extraction; and (6) RNA extraction.

## 3.1. LMM Performed Using Disposable Needles

### 3.1.1. Preparation of Membrane-Mounted Glass Slides

1. The polyethylene membrane (1.5 µm, MMI) is layered between two sheets of A4 paper and cut to size (~2 × 3 cm) using a paper guillotine. The noncharged Superfrost glass slides (Menzel-Glaser) are dipped in 70% ethanol to make wrinkle-free application of the membranes easier.
2. After removing one of the papers, the precut membrane is applied to the wet glass with the help of the remaining paper support *(10)*. This is then removed, and the membranes are smoothed out using sterile cotton swabs.
3. Subsequently, the membrane is attached to the glass slide using Fixogum rubber cement along the two short opposing sides (*see* **Notes 3** and **4**). Gloves should be worn throughout the whole procedure.
4. The membrane-mounted slides are then put under UV light overnight (*see* **Note 5**).

## 3.1.2. Instructions for Improving Quality of Morphology

Improvement of morphology can be achieved using several different volatile fluids, applied using a simple plastic spray bottle. We use Cytotec, which is a commercially available fixative for cytology specimens (*see* **Note 6** for alternatives).

### 3.2. Single-Step Collection

Commercially available metal slides are placed under UV light for 30 min (*see* **Note 5**). Each metal slide is used together with a single Menzel glass slide.

### 3.3. Preparations Common to Both Methods

## 3.3.1. Preparation of Sections From FFPE Tissues

1. Gloves should be worn and changed regularly.
2. The microtome is cleaned with ethanol before each new case.
3. Serial 5-μm sections are cut with a fresh knife, floated out in a hot water bath for 30 s, and finally mounted on the prepared slides.
4. Subsequently, they are incubated at 60°C for 15 min to adhere the tissue to the membrane (*see* **Note 7**).
5. Removal of paraffin is carried out for 2 × 2 min in xylene, then 2 min each in 99%, 96%, and 70% ethanol.
6. The sections are stained for either 3 min in Mayer's hematoxylin, or for 10 s in methyl green or toluidine blue (*see* **Note 8**), rinsed in distilled water for 1 min, and finally dehydrated in 99% ethanol for 1 min.
7. When the membrane is glued on to the glass slide, the sections are dried at RT for approx 30 min (*see* **Note 9**).
8. The metal slides for single-step collection should be dried for approx 5 min at RT.

## 3.3.2. Preparation of Frozen Sections

1. The procedure for cleaning and cutting is essentially the same as for FFPE sections.
2. Frozen sections are fixed in acetone for 3 min immediately after being cut, air-dried for 15 s, stained with either Mayer's hematoxylin for 3 min, or methyl green or toluidine blue for 10 s (*see* **Note 8**).
3. Solutions are washed in distilled water for 1 min, and dehydrated consecutively in 96% and 99% ethanol for 1 min each.
4. Finally, the sections are dried at RT for 30 min or 5 min, depending on the type of slide used (*see* **Note 9**).

### 3.4. Laser-Assisted Microdissection

## 3.4.1. LMM Using Disposable Needles

1. We use the SLμCUT UV laser system for microdissection.
2. Before each new sample, the robot-stage is cleaned with ethanol.

3. When working with single cells, we irradiate the same working area with a portable UV lamp.
4. The power, focus, and speed of the laser should be adjusted for each new specimen.
5. After visual identification of the target, the computer mouse and software are used to select and cut out the chosen area.
6. This procedure and the transfer of the cut fragment will differ between the various available systems (*see* **Note 2** for alternatives).
7. We use a 31G needle (Becton Dickinson) clamped in a holder on the microscope to collect the specimen.
8. The computer mouse and software position the microdissected fragment under the needle point.
9. The needle is then lowered to pick up the cut fragment, which is transferred to a PCR tube (*see* **Note 10**; **Fig. 1A** in this chapter; **Fig. 1A–D** in Chapter 12).

### 3.4.2. Single-Step Collection

1. The same UV laser system and software are used, and the preparation of slides and tissues is essentially the same.
2. The metal slide containing the membrane and cut tissue is inverted and put on top of a Menzel glass slide.
3. The tissue now lies between the membrane and glass slide (*see* **Note 11**, **Fig. 1B** in this chapter and **Fig. 2A–D** in Chapter 12).
4. This sandwich is placed on the robot stage with the metal slide on top, and clamped in a holder.
5. The adhesive cap is placed in the cap-holder, and centered over the light beam.
6. The lid is then lowered onto the membrane surface.
7. The procedure of adjusting the laser and cutting the target is carried out as described above.
8. After cutting one or more samples, the cap (with adherent microdissected fragments) is lifted away from the stage, and the tube closed for transportation.
9. Digestion buffer is added to the cap, and incubation takes place in a hot-air oven with the tubes inverted (*see* **Note 11**).

### 3.5. DNA Extraction

1. The extraction method is essentially the same, regardless of which sampling method is used.
2. Using a disposable needle, the dissected tissue is transferred into a 0.2-mL MicroAmp tube containing 28.6 µL of DNA extraction buffer and 1.4 µL of 5 mg/mL proteinase K solution (*see* **Note 12**).
3. This is incubated overnight at 56°C.
4. Following single-step collection, 10 µL of digestion buffer is added to the cap.
5. The digestion buffer for single-cell microdissection contains, in addition, 1 µL of 20 mg/mL glycogen (*see* **Note 11**).
6. After inversion, the tubes are placed in a hot air oven at 48°C overnight.

7. In both protocols, the enzyme is inactivated at 95°C for 5 min before subsequent PCR analysis is performed.
8. The samples can be stored at –20°C for months.
9. For larger dissected samples, an ethanol precipitation may be advisable (*see* **Note 13**).

### *3.6. RNA Extraction*

1. The same RNA extraction protocol is used with both methods.
2. Microdissected tissue samples are either placed in a reagent tube containing 200 µL of digestion buffer and 5 µL of 20 mg/mL proteinase K, and are subsequently allowed to incubate at 60°C overnight *(11)* (*see* **Note 12**), or the inverted tube containing the tissue and same amount of digestion buffer are incubated overnight in an hot air oven.
3. The enzyme is inactivated at 95°C for 5 min.
4. The RNA is purified using phase separation by 500 µL TRIzol and 100 µL chloroform (*see* **Note 14**).
5. The samples are left for 15 min on ice, after which they are centrifuged at 12,200$g$ for 15 min.
6. The aqueous phase is removed to a new tube, and RNA is precipitated by an equal volume of isopropanol, 0.1 vol 3 $M$ sodium acetate and 1 µL of glycogen.
7. The samples are now kept overnight at –20°C and can be stored for several weeks.
8. When required, the samples are centrifuged at 12,200$g$ for 30 min, the supernatant is removed, and the pellet is washed once with 500 µL 70% ethanol.
9. After centrifugation at 7500$g$ for 5 min, the supernatant is removed, and the pellet is left to air-dry with the tube open on a thermomixer at 65°C for 5–15 min.
10. The pellet is resuspended in 10 µL of sterile water, and can be stored at –80°C for several weeks.

## 4. Notes

1. There are two main systems for performing laser-assisted microdissection. In laser capture microdissection (LCM), an infrared laser is used to melt a special transfer film, which binds to cells identified microscopically. The transfer film with the attached cells is then lifted off and can be used for subsequent analysis. This technology is described extensively in other chapters. In contrast, laser microbeam microdissection (LMM) uses a UV laser to cut selected cells out of a cytological or histological preparation. These two systems each have advantages and disadvantages and the choice of technology will be decided by the purpose of the study and, of course, by the availability of equipment. For example, LCM is particularly suited to projects in which many samples are to be rapidly collected from large tissue areas. In contrast, the superior resolution of the laser cutting in LMM means that this technique is better suited for precise microdissection, particularly of single cells. By mounting tissue sections on support membranes, large tissue fragments can also be dissected intact using LMM.
2. At least three UV laser-based systems for histological microdissection (LMM) are currently commercially available: (1) PALM MicroBeam system (PALM

Microlaser Technologies, Bernried, Germany); (2) SLμCUT (Molecular Machines & Industries, Glattbrugg, Switzerland [formerly SL Microtest]); (3) Leica Laser Microdissection—AS LMD (Leica, Wetzlar, Germany).

These are essentially similar in principle and use similar types of UV lasers to microdissect target cells and ablate adjacent unwanted tissue. They differ in the specifics of the technology used, particularly with regard to the mode of collection and transfer of the microdissected specimen for downstream analysis. Thus, in the PALM system, the laser itself can be used to catapult the microdissected cells into the lid of a microfuge tube in a "no touch" technique—laser pressure catapulting (LPC) *(7)*. In the SLμCUT MMI system, transfer is by needle or glass pipet, or by single-step collection (either using a version of LPC or by picking the cells up directly using an adhesive film in the lid of a reaction tube). In the Leica AS LMD system, the section is inverted so that after cutting, the microdissected cells fall into the reaction tube under the influence of gravity. The latest versions of these systems have improved computer software and a number of new features that together allow more sophisticated microdissection techniques to be applied. These include (though not in all systems) automatic dissection of preselected fields with sorting into multiple reagent tubes, automatic multiple sampling from within a larger preselected area, and variable cut techniques. Each system appears to be adequate for most basic microdissection tasks, although each has its proponents and opponents.

3. Ready-made MOMeNT slides can be purchased from MMI, PALM, or Leica. We also make membrane-mounted slides ourselves; it is not particularly laborious and it allows us to prepare the exact number required each time. This avoids a possible source of contamination and minimizes the risk of introducing additional RNase/DNase. It is possible to keep the mounted slides in a dust-free, light-tight glass-slide storage box at RT, and tissue sections on membrane-mounted slides can be stored in a similar way. We have been able to extract and PCR-amplify DNA from sections stored in this way (so far for at least 2 yr). Similarly, we can retrieve RNA for PCR-based analysis from slides stored for up to at least 3 mo.

4. We have used different kinds of glue, including nail polish, but Fixogum rubber cement works best in our hands, especially when MOMeNT is used in combination with immunohistochemistry or *in situ* hybridization (*see* Chapter 12).

5. There are different protocols for overcoming the hydrophobic nature of the membranes. Both PALM and Leica recommend irradiating the membranes with UV light for 30 min. Coating the membranes with 0.1% poly-L-lysine may also be used. We treat the membrane-mounted slides under UV light overnight to eliminate any contaminating nucleic acids. To increase adherence of the tissue section to the membranes it is also possible to combine UV light with poly-L-lysine, or when extra adhesion is required, with 2% or 8% 3-aminopropyltriethoxysilane (APES) *(10)* (*see* Chapter 12). Extra precautions must be taken when applying these fluids, as they may interfere with LPC and transfer of the cut segment in cases where the fluids have leaked under the membrane. The commercially avail-

able metal slides for single-step collection, are purchased individually wrapped to ensure a DNase- and RNase-free environment, as well as to prevent any contamination with foreign nucleic acids. In this situation, we find treatment with UV light for 30 min to be sufficient.

6. A disadvantage of LAM is inferior tissue morphology during microdissection caused by not being able to use cover slips. This can make it difficult to identify targets by morphology alone (*see* Chapter 12). We have tried several different volatile fluids as "optical media" to compensate for this. Xylene is best for giving optimal morphological details, but is more difficult to distribute than Cytotec, and its use should be restricted on safety grounds. Cytotec is an alcohol-based fixative used in the preparation of cytological smears. We use Cytotec in routine LMM (*see* **Fig. 1E–F** in Chapter 12) and xylene when optimal morphology is important, e.g., when dissecting single cells.

The fluids are applied using a plastic spray bottle. After identifying and marking the target cells, the fluid is left to evaporate before laser microdissection is performed. The specially designed tubes for single-step collection may contain a light diffuser in the cap, improving the morphology. Similarly, the glass slide below the metal slides also helps to improve morphology. Since this described method is one in which the membrane is stuck on the adhesive cap, a volatile medium cannot be used.

7. In most cases, we find it adequate to incubate the slides at 60°C for 15 min to secure adhesion of tissue to the membrane-mounted slides, but in some cases prolonged incubation (up to 2 h) is required in combination with coating of the membranes (*see* **Note 3**).

8. Mayer's hematoxylin binds to DNA and several papers have reported that this dye may interfere with PCR-based amplification of DNA *(12–14)*. In contrast, other groups have not found this to be a problem *(15)* and single-cell microdissection of hematoxylin-stained sections with subsequent PCR-based analysis of DNA is possible *(16)*. It appears that RNA is not affected in the same way; several papers have reported successful RT-PCR on single cells microdissected from hematoxylin-stained FFPE *(7,8)*. We find that Mayer's hematoxylin does not affect RNA quantification, and it is our preferred histological stain because it gives improved morphological detail. However, when working with small numbers of (<50) microdissected cells or single cells, it might be advisable to use a different stain, such as methyl green or toluidine blue, when amplifying DNA or when performing quantitative analysis of DNA. Eosin is frequently used together with hematoxylin, but we find that this quite often results in a very dark stain with inferior morphology, especially when staining lymph nodes.

9. A major obstacle to successful LMM is wet sections, or fluid leaking under the membranes, as this interferes with the cutting efficiency of the laser. After the slides have been prepared with FFPE or frozen tissue sections, they should be allowed to dry at RT for approx 30 min. We then normally perform LMM immediately. In special cases, the slides can be dried at 37°C for 30 min. Some authors *(17)* and PALM recommend that sections be dried at 37°C overnight, but we

believe that the hands-on time should be as short as possible to diminish problems with external and internal RNase/DNase. The metal slides for single-step
collection take less time to dry, and only 5–10 min is required.

10. The needle with the microdissected fragment can be transferred in different ways.
We previously cut off the entire needle together with the microdissected cells,
and placed everything in the reagent tube. Now we prefer to use the needle to
place the cut fragment into a drop of fluid on the wall of the PCR tube. If the
sample is placed directly onto the wall, static electrical forces can make it very
difficult to get the microdissected sample into the digestion buffer.

11. In single-step collection, the tissue lies protected between the membrane and
the opposing clean Menzel glass slide. The adhesive film in the cap is not in
contact with the tissue, only with the membrane, to which the tissue is adherent. Single-step collection allows for sampling of several segments or single
cells in one cap without removing the cap in between. We prefer to use single
step collection for microdissection of single cells. Sampling of 10 cut cells
from paraffin-embedded tissue, with subsequent I-PEP-PCR, is sufficient for
PCR of a housekeeping gene (*see* **Fig. 2E** in Chapter 12). Digestion of sampled
single cells or small tissue fragments (<100 cells), can preferentially be accomplished by adding 10 µL of buffer directly onto the cells. Since single cells are
not visible to the eye, the cells should be sampled centrally in the cap. Similarly, the digestion buffer should also be placed centrally. Digestion of single
cells for DNA analysis is accomplished using our DNA digestion buffer with
glycogen, and incubation is performed overnight. Glycogen is used as a carrier
to prevent nucleic acid binding to the plastic wall of the reaction tube. When
dissecting larger fragments for subsequent RNA extraction, we prefer to use
our standard total RNA extraction protocol.

12. We usually perform proteolytic digestion overnight. Shorter incubation times
result in products suitable for PCR. We have obtained amplifiable DNA from
pooled single cells after 2 h incubation; however, further incubation increases the
DNA yield and the chances for successful amplification. Using the single-step
collection system, it is possible to check that all cells present have been totally
digested. To do this, the cap is replaced on the cap-holder, lowered onto a glass
slide, and the digestion status is controlled visually. Proteinase K digestion can
be performed using different concentrations of the enzyme. For DNA and RNA
extraction from larger cut areas, we use 0.3 mg/mL. When digesting single cells,
we use 4 mg/mL *(18)*.

13. When microdissecting 1500 cells or less for DNA analysis, we use the template
in the digestion buffer, without any precipitation steps. For larger samples, an
ethanol precipitation should be considered.

14. We have tested different commercially available kits for total RNA extraction,
and we prefer the TRIzol reagent for this procedure. However, new kits are being
produced on a regular basis and readers should choose the method with which
they are most familiar.

## References

1. Bohm, M., Wieland, I., Schutze, K., and Rubben, H. (1997) Microbeam MOMeNT: non-contact laser microdissection of membrane-mounted native tissue. *Am. J. Pathol.* **151,** 63–67.
2. Aubele, M., Zitzelsberger, H., Schenck, U., Walch, A., Hofler, H., and Werner, M. (1998) Distinct cytogenetic alterations in squamous intraepithelial lesions of the cervix revealed by laser-assisted microdissection and comparative genomic hybridization. *Cancer* **84,** 375–379.
3. Luo, L., Salunga, R. C., Guo, H., et al. (1999) Gene expression profiles of laser-captured adjacent neuronal subtypes. *Nat. Med.* **5,** 117–122.
4. Weber, R. G., Scheer, M., Born, I. A., et al. (1998) Recurrent chromosomal imbalances detected in biopsy material from oral premalignant and malignant lesions by combined tissue microdissection, universal DNA amplification, and comparative genomic hybridization. *Am. J. Pathol.* **153,** 295–303.
5. Bohm, M. and Wieland, I. (1997) Analysis of tumour-specific alterations in native specimens by PCR: How to procure the tumour cells! (review). *Int. J. Oncol.* **10,** 131–139.
6. Emmert-Buck, M. R., Bonner, R. F., Smith, P. D., et al. (1996) Laser capture microdissection. *Science* **274,** 998–1001.
7. Schutze, K. and Lahr, G. (1998) Identification of expressed genes by laser-mediated manipulation of single cells. *Nat. Biotechnol.* **16,** 737–742.
8. Lahr, G. (2000) RT-PCR from archival single cells is a suitable method to analyze specific gene expression. *Lab. Invest.* **80,** 1477–1479.
9. Banks, R. E., Dunn, M. J., Forbes, M. A., et al. (1999) The potential use of laser capture microdissection to selectively obtain distinct populations of cells for proteomic analysis—preliminary findings. *Electrophoresis* **20,** 689–700.
10. Gjerdrum, L. M., Lielpetere, I., Rasmussen, L. M., Bendix, K., and Hamilton-Dutoit, S. (2001) Laser-assisted microdissection of membrane-mounted paraffin sections for polymerase chain reaction analysis: identification of cell populations using immunohistochemistry and in situ hybridization. *J. Mol. Diagn.* **3,** 105–110.
11. Specht, K., Richter, T., Muller, U., Walch, A., Werner, M., and Hofler, H. (2001) Quantitative gene expression analysis in microdissected archival formalin-fixed and paraffin-embedded tumor tissue. *Am. J. Pathol.* **158,** 419–429.
12. Burton, M. P., Schneider, B. G., Brown, R., Escamilla-Ponce, N., and Gulley, M. L. (1998) Comparison of histologic stains for use in PCR analysis of microdissected, paraffin-embedded tissues. *Biotechniques* **24,** 86–92.
13. Murase, T., Inagaki, H., and Eimoto, T. (2000) Influence of histochemical and immunohistochemical stains on polymerase chain reaction. *Mod. Pathol.* **13,** 147–151.
14. Serth, J., Kuczyk, M. A., Paeslack, U., Lichtinghagen, R., and Jonas, U. (2000) Quantitation of DNA extracted after micropreparation of cells from frozen and formalin-fixed tissue sections. *Am. J. Pathol.* **156,** 1189–1196.

15. Ehrig, T., Abdulkadir, S. A., Dintzis, S. M., Milbrandt, J., and Watson, M. A. (2001) Quantitative amplification of genomic DNA from histological tissue sections after staining with nuclear dyes and laser capture microdissection. *J. Mol. Diagn.* **3,** 22–25.

16. Kuppers, R., Zhao, M., Hansmann, M. L., and Rajewsky, K. (1993) Tracing B cell development in human germinal centres by molecular analysis of single cells picked from histological sections. *EMBO J.* **12,** 4955–4967.

17. Lehmann, U. and Kreipe, H. (2001) Real-time PCR analysis of DNA and RNA extracted from formalin-fixed and paraffin-embedded biopsies. *Methods* **25,** 409–418.

18. Dietmaier, W., Hartmann, A., Wallinger, S., et al. (1999) Multiple mutation analyses in single tumor cells with improved whole genome amplification. *Am. J. Pathol.* **154,** 83–95.

# 12

## Laser-Assisted Microdissection of Membrane-Mounted Sections Following Immunohistochemistry and *In Situ* Hybridization

### Lise Mette Gjerdrum and Stephen Hamilton-Dutoit

### Summary

Laser microbeam microdissection (LMM) is an increasingly important histological technique for obtaining homogeneous cell populations and tissue components in order to analyze target-specific changes in genes, gene expression, and proteins. The quality of data obtained with LMM is heavily dependent on the precision with which the target for microdissection can be identified. Since no cover slip is used during LMM, tissue morphology is poor compared with traditional light microscopy. This hampers morphological recognition of targets for microdissection in routinely stained sections and can be a limiting factor in the use of this technique. Immunohistochemistry (IHC) and *in situ* hybridization (ISH) can improve the identification of specific cell populations *in situ* in tissue sections, but there are a number of problems in applying these methods to slides prepared for LMM. In this chapter, we present optimized protocols that allow IHC to be performed for detecting a wide range of antigens in conjunction with LMM, both on formalin-fixed paraffin-embedded and on frozen sections. In addition, we present a quick, versatile protocol for performing ISH on archival material suitable for LMM.

**Key Words:** Laser-assisted microdissection; membrane-mounted tissue; MOMeNT; frozen tissue; formalin-fixed paraffin-embedded tissue; immunohistochemistry; *in situ* hybridization.

## 1. Introduction

This chapter focuses on methods for improved identification of cell populations in membrane-mounted sections intended for laser microbeam microdissection (LMM). Pathologically altered tissues are often markedly heterogeneous-malignant tumors, for example containing a variable (and often significant) admixture of non-neoplastic cells. Some tumors may be made up

From: *Methods in Molecular Biology, vol. 293: Laser Capture Microdissection: Methods and Protocols*
Edited by: G. I. Murray and S. Curran © Humana Press Inc., Totowa, NJ

of neoplastic cells that are morphologically relatively similar to normal cell types (e.g., many lymphomas), making reliable identification of target cells in paraffin-embedded sections difficult and in frozen sections often impossible. Furthermore, since cover slips are not used during LMM, tissue morphology is poor compared with traditional light microscopy, hampering the morphological recognition of targets for microdissection in routinely stained sections. One approach to avoiding the problems associated with poor morphology is to immunostain target cells with antibodies against specific antigens not expressed in other cells in the tissue *(1–6)*, thus increasing the specificity, precision, and speed of microdissection. LMM may be performed in combination with MOMeNT (microdissection of membrane-mounted native tissue), thus facilitating dissection and transfer of large intact specimens and reducing the risk of contamination. However, this technique is difficult to use in combination with traditional immunohistochemistry (IHC) and *in situ* hybridization (ISH) staining methods. Both the tissue section and the membranes are easily damaged using standard staining protocols. In particular, the use of microwave superheating for heat-induced epitope retrieval (HIER)—an often essential step if sufficiently sensitive staining of paraffin sections is to be achieved—may result in detachment of both the membranes and sections, and it is potentially damaging to target nucleic acids *(7)*.

In response to these problems, we have optimized protocols for IHC and ISH using "low"- temperature HIER, which can be used reliably with membrane-mounted tissue sections. There are many different staining protocols for IHC and we have focused on the detection systems used routinely in our laboratory. The protocols presented here can be performed on formalin-fixed paraffin-embedded (FFPE) and frozen tissue, and are suitable for detection of a wide range of antibodies and different nucleic acids *(8)*. The protocols are easy to perform, and have been tested in various different settings and organ-systems.

## 2. Materials

### 2.1. Immunohistochemistry for MOMeNT on Formalin-Fixed Paraffin-Embedded Slides

1. Humid chamber.
2. Disposable plastic pipets.
3. Membrane-mounted slides—two types as described in Chapter 11.
4. Xylene stored at room temperature (RT).
5. Graded alcohols (99%, 96%) at RT.
6. 0.5% (v/v) $H_2O_2$ in methanol: 100 mL methanol + 1.5 mL 35% w/w $H_2O_2$.
7. TEG buffer: Tris-EGTA, pH 9.0, 0.1% diethylpyrocarbonate (DEPC) at RT.

8. Primary antibody: Melan-A (M7196, DakoCytomation, Copenhagen, Denmark) or Ki-67 (M7240, DakoCytomation), both used 1:10 in TBS, stored at 4°C.
9. TBS buffer: 50 m$M$ Tris-HCl, 150 m$M$ NaCl$_2$, pH 7.6 , 0.1% DEPC at RT.
10. EnVision™ + (peroxidase conjugated mouse; DakoCytomation K4001) at 4°C.
11. Chromogen: Diaminobenzidine (DAB) 4170 (Kem-En-Tec Diagnostics A/S, Copenhagen, Denmark).
12. Hydrogen peroxide (H$_2$O$_2$).
13. Histological counterstain: Mayer's hematoxylin, toluidine blue, or methyl green.

## 2.2. Immunohistochemistry for MOMeNT on Frozen Sections

1. Humid chamber.
2. Disposable plastic pipets
3. Membrane-mounted slides—two types as described in Chapter 11.
4. Absolute acetone at 4°C.
5. Coons buffer: 0.9% NaCl, 0.45 g Na$_2$HPO$_4$·2H$_2$O, 1.68 g NaH$_2$PO$_4$·H$_2$O at pH 7.1 per liter with 1% bovine serum albumin (BSA) at 4°C.
6. Primary antibody: Melan-A (M7196, DakoCytomation) 1:10 or Ki-67 (M7240, DakoCytomation) 1:5, both in Coons buffer at 4°C.
7. Secondary antibody (biotinylated swine anti-mouse immunoglobulins; Dako-Cytomation, E 0453) and avidin-biotinylated horseradish peroxidase complex (ABC, K0355, DakoCytomation) or EnVision™ + (peroxidase conjugated mouse; DakoCytomation K4001) at 4°C.
8. Tris buffer: 100 m$M$ Tris-HCl, pH 7.6, 150 m$M$ NaCl$_2$ at RT.
9. Chromogen: DAB (3,3'-diaminobenzidine tetrahydrochloride) 4170, Kem-En-Tec Diagnostics A/S).
10. H$_2$O$_2$.
11. Histological counterstain: Mayer's hematoxylin, toluidine blue, or methyl green.
12. Graded alcohols (99%, 96%) at RT.

## 2.3. In Situ Hybridization for MOMeNT on Formalin-Fixed Paraffin-Embedded Slides

1. Humid chamber.
2. Disposable plastic pipets.
3. Membrane-mounted slides—two types as described in Chapter 11.
4. Xylene at RT.
5. Graded alcohols (99%, 96%) at RT.
6. 0.1% DEPC/H$_2$O at RT.
7. 0.4% Paraformaldehyde/10x PBS (pH 7.4) at RT.
8. Proteinase K (S3004, 1:10 in TBS, DakoCytomation) at 4°C .
9. Cover slip (nonsiliconized).
10. EBER-specific probe (developed in our laboratory): hybridization mixture consists of 2 ng/µL probe, 50% deionized formamide, 2x standard saline citrate (SSC), 10% dextran sulfate, and 200 µg/mL tRNA. Store at 4°C until use.

11. Monoclonal anti-digoxin mouse primary antibody (D8156 [clone D1-22], 1:2500; Sigma-Aldrich Denmark A/S, Copenhagen, Denmark).
12. 0.5% (v/v) $H_2O_2$ in methanol: 100 mL methanol + 1.5 mL 35% w/w $H_2O_2$.
13. 2x SSC: 3 $M$ $NaCl_2$, 0.3 $M$ Na citrate, pH 7.0.
14. EnVision™+ (peroxidase conjugated mouse; DakoCytomation, K4001).
15. TBS buffer at RT.
16. Chromogen (DAB; 4170 Kem-En-Tec Diagnostics A/S).
17. Counterstain: Mayer's hematoxylin or toluidine blue.

## 3. Methods

In the following section, we present brief staining protocols for IHC on FFPE and frozen tissues, and for ISH on FFPE. As a model for IHC protocols, we stained sections of malignant melanoma lymph node metastases for Melan-A and Ki-67 (*see* **Fig. 1G–H**), on both FFPE and frozen tissues. ISH was performed on paraffin sections from a post-transplantation B-cell lymphoma. This tumor was Epstein-Barr virus (EBV)-positive, as shown by nuclear positivity for EBERs (small EBV encoded early RNAs) detected using RNA-ISH with a digoxigenin-labeled polymerase chain reaction (PCR)-generated single-stranded DNA probe developed in our laboratory *see* **Fig. 2A–D**).

## 3.1. Immunohistochemical Stains for Melan-A and Ki-67 on FFPE Tissues Using EnVision + Staining Protocol

### 3.1.1. Removal of Paraffin and Blocking of Endogenous Peroxidase

1. Preparation of membrane-mounted slides and cutting of FFPE tissues has been described in detail in Chapter 11.
2. If the sections are handled carefully, no extra adhesion (e.g., with poly-L-lysine or APES) of the tissue is normally needed (*see* **Note 1**).
3. Removal of paraffin is carried out in xylene for $2 \times 2$ min, followed by 2 min in 99% and 96% alcohol each.
4. Endogenous peroxidase is blocked with 0.5% $H_2O_2$ in methanol (*see* **Note 2**) and sections are briefly washed in distilled water for 1 min.

---

Fig. 1. (*opposite page*) UV laser-assisted microdissection of membrane-mounted immunostained paraffin sections. (**A–D**) Membrane-mounted sections from Epstein-Barr-infected, undifferentiated nasopharyngeal carcinoma stained with pancytokeratin AE1/3. (**A**) Positive staining helps identify target carcinoma areas before microdissection. (**B**) Targeted area after cutting. (**C**) Same area after removal of target fragment. (**D**) Microdissected fragment on the needle tip. (**E–F**) Skin stained with H & E prior to microdissection. Before (**E**) and after (**F**) using optical medium (Cytotec) to improve morphology. (**G–H**) Membrane-mounted sections from lymph node metastasis from malignant melanoma stained for cytoplasmic Melan-A (**G**) and nuclear Ki-67 antigen (**H**).

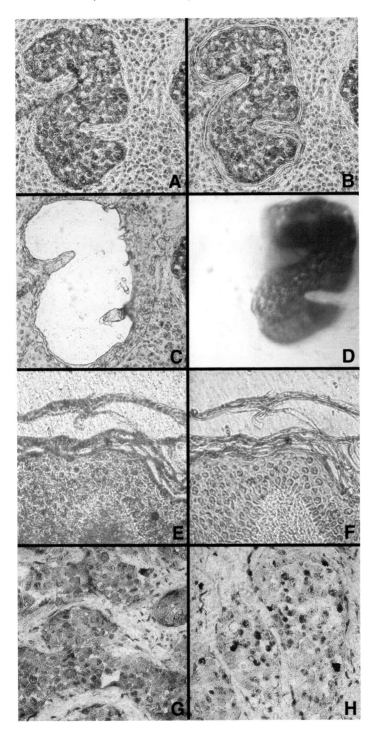

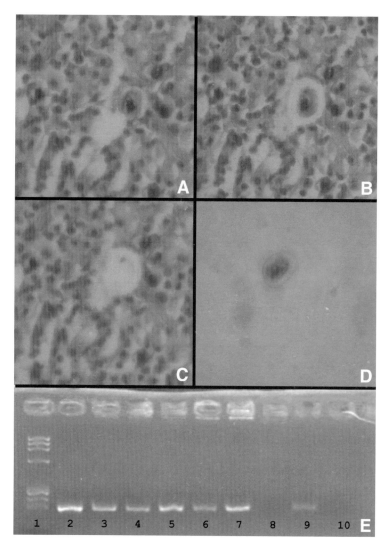

Fig. 2. UV laser-assisted single-cell microdissection of EBV-positive Hodgkin's lymphoma in paraffin sections (**A–D**). Hodgkin/Reed-Sternberg (HRS) cells show strong nuclear staining after *in situ* hybridization for EBERs. A single HRS cell is seen before microdissection (**A**) and after isolation by laser cutting (**B**). (**C**) The same field after the cell is removed. (**D**) High-power view of the adhesive cap in the PCR tube containing a single microdissected HRS cell. (**E**) Gel analysis of PCR products after amplification for the housekeeping gene NPM. Lane 1: size marker; lane 2: positive control consisting of 50 microdissected neoplastic and non-neoplastic cells; lane 3: 25 individually dissected HRS cells; lanes 4 and 5: 15 individually dissected HRS cells; lanes 6 and 7: 10 individually dissected HRS cells; lanes 8 and 10: negative controls; lane 9: positive whole section Hodgkin's lymphoma control.

## 3.1.2. "Low"-Temperature Heat-Induced Epitope Retrieval

1. Unmasking of epitopes is achieved by prolonged "low"-temperature incubation in TEG buffer (pH 9.0) at 60°C in a hot-air oven *(8)* (*see* **Note 3**).
2. The sections are placed in a glass Coplin jar, covered with ample buffer (to avoid drying out), and incubated overnight (approx 18 h).
3. The next day, the sections are allowed to cool down for 20 min in the same buffer.
4. The slides are then placed in a humid chamber and washed briefly in TBS buffer using disposable plastic pipets.

## 3.1.3. Immunostaining and Visualization

1. Sections are covered with primary antibody solution (200 µL of Melan-A or Ki-67; both diluted 1:10 in TBS), incubated for 5 min, then washed in TBS using disposable plastic pipets for approx 1 min (*see* **Note 2**).
2. They are then incubated in 200 µL undiluted EnVision + for 10 min, before again being washed in TBS.
3. One tablet of DAB is dissolved in 10 mL distilled water approx 30 min before use.
4. Immediately before application, 10 µL $H_2O_2$ is added.
5. Slides are incubated in DAB for 10 min and washed in TBS.
6. The sections are counterstained with Mayer's hematoxylin for 2 min (*see* **Note 4**), washed in distilled water for 1 min, and dehydrated through graded alcohols (99%, 96%, 1 min each).
7. The sections are allowed to dry in a fume cupboard before microdissection (*see* **Note 5**).

## 3.2. Melan-A and Ki-67 Staining Protocols for Frozen Sections Using EnVision + and ABC Detection

### 3.2.1. Fixation

1. The cut frozen sections are placed on membrane-mounted sections as described in Chapter 11, before being fixed in cold absolute acetone for 3 min, and then dried for 15 s.
2. The slides are placed in a humid chamber and washed with Coons buffer for 1 min using a disposable plastic pipet.

### 3.2.2. Immunostaining and Visualization

1. Sections are covered with the primary antibody solution (Mela-A [1:10] or Ki-67 [1:5] in Coons buffer, 200 µL each), incubated for 5 min, and then washed in Coons buffer.
2. For Melan-A staining, 200 µL of undiluted EnVision™ + is applied for 10 min and then washed in Tris-HCl.
3. Alternatively for Ki-67 staining, sections are first incubated for 5 min in 200 µL of secondary biotinylated anti-mouse immunoglobulin antibody, washed in TRIS, incubated for 5 min in the tertiary layer avidin-biotinylated horseradish peroxidase complex (200 µL), and finally washed in Tris-HCl.

4. Signal detection using DAB, counterstaining, dehydration and drying are carried out as described for paraffin section immunohistochemistry.

### 3.3. In Situ *Hybridization for EBER on FFPE Tissues Using EnVision + Visualization System*

#### 3.3.1. Preparation of Slides and Sections

1. The preparation of the membrane-mounted glass slides and cutting of the tissue is described in detail in Chapter 11.
2. Adhesion of the tissue is achieved by heating for 15–30 min (*see* **Note 1**) in a hot-air oven (60°C).
3. Sections are deparaffined in xylene for 2 × 2 min, followed by 99% and 96% ethanol, for 2 min each.

#### 3.3.2. Tissue Digestion and Fixation

1. The sections are immersed in 0.1% DEPC/$H_2O$ for 2 × 2 min, and then digested with proteinase K (1:10 in TBS) for 4–10 min.
2. After washing in 0.1% DEPC/$H_2O$ for 1 min (*see* **Note 2**), the sections are post-fixed for 15 min in 0.4% paraformaldehyde in PBS (*see* **Note 6**).
3. After washing in 0.1% DEPC/H2O for 1 min the slides are dehydrated by dipping 5 times in 99% ethanol, and then air-dried for 2 min.

#### 3.3.3. Hybridization and Visualization

1. Sections are hybridized with the probe for 60–90 min at 55°C (*see* **Note 7**).
2. Sections are covered with a standard nonsiliconized glass cover slip.
3. PAP pens (or similar hydrophobic slide markers) should not be used (*see* **Note 8**).
4. The cover slip is picked up carefully with forceps and excess probe is removed by sequential washing steps in graded SSC: 2x SSC at 55°C for 8 min; 0.2x SSC at 40°C for a further 8 min; and finally 0.1x SSC at RT for 4 min (*see* **Note 9**).
5. The slides are washed in TBS for 2 × 2 min, followed by labeling with anti-digoxin for 8 min.
6. After immunolabeling, the sections are blocked for endogeneous peroxidase using 0.5% $H_2O_2$ in methanol for 5 min, immersed in TBS for 2 × 3 min (*see* **Note 2**), and EnVision + is applied undiluted for 8 min.
7. Sections are rinsed in TBS for 2 × 3 min, followed by DAB for 5 min.
8. One tablet of DAB is dissolved in 10 mL of distilled water approx 30 min before use, and immediately before application, 10 μL $H_2O_2$ is added.
9. The slides are immersed in distilled water for 1 min, and then counterstained with 0.1% toluidine blue for 30 s (*see* **Note 4**).
10. The sections are dehydrated in graded ethanol (70%, 96%, and 99% ethanol for 2 min each) before being air-dried.
11. The slides are now ready for microdissection (*see* **Note 10**, **Fig. 1A–D**).

## 4. Notes

1. Adhesion of the cut sections to the membranes varies according to the tissue. Some tissues require firm adhesion, which can be achieved by coating the membrane with 8% APES *(8)*. However, if the sections are handled gently during the washing steps, we find that most tissues stay on the membrane during the procedure. We normally place the MOMeNT slides in a hot-air oven at 60°C for 15 min to help ensure tissue adhesion. For sections to be used for ISH, we prefer to incubate the slides for 30 min as they undergo treatment in proteinase K.

2. In our standard protocols, we block endogenous peroxidase using methanol and $H_2O_2$ for 10 min, but in this short staining procedure, we find 5 min to be sufficient. All washing steps are reduced by several minutes, without any significant impact on the staining result. This is important, as it is deleterious to mRNA preservation for the slides to be immersed in aqueous media, in part because this activates tissue ribonucleases (RNAses). This is particularly important when dealing with cryostat sections *(1,3)*. In FFPE sections, the internal RNAses are inactivated during formalin fixation. In these cases, the major threat to RNA integrity comes from external RNAses introduced from the outside environment (e.g., in wash buffers). We use DEPC-treated buffers and equipment to avoid this potential problem.

3. Formalin fixation results in extensive cross-linking of proteins that will often mask target epitopes from immunohistochemical detection. Optimal detection of many antigens in paraffin sections requires the use of short periods of superheating (e.g., to >100°C for 20 min) in buffers of variable pH to unmask these hidden epitopes. The membranes used in MOMeNT can rarely survive this treatment. As an alternative, the protocols described here make use of prolonged heating at a lower temperature (60°C) to achieve similar antigen retrieval without damaging the sections or membranes. "Low"-temperature HIER is performed in a hot-air oven at 60°C. The sections are placed in an ample volume of buffer (TEG, pH 9.0) for different retrieval times depending on the antibody used. For Melan-A and Ki-67, 18 h is sufficient to give clear positive staining with low background (*see* **Fig.1G–H**), while 24 h is recommended for staining cytokeratins (e.g. pan-cytokeratins AE1/AE3 [DakoCytomation; *see* **Fig. 1A–D**], KL1, and high molecular weight cytokeratin) *(8)*. For detection of antigens with other antibodies (e.g., CD30 antigen and EBV-encoded LMP-1, DakoCytomation) 48 h of retrieval is required to achieve strong positive staining with low background. During "low"-temperature HIER, it is important that the sections are completely immersed in TEG buffer, so that they do not dry out. The glue used for attaching the membranes to the slides is silicone-based and in some cases this can become detached. To overcome this problem, we apply a second layer of glue to the edge of the membrane, after UV light treatment. The metal slides for single step collection can undergo "low"-temperature HIER without any difficulties. Pronase digestion for antigen retrieval does not work well in our hands, resulting in either

weak stains with low background, or strong, false-positive stains with high background.

4.  Mayer's hematoxylin binds to DNA; several papers have reported that this dye may interfere with PCR-based amplification of DNA *(7,9,10)*. In contrast, other groups have not found this to be a problem *(11)* and single cell microdissection of hematoxylin-stained sections with subsequent PCR-based analysis of DNA is possible *(12)*. It appears that RNA is not affected in the same way, and several papers have reported successful RT-PCR on single-cells microdissected from hematoxylin-stained FFPE *(13,14)*. However, when working with small numbers of microdissected cells (<50) or single cells, it may be advisable to use methyl green or toluidine blue instead. In these circumstances we prefer to use toluidine blue as the counterstain.

5.  It is difficult, and often impossible, to microdissect wet sections, so the slides must be allowed to dry. *See* **Note 9** in Chapter 11 for further details.

6.  Paraformaldehyde (0.4%) is made by dissolving 8 g of paraformaldehyde (Sigma P-6148) in 100 mL of 20x PBS and 100 mL of 0.1% DEPC/$H_2O$. The solution is heated to a temperature of 70°C whilst mixing until it is clear (approx 60 min). Prolonged storage of this fixative is possible at –20°C. Thawed fixative should be replaced each week.

7.  We use a digoxigenin-labeled PCR-generated single-stranded DNA probe developed within our laboratory for EBER-ISH. A number of alternative probes for EBER-ISH are available commercially. For example, fluoresceinated EBER-specific oligonucleotide probe (NCL-EBV, Novocastra, Newcastle, UK) can be used as recommended by the manufacturer. Hybridized probe can be detected in various ways, for example using monoclonal mouse anti-FITC antibody (M0878, DakoCytomation), followed by biotinylated second-stage antibody and ABC.

8.  We do not normally use PAP pens (or similar hydrophobic slide markers) to encircle the tissue prior to immunostaining or hybridization, as these may cause either damage or wrinkling to the membranes.

9.  This wash step is crucial. We have used different stringent washing solutions and some result in loss of both the hybrids and/or the tissue. For example, in our routine ISH protocol, we use stringent wash concentrate (DakoCytomation S3500, 1:50 for 25 min). However, this results in unacceptable section loss when used with membrane-mounted slides.

10. Extraction of DNA and RNA is described in detail in Chapter 11. No special kits or precautions are needed. These protocols all allow for extraction and amplification of standard quality template. Standard methods are used to visualize amplified DNA on ethidium bromide-tained agarose gels *(8)* (*see* **Fig. 2E**), and quantification of RNA is by real-time quantitative RT-PCR. Immunostaining in frozen tissues does decrease the overall yield of available RNA, compared to hematoxylin stain alone. The choice of antibody and staining system might also have an impact on the yield, but in our hands, quantification of RNA is still possible.

## References

1. Murakami, H., Liotta, L., and Star, R. A. (2000) IF-LCM: laser capture microdissection of immunofluorescently defined cells for mRNA analysis rapid communication. *Kidney Int.* **58,** 1346–1353.
2. Fink, L., Kinfe, T., Stein, M. M., Ermert, L., Hanze, J., Kummer, W., et al. (2000) Immunostaining and laser-assisted cell picking for mRNA analysis. *Lab Invest.* **80,** 327–333.
3. Fend, F., Emmert-Buck, M. R., Chuaqui, R., Cole, K., Lee, J., Liotta, L. A., and Raffeld, M. (1999) Immuno-LCM: laser capture microdissection of immunostained frozen sections for mRNA analysis. *Am. J. Pathol.* **154,** 61–66.
4. Fink, L., Kinfe, T., Seeger, W., Ermert, L., Kummer, W., and Bohle, R. M. (2000). Immunostaining for cell picking and real-time mRNA quantitation. *Am. J. Pathol.* **157,** 1459–1466.
5. Jin, L., Thompson, C. A., Qian, X., Kuecker, S. J., Kulig, E., and Lloyd, R. V. (1999) Analysis of anterior pituitary hormone mRNA expression in immunophenotypically characterized single cells after laser capture microdissection. *Lab. Invest.* **79,** 511–512.
6. d'Amore, F., Stribley, J. A., Ohno, T., Wu, G., Wickert, R. S., Delabie, J., et al. (1997) Molecular studies on single cells harvested by micromanipulation from archival tissue sections previously stained by immunohistochemistry or nonisotopic *in situ* hybridization. *Lab. Invest.* **76,** 219–224.
7. Murase, T., Inagaki, H., and Eimoto, T. (2000) Influence of histochemical and immunohistochemical stains on polymerase chain reaction. *Mod. Pathol.* **13,** 147–151.
8. Gjerdrum, L. M., Lielpetere, I., Rasmussen, L. M., Bendix, K., and Hamilton-Dutoit, S. (2001) Laser-assisted microdissection of membrane-mounted paraffin sections for polymerase chain reaction analysis: identification of cell populations using immunohistochemistry and *in situ* hybridization. *J. Mol. Diagn.* **3,** 105–110.
9. Burton, M. P., Schneider, B. G., Brown, R., Escamilla-Ponce, N., and Gulley, M. L. (1998) Comparison of histologic stains for use in PCR analysis of microdissected, paraffin-embedded tissues. *Biotechniques* **24,** 86–92.
10. Serth, J., Kuczyk, M. A., Paeslack, U., Lichtinghagen, R., and Jonas, U. (2000) Quantitation of DNA extracted after micropreparation of cells from frozen and formalin-fixed tissue sections. *Am. J. Pathol.* **156,** 1189–1196.
11. Kuppers, R., Zhao, M., Hansmann, M. L., and Rajewsky, K. (1993) Tracing B cell development in human germinal centres by molecular analysis of single cells picked from histological sections. *EMBO J.* **12,** 4955–4967.
12. Lehmann, U. and Kreipe, H. (2001) Real-time PCR analysis of DNA and RNA extracted from formalin-fixed and paraffin-embedded biopsies. *Methods* **25,** 409–418.
13. Schutze, K. and Lahr, G. (1998) Identification of expressed genes by laser-mediated manipulation of single cells. *Nat. Biotechnol.* **16,** 737–742.
14. Lahr, G. (2000) RT-PCR from archival single cells is a suitable method to analyze specific gene expression. *Lab. Invest.* **80,** 1477–1479.

# 13

## Laser-Assisted Cell Microdissection Using the PALM System

Patrick Micke, Arne Östman, Joakim Lundeberg, and Fredrik Ponten

### Summary

Laser-assisted microdissection has enabled the collection of morphologically defined cell populations from a tissue section. The PALM® Robot MicroBeam laser microdissection system provides a robust system for the retrieval of specified cells (including single cells). Due to the fragile nature of DNA, and in particular RNA, robust protocols are required to obtain reliable data from a limited number of cells (1–10.000 cells). This chapter describes the application of the PALM MicroBeam system to isolate RNA and DNA from cells in a complex tissue for subsequent molecular analysis. Protocols for successful analysis of RNA from 500 to 1000 cells, including steps to produce cDNA for subsequent polymerase chain reaction analysis, are given. The cDNA could also be used as a template for linear amplification in order to perform gene array analysis. Furthermore, a protocol for genomic analysis of p53 mutations from single cells is given. The described procedures emphasize preparation of tissue, laser microdissection including catapulting of cells, and extraction of RNA and DNA. Downstream experiments for validation are also shown.

**Key Words:** RNA isolation; real-time quantitative PCR; p53; PALM; microdissection.

## 1. Introduction

The PALM® Robot Microbeam laser microdissection system (P.A.L.M. GmbH, Bernried, Germany) provides a valuable tool for laser microdissection of selected cell populations and single cells from tissue sections. A need for large number of cells often limits the possibility of analyzing defined cell populations due to the mixture of cell types in complex tissues. However, recent technical developments to analyze nucleic acids and proteins from only limited amounts of cells has created a prerequisite for precise microdissection. Microdissection of cells defined under the microscope ensures a relevant

From: *Methods in Molecular Biology, vol. 293: Laser Capture Microdissection: Methods and Protocols*
Edited by: G. I. Murray and S. Curran © Humana Press Inc., Totowa, NJ

selection of templates for subsequent molecular analysis. Microscopical evaluation remains the gold standard to determine morphological characteristics that define normal and abnormal cell populations in a tissue; thus, tools for noncontact microdissection are required to define genetic and transcriptional events underlying phenotypic differences between normal and abnormal cells.

The principle of the PALM system is based on a pulsed UVA laser that is focused through the microscope to allow laser ablation of cells and tissue on a tissue section. The physical mechanism for such laser beam function is photofragmentation, a photochemical process that "transforms" biological material into atoms that are blown away at supersonic velocities. This force is restricted to a minute laser focal spot (<1 μm), leaving adjacent material (cells, nucleic acids, proteins, etc.) fully intact *(1,2)*. This chapter describes the application of the PALM MicroBeam laser microdissection system to isolate RNA and DNA for downstream molecular biological analysis.

The original quality and subsequent handling of tissue used as a template source is of fundamental importance. This is especially true for RNA, which is rapidly degraded by the widespread presence of ribonuclease. We have focused on using fresh, unfixed tissues to develop a robust methodology. Molecular analysis of material fixed in various fixatives is also possible. Formalin fixation degrades genomic DNA and may also create artifactual mutations *(3)*. Although selected DNA gene fragments and to a lesser extent mRNA, can be amplified from only few or single fixed cells, more global approaches, e.g., transcription profiling using microarray technology, may require substantially more cells.

## 2. Materials

1. RNaseZap (Sigma, St. Louis, MO).
2. Superfrost plus charged glass slides (Menzel Gläser, Braunschweig, Germany).
3. PALM LiquidCoverglass (P.A.L.M. AG, Bernried, Germany).
4. Polyethylene membrane covered slides (P.A.L.M. AG.).
5. Zincfix buffer: 5 g $ZnCl_2$, 6 g $ZnAc_2 \cdot 2 H_2O$, 0.1 g $CaAc_2$, in 1 L 0.1 $M$ Tris-HCl, pH 7.4, make fresh if required, stable at room temperature (RT) for approx 1 wk.
6. 70% Ethanol.
7. 95% Ethanol.
8. Hematoxylin solution.
9. PALM Laser-MicroBeam System (P.A.L.M. AG.).
10. RNase free water (Ambion, Austin, TX).
11. RNasin (Promega UK, Southampton, UK).
12. Zymogen micro RNA Isolation kit (Zymo Research, Orange, CA).
13. Agilent 2100 Bioanalyser (Agilent Biotechnologies, Palo Alto, CA).
14. RNA 6000 Pico Labchip kit (Agilent Biotechnologies).
15. RNA 6000 ladder (Ambion).

16. Oligo-dT primer.
17. Primer for real-time polymerase chain reaction (PCR) analysis of GAPDH.
18. Linear acrylamide (Ambion Ltd., Cambridgeshire, UK).
19. Ultrapure dNTPs (Clontech, Palo Alto, CA).
20. Reverse transcriptase 5X buffer (included in Superscript II kit, Invitrogen, Lidingö, Sweden).
21. DTT (included in Superscript II kit, Invitrogen).
22. Superscript II (included in Superscript II kit, Invitrogen).
23. Microcon 100 columns (Millipore AB, Sundbyberg, Sweden).
24. TaqMan Universal PCR Master Mix (Applied Biosystems, Foster City, CA)
25. PCR buffer: 10 m$M$ Tris-HCl, pH 8.3, 50 m$M$ KCl.
26. Glass capillary (Femtotips, Eppendorf).

## 3. Methods

The methods described below outline (1) preparation of tissue samples, (2) staining of tissue sections for microdissection, (3) microchip gel electrophoresis, 4) laser microdissection and laser pressure catapulting, (5) RNA extraction, (6) cDNA synthesis, (7) quantitative real-time PCR, and (8) genetic analysis from single cells. Special care should be taken when handling RNA samples, especially when nanogram amounts are processed (*see* **Note 1**).

### 3.1. Preparation of Tissue Samples

After surgical resection, samples are dissected and snap-frozen in liquid nitrogen. Frozen tissue is stored at –70°C until further processing (*see* **Note 2**). Frozen tissue is cut into 10-μm-thick cryosections and mounted either on glass slides or slides covered with a thin polyethylene membrane (*see* **Note 3**). The slides are immediately stored at –70°C until further use.

### 3.2. Staining of Tissue Sections for Laser Microdissection

1. Add to the tissue section mounted on the slide 50 μL hematoxylin mixed with 1 μL RNasin (5000 U/L) on (*see* **Note 4**).
2. Incubate slide in a cuvet filled with 70 mL Zincfix for 30 s.
3. Incubate in 70% and 95% ethanol for 30 s consecutively.
4. Place the section directly onto the PALM device; after 3 min of air-drying the tissue is ready for laser microbeam microdissection (**Fig. 1**).
5. To improve the poor morphology in air-dried tissue sections, 50 μL of a newly developed PALM LiquidCoverglass diluted in isopropanol (1:3) can be applied directly on the tissue section (**Fig. 1B**). Laser microdissection can be performed after 5 min drying of the section (*4*; *see* **Note 5**).

### 3.3. Laser Microdissection and Laser Pressure Catapulting

Sections are laser microdissected following the manufacturer's protocol for the PALM Laser-MicroBeam System. The UV laser microbeam is coupled to

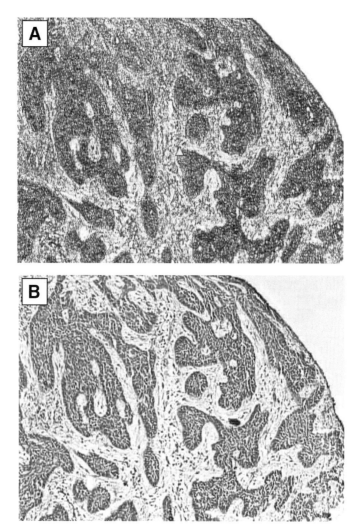

Fig. 1. Morphology of sections for microdissection. Sections of snap-frozen basal
cell cancer were stained with hematoxylin. Pictures were taken without cover (**A**) and
with PALM LiquidCoverglass (**B**).

the epifluorescence illumination port of the microscope. A motorized
controlled microscope stage is attached to the microscope and a frame grabber
enables the observation of the microscopic image on a computer screen (**Fig. 3A**).
The image is overlaid with a graphical user interface enabling the user to perform
laser manipulation of tissue directly on the screen (**Fig. 3B**). A microfuge cap
moistened with 1 µL mineral oil and 1 µL RNAsin (*see* **Note 7**) is mounted

**A   B**

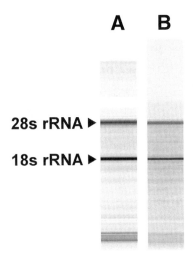

28s rRNA ▶

18s rRNA ▶

Fig. 2. RNA analysis of unstained and stained tissue. The RNA was extracted from tissue sections and separated with a Pico Labchip in an Agilent 2100 Bioanalyzer. The size range of RNA transcripts is estimated by the ladder marker. The electrophero-grams were converted to a gel-like image and showed total RNA of the original tissue (**A**) and from tissue after staining and storage at –70°C (**B**).

upside down just above the tissue section. To select and isolate areas of inter-est, microdissection is performed by cutting with a fine focused laser beam producing a gap of 0.5 to 1.2 μm (**Fig. 4A**). Single unwanted cells can selec-tively be eliminated with single "shots" of pulsed laser. Following isolation of cells, a high-energy pulse of the focused laser beam just below the focal plane of the tissue specimen is used to create a pressure wave separating the targeted tissue and catapulting it into the microfuge cap. When membrane-mounted slides are used, one single shot is sufficient to catapult the target tissue into the microfuge cap (**Fig. 4**). When normal glass slides are used multiple pulses are necessary, each catapulting only a fragment of the isolated tissue cells (**Fig. 4**). In each session approx 500–5000 cells are collected within 30–60 min. Extrac-tion buffer is added into the microfuge tube, vortexed briefly, and centrifuged for 5 min to spin down cells from the lid. The samples are stored on ice or frozen at –70°C.

For genomic analysis of single cells a small glass capillary (Femtotips) are used to transfer the isolated cell to the microfuge tube. The tip of the glass capillary with the attached cell is broken off against the bottom of the microfuge tube containing 10 μL of 1X PCR buffer and the samples are covered with 50 μL of mineral oil.

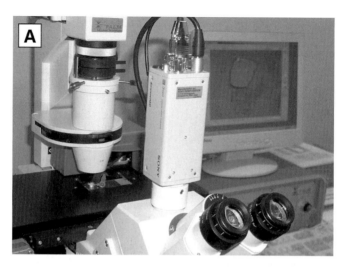

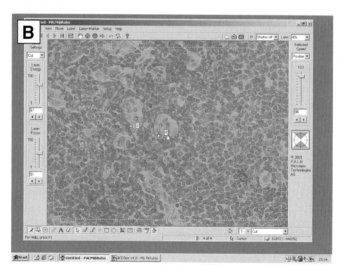

Fig. 3. Image of the PALM microbeam system. The laser system is coupled to a Zeiss light microscope (**A**). The slides are positioned on a stage, that can be computer controlled (**B**).

---

Fig. 4. (*opposite page*) Laser microdissection of basal cell cancer. When glass slides without membrane were used (**A**), areas of interest were circumscribed and separated from unwanted cells (**B**). Multiple laser shots are necessary to catapult the tissue into a lid of a PCR tube (**C**). Cell fragments are visible in the cap (**D**). If membrane mounted slides are processed (**E**) and separated (**F**), the target cells can be catapulted often with a single shot (**G**). The membrane avoids fragmentation and preserves morphology after catapulting (**H**).

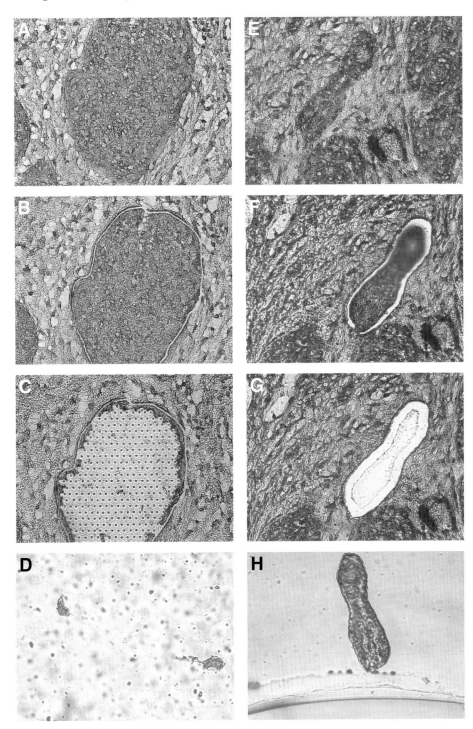

### 3.4. RNA Extraction

RNA isolation is performed using a column-based RNA extraction method according to the manufacturer's instructions (Mini RNA isolation Kit, *see* **Notes 8 and 9**). In brief:

1. Dissolve collected cells in 200 µL RNA extraction buffer.
2. Incubate on ice for 20 min; vortex briefly every 10 min.
3. Add 200 µL 100% ethanol, mix briefly.
4. Incubate on ice 10 for min.
5. Transfer mixture to spin column and spin at maximum speed for 1 min.
6. Wash twice with 200 µL wash buffer and repeat 1 min centrifugation.
7. Add 8 µL RNAse-free water directly onto the dry filter and wait for 2 min.
8. Elute in a new collection tube by centrifugation at maximum speed for 1 min.
9. Add 1 µL RNAsin and store at –70°C.

The purified RNA is used for quantitative real-time PCR analysis or, after linear amplification, for global expression analysis on gene arrays. Since these methods are detailed in other chapters, they are only presented brief in this chapter (*see also* **ref. 4**).

### 3.5. Microchip Gel Electrophoresis

The Agilent 2100 Bioanalyser and a RNA 6000 Pico Labchip kit are utilized to evaluate RNA quality of cells from tissue sections (*see* **Note 6**). One µL corresponding to 1% (v/v) of each RNA probe is transferred to the Pico Labchip, together with 1 µL of RNA 6000 ladder. The analysis is performed according to the manufacturer's instructions and an example of results are shown as a gel-like image (**Fig. 2**).

### 3.6. cDNA Synthesis From Microdissected Cells

cDNA synthesis is performed according to a method that is part of a T7-based linear RNA amplification protocol of Scheidl et al. (*5*). An oligo-dT primer extended with the T7 promoter sequence is used (*see* **Note 10**).

1. Add 1 µL (1 µg/µL) oligo-dT primers (5'-AAA-CGACGGCCAGTGAATTGT AATACGACTCACTATAGGCGCTTTTTTTTTTTTTTTT–3') (*5*) and 0.5 µL (5 µg/µL) of linear acrylamide to the eluted 9 µL of total RNA.
2. Heat sample to 70°C for 5 min and rapidly cool the sample on ice.
3. Add 4 µL reverse transcriptase buffer, 2 µL DTT, 2 µL ultrapure dNTPs, 1 µL RNAsin, and 2 µL Superscript II reverse transcriptase.
4. Perform reverse transcription at 42°C for 1 h.
5. When small amounts (<1 ng) of cDNA are expected, purification of cDNA should be considered. This can be performed by phenol extraction followed by three times washing with water on Microcon 100 columns. The volume of the purified cDNA is adjusted to 20 µL (*see* **Note 11**).

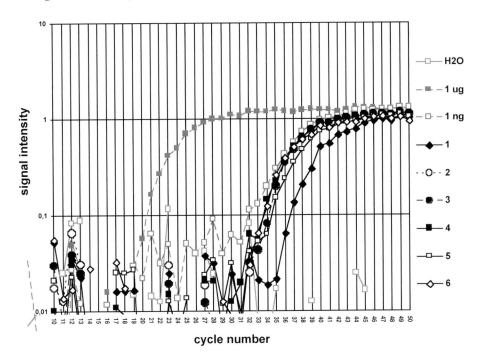

cycle number

Fig. 5. Real-time PCR of cDNA derived from microdissected tissue. Extracted RNA from microdissected tissue samples (Sample 1: 800 cells; 2: 500 cells; 3: 300 cells; 4: 300 cells; 5: 300 cells; 6: 400 cells) was used for cDNA synthesis. 10% of cDNA was used for TaqMan real-time PCR (A). Amplificaton curves represent the mean from duplicates. For positive controls 1 µg or 1 ng of total RNA extracted from a cell line was used. As negative control ($H_2O$) a sample of water was processed exactly in the same way as the tissue (i.e., RNA extraction, cDNA synthesis, real-time PCR). Please note that sample 2 failed to give an adequate amplification curve.

### 3.7. Quantitative RT-PCR

In order to validate the RNA extraction procedure and to test its suitability for RT-PCR analysis, primers for the housekeeping gene human glyceralde-hyde-3-phosphate dehydrogenase (GAPDH) are used in TaqMan real-time PCR (**Fig. 5**). Two microliter (10%) is added to a 50-cycle TaqMan PCR assay using the TaqMan Universal PCR Master Mix with following primers for GAPDH mRNA sequence: Forward primer, 5'-CCCATGTTCGTCATGGGTGT (200 n*M*); Reverse primer, 5'-TGGT CATGAGT CCTTCCACGATA (200 n*M*); Probe, 5'-FAM-CTGCACCAC CAACT GCTTAGCACCC-TAMRA (150 n*M*). The reaction is performed with the ABI PRISM 7000HT real-time PCR cycler (Applied Biosystems, Foster City, CA) under conditions recom-mended by the manufacturer (**Fig. 5**).

## 3.8. Genetic Analysis of Single Cells

Genomic analysis of the *p53* gene (exons 4–11) can be performed from microdissected single cells by amplification in a multiplex/nested configuration *(Fig. 6, 6,7)*. Primers for the mitochondrial sequence are incorporated in the multiplex PCR to generate a specific sequence for each individual (*see* **Note 13**). The outer multiplex amplification is performed in one tube with 18 primers for 30 cycles. After a 100-fold dilution, inner region specific amplifications for 30 cycles are performed. All primers used are listed in **Table 1**.

1. For the outer multiplex amplification prepare the reagent mix on ice. The volume of each sample is 10 µl consisting of:
   (a) 2 µL 10X Pfu buffer; (b) 2 µL (2 m*M*), dNTP; (c) 1 µL outer primer mix (consisting of 5-pmol/µL of each of the 18 outer primers in water); (d) 0.7 µL Pfu Turbo Polymerase 2.5 U/µL and (e) 4.3 µL Millipore water. Always prepare sufficient mix for a couple of additional samples, since a small amount is lost during multiple pipetting. Also include some negative controls without DNA—at least 3 for every 10 samples is appropriate, in each run.
2. Add 10 µL of the mix to each single cell sample, just letting the tip touch the liquid surface beneath the oil. Keep samples on ice.
3. Initiate the PCR by denaturation at 98°C for 2 min. Then amplify the samples in two steps. First run 4 cycles of denaturation at 98°C for 15 s; annealing at 55°C for 4 min, and extension at 72°C for 30 min. This is followed by 26 cycles of denaturation at 98°C for 15 s; annealing at 55°C for 30 s, and extension at 72°C for 1 min. End the program with an extension step at 72°C for 10 min and a hold step at 4°C.

Each outer PCR product is used as a template in nine inner amplifications, one for each exon and one from the mitochondrial sequence.

4. For nested amplification a 50-µL reaction is used. Prepare reagent mixes (on ice), one for each fragment. For *p53* exons 4, 6–11 and Mito, the mix per sample consists of:
   (a) 5 µL 10X PCR buffer II; (b) 5 µl dNTPs (2 m*M*); (c) 4 µL MgCl$_2$ (25 m*M*); (d) 1 µL inner primer (forward)(10 pmol/µL); (e) 1 µL inner primer (reverse) (10 pmol/µL); (f) 0.2 µL AmpliTaq DNA polymerase 5 U/µL; and (g) 33.3 µL Millipore water. For p53 exon 5 the mix consists of: (a) 5 µL 10X PCR buffer II; (b) 5 µL dNTPs (2 m*M*); (c) 5 µL MgCl$_2$ (25 m*M*); (d) 1 µL inner primer (forward)(10 pmol/µL); (e) 1 µL inner primer (reverse)(10 pmol/µL); (f) 0.7 µL Pfu Turbo polymerase (2.5 U/µL); and (g) 31.8 µL Millipore water.
5. For each sample dispense 49.5 µL of the reagent mix in an Eppendorf tube (0.5 mL) and cover with 30–50 µL mineral oil.
6. Add 0.5 µL of the outer reaction product as template. Make sure that the oil layer is penetrated when dispensing.

## laser-assisted single cell microdissection

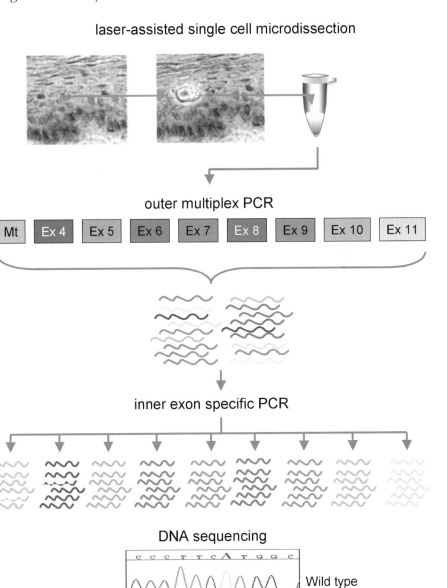

### outer multiplex PCR

| Mt | Ex 4 | Ex 5 | Ex 6 | Ex 7 | Ex 8 | Ex 9 | Ex 10 | Ex 11 |

### inner exon specific PCR

### DNA sequencing

c c c T T c A T G G c — Wild type

c c c T T c C T G G c — Mutant

Fig. 6. Schematic illustration of the different steps involved in genetic analysis of the p53 gene from single cells. The strategy of multiplex/nested PCR followed by direct DNA sequencing is depicted.

**Table 1**
**Primers Used for Multiplex PCR of Genomic DNA From Single Cells**

| Target | Amplification | Primer sequence | |
|---|---|---|---|
| | | Forward | Reverse |
| Exon 4 | Outer | 5'-CTGGGACCTGGAGGGCTGGG | 5'-AGAGGAATCCCAAAGTTCCA |
| | Inner | 5'-CTGAGGACCTGGTCCTCTGAC | 5'-ATACGGCCAGGCATTGAAGT |
| Exon 5 | Outer | 5'-TGCTGCCGTGTTCCAGTTGC | 5'-CAATCAGTGAGGAATCAGAGG |
| | Inner | 5'-TTCACTTGTGCCCTGACTT | 5'-ACCAGCCCTGTGTCTCTCC |
| Exon 6 | Outer | 5'-GGCTGGAGAGACGACAGGGC | 5'-CGGAGGGCCACTGACAACCA |
| | Inner | 5'-TTGCCCAGGGTCCCCAGGCC | 5'-CTTAACCCCTCCTCCCAGAG |
| Exon 7 | Outer | 5'-CCTCCCCTGCTTGCCACAGG | 5'-GGAAGAAATCGGTAAGAGGTGG |
| | Inner | 5'-CGCACTGGCCTCATCTTGGG | 5'-CAGCAGGCCAGTGTGCAGGG |
| Exon 8 | Outer | 5'-ACAGGTAGGACCTGATTTCC | 5'-TGAATCTGAGGCATAACTGC |
| | Inner | 5'-GCCTCTTGCTTCTTTTCC | 5'-CCCTTGGTCTCCTCCACCGC |
| Exon 9 | Outer | 5'-AGCAAGCAGGACAAGAAGCG | 5'-GTTAGCTACAACCAGGAGCC |
| | Inner | 5'-GCCTCAGATTCACTTTTATCACC | 5'-CTGGAAACTTTCCACTTGAT |
| Exon 10 | Outer | 5'-GATCCGTCATAAAGTCAAAC | 5'-TTGACCATGAAGGCAGGATG |
| | Inner | 5'-CTTGAACCATCTTTTAACTCAGG | 5'-AATCCTATGGCTTTCCAACCTAGG |
| Exon 11 | Outer | 5'-CTTCAAAGCATTGGTCAGGG | 5'-GGGTTCAAAGACCCAAAACC |
| | Inner | 5'-CACAGACCCTCTCACTCATG | 5'-GCAGGGGAGGGAGGAGAGATGGG |
| Mito. | Outer | 5'-CCTGAAGTAGGAACCAGATG | 5'-ACACCAGTCTTGTAAACCGG |
| | Inner | 5'-CTCCACCATTAGCACCCAAAG | 5'-TGATTTCACGGAGGATGGTGG |

162

7. Initiate PCR by denaturation at 94°C for 5 min (Taq Polymerase)/98°C for 2 min (Pfu polymerase). Amplify the samples for 30 cycles by denaturation at 94°C for 30 s/ 98°C for 15 s, annealing at 63°C for 30 s and extension at 72°C for 1 min. The program is ended with an extension step at 72°C for 10 min and a hold step at 4°C.
8. The resulting PCR products are analyzed on a 1% agarose gel.

## 4. Notes

Since we started to work with the PALM system many modifications and variations have been made to facilitate techniques and improve protocols.

1. The work with RNA requires lab work under RNase-free conditions. The working place and pipets should be decontaminated with RNaseZap. Gloves should be worn at all times and be changed frequently. Only RNase-free filter pipette tips should be used. If possible, all reagents used should be RNase-free.
2. The quality of the original material is most important for all kinds of RNA analyses. Best results were obtained with tissue specimens, which were immediately frozen after resection. Other possibilities that were suggested involved storage in Zincfix *(8)* or commercially available buffers (e.g., RNAlater from Ambion). Formalin fixation and paraffin embedding leads to rapid RNA degradation, making downstream applications very difficult *(8,9)*. Although there are several descriptions of successful RNA amplification from formalin-fixed and paraffin-embedded tissue using PCR methods *(10–12)*, this option should be considered only if no other materials are available. If OCT or other freezing media are used it is important to remove them before laser microdissection (e.g., incubation in water). If the tissue specimens are embedded the OCT is usually removed by incubation in staining and dehydrating solutions.
3. In order to shorten and facilitate the staining procedure the tissue specimens can also be stained directly after cutting in the cryostat and then stored, after dehydration, at –70°C.
4. In general the use of membrane-coated slides is preferable for several reasons: (1) The membrane serves as a backbone underlying the tissue and facilitates the laser catapulting. (2) It protects the sample from contaminating tissue fragments. (3) After catapulting the complete tissue fragment can be found with excellent preserved morphology in the lid. (4) Our experience indicates higher amounts of extracted RNA and more reproducible results when membrane-coated slides were used. However, some users state that no differences were observed when both slide types were compared (membrane coated vs non-membrane-coated).
5. Instead of hematoxylin, other staining reagents could be considered. A recent study suggested Nuclear Fast Red to be best in terms of preserving RNA integrity *(13)*. In general incubation times should be kept as short as possible in any staining procedure.
6. Analysis of quantity and integrity of RNA is an important and necessary step prior to later applications. This avoids laborious and expensive processing of RNA probes of low quality or RNA subjected to degradation during handling.

Furthermore it circumvents misinterpretation of results. A very useful tool in this respect is the Bioanalyzer (Agilent). The combination of capillary electrophoresis, microfluidics, and nucleic acid binding dyes allows for evaluation of both concentration and quality of RNA. However, since the RNA has to be diluted in pure water, salt and proteins remaining in the RNA sample could lead to false negative results. Moreover the detection limit is around 1–0.1 ng/μL. Therefore the application for microdissected cells is most often limited to tissues with very high RNA content (e.g., liver) or high cell numbers (>5000 cells).

7. The morphology of air-dried uncovered sections under the light microscope is poor, which limits the repertoire of cell types that can be isolated without contamination. Single cells are hardly distinguished, and counter staining or other forms of labeling are less effective. A method to overcome some of these problems and to improve morphology is to apply PALM LiquidCoverglass, a fluid that was developed by Tone Bjørnsen and has been shown not to affect RNA integrity *(4)*.

8. When normal glass slides (without membrane covering) are used for laser microdissection, it is recommended to apply at least 30 μL of fluid in the lid for laser catapulting. Preferentially the buffer for the later RNA extraction should be applied (e.g., Zymogen RNA extraction buffer, TRIzol, etc.).

9. The method of RNA extraction can be crucial. We tested different microcolumn-based RNA isolation kits and saw significant differences in RNA quantity and quality. Inexpensive and effective alternatives are phenol-based extraction methods (e.g., TRIzol), followed by an alcohol precipitation with a coprecipitant (e.g., linear acryl amide or Pellet Paint). However, the recovery rate of RNA samples of known concentration should be tested before including a new method or kit in the RNA extraction process.

10 The amount of isolated RNA depends not only on the amount of microdissected cells but also on the type of cells that were isolated. For instance, metabolic active liver cells contain much more RNA than inactive fibroblasts or osteoblasts (Ambion, technical notes 8/2003).

11. The cDNA synthesis was performed according to a method, that is part of a T7-based linear RNA amplification protocol of Scheidl et al. *(5)*. It includes an oligo-dT primer with the T7 promoter sequence. This enabled us not only to perform quantitative real-time PCR, but also to amplify good quality preparations to μg amounts of Cy-labeled RNA for gene array analysis. If this application is not appropriate, a common poly-dT primer can be used instead. The sensitivity might improve by using random hexamers. However, this could also result in shorter transcripts and increase the risk of artifacts caused by DNA contaminations.

12. If very small amounts of RNA (<10 ng) are reverse transcribed and analyzed by real-time PCR, purification of cDNA before amplification could be advantageous. The unfavorable ratio between cDNA on one hand and enzymes, poly dT-primers, and nucleotides on the other hand can negatively affect the polymerase chain reaction.

13. Mitochondrial DNA can be used as a specific signature for each sample and thus serves as a control for contamination between samples. Since there are approx

1000 copies of mitochondrial DNA in each cell compared to two copies of the *p53* gene, amplification of mitochondrial DNA can also be used to distinguish between loss of the cell (no amplified fragments) and degradation of the template (amplification only of mitochondrial DNA) in cases of amplification failure.

14. Two potential problems with single-cell PCR include risk of contamination and random amplification failure of one allele, i.e., allele dropout (ADO) *(6,7)*. Preparation of PCR reagent mixture, adding PCR reagent mixture to the single cell sample and adding outer PCR product to the reagent mix for the inner PCR should be performed in separate rooms to avoid contamination. Coats and gloves should be used at all times. All pipetting should be performed with filtertips and the bench area where the template is added should be decontaminated using UV light. Negative controls without DNA should be included in abundance.

15. The random frequency of ADO, using the outlined procedure with frozen sections, has been approx 50%. ADO is important to consider when interpreting data. The ADO rate in cultured cells was lower compared to cells retrieved from tissue sections. The reported ADO rates, usually based on analysis of cultured cells, are between 10 and 80% and are known to be affected by fragment length and number of fragments amplified in parallel.

## Useful Links

16. www.ambion.com/techlib/index.html: Very useful technical notes about the work with RNA.
17. http://www.palm-mikrolaser.com: Technical notes and protocols for the use of the PALM laser systems.

## Acknowledgments

The authors thank Anna Asplund, Christina Sundquist, and Åsa Sivertsson for expertise work with the development of microdissection strategies using the PALM system and subsequent technologies. This work was supported by grants from the Swedish Cancer Foundation to Fredrik Ponten, Joakim Lundeberg, and Arne Östman.

## References

1. Schutze, K., Burgemeister, R., Clemment-Sengewald, A., Ehnle, S., Friedemann, G., Lahr, G., et al. (2003) Non-contact live cell laser micromanipulation using PALM microlaser systems. *P.A.L.M Scientific Edition*, 11.
2. Schutze, K., Posl, H., and Lahr, G. (1998) Laser micromanipulation systems as universal tools in cellular and molecular biology and in medicine. *Cell. Mol. Biol. (Noisy-le-grand)* **44**, 735–746.
3. Williams, C., Ponten, F., Moberg, C., Soderkvist, P., Uhlen, M., Ponten, J, et al. (1999) A high frequency of sequence alterations is due to formalin fixation of archival specimens. *Am. J. Pathol.* **155**, 1467–1471.
4. Micke, P., Bjørnsen, T., Scheidl, S., Strömberg, S., Demoulin, J. B., Ponten, F., et al. (2004) A fluid cover medium provides superior morphology and preserves

RNA integrity in tissue sections for laser microdissection and pressure catapulting. *J. Pathol.* **202,** 130–138.

5.  Scheidl, S. J., Nilsson, S., Kalen, M., Hellstrom, M., Takemoto, M., Hakansson, J.,
    and Lindahl, P. (2002) mRNA expression profiling of laser microbeam micro-
    dissected cells from slender embryonic structures. *Am. J. Pathol.* **160,** 801–813.

6.  Persson, Å., Ling, G., Williams, C., Bäckvall, H., Ponten, J., Ponten, F., and
    Lundeberg, J. (2000) Efficient analysis of single cells obtained from histological
    tissue sections. *Anal. Biochem.* **287,** 25–31.

7.  Ling, G., Persson, Å., Berne, B., Uhlen, M., Lundeberg, J., and Ponten, F. (2001)
    Persistent p53 mutations in single cells from normal human skin. *Am. J. Pathol.*
    **159,** 1247–1253.

8.  Wester, K., Asplund, A., Backvall, H., Micke, P., Derveniece, A., Hartmane, I.,
    et al. (2003) Zinc-based fixative improves preservation of genomic DNA and pro-
    teins in histoprocessing of human tissues. *Lab. Invest.* **83,** 889–899.

9.  Vincek, V., Nassiri, M., Knowles, J,. Nadji, M., and Morales, A. R. (2003) Pres-
    ervation of tissue RNA in normal saline. *Lab. Invest.* **83,** 137–138.

10. Lehmann, U. and Kreipe, H. (2001) Real-time PCR analysis of DNA and RNA
    extracted from formalin-fixed and paraffin-embedded biopsies. *Methods* **25,**
    409–418.

11. Becker, I., Becker, K. F., Rohrl, M. H., and Hofler, H. (1997) Laser-assisted prepa-
    ration of single cells from stained histological slides for gene analysis. *Histochem.
    Cell Biol.* **108,** 447–451.

12. Specht, K., Richter, T., Muller, U., Walch, A., Werner, M., and Hofler, H. (2001)
    Quantitative gene expression analysis in microdissected archival formalin-fixed
    and paraffin-embedded tumor tissue. *Am. J. Pathol.* **158,** 419–429.

13. Burgemeister, R., Gangnus, R., Haar, B., Schutze, K., and Sauer, U. (2003) High
    quality RNA retrieved from samples obtained by using LMPC (laser microdissec-
    tion and pressure catapulting) technology. *Pathol. Res. Pract.* **199,** 431–436.

# 14

## Laser Microdissection and RNA Analysis

### Ludger Fink and Rainer Maria Bohle

#### Summary

Microdissection techniques have become an important tool to link histomorphology and pathophysiological events using modern methods of molecular biology. They allow isolation of cell clusters or even single cells precisely under optical control from complex tissue structures for further analysis of DNA, RNA, and proteins. In particular, the fragile RNA molecules can be preserved during microdissection so that gene expression and regulation measurement become feasible in a cell type-specific manner within complex tissues. This report focuses on and outlines the procedures for RNA investigation, from tissue fixation, sectioning, and staining to downstream applications (RT-PCR, mRNA quantification, and mRNA preamplification). Standards for the preparation of RNA from frozen and formalin-fixed tissues are presented. Specific protocols are given for both the isolation of RNA from small numbers of cells ( 50 cells) as well as for larger cell numbers. While most of the procedures are identical for the microdissection systems, special features of each technique are mentioned.

**Key Words:** Laser microdissection; gene expression, formalin fixation; immunohistochemistry; immunofluorescence; mRNA quantification; real-time PCR; RNA preamplification; LCM.

## 1. Introduction

The analysis of commonly used tissue homogenates leads inevitably to an average measurement of the biochemical components (such as, nucleic acids and proteins) from various cell types. There is a high risk that changes in mRNA expression of an individual cell type may be masked by the surrounding cells. To overcome tissue heterogeneity, cells and cell types must be isolated selectively for further analysis. Microdissection and micromanipulation techniques were consequently developed (*1–3*) and, particularly, laser-based

From: *Methods in Molecular Biology, vol. 293: Laser Capture Microdissection: Methods and Protocols*
Edited by: G. I. Murray and S. Curran © Humana Press Inc., Totowa, NJ

microdissection systems proved to allow simple, rapid and precise retrieval of target cells *(4)*.

When planning a microdissection study from tissue harvesting to the analytic steps the isolation of nucleic acids comprise a number of steps that may cause fragmentation and degradation of RNA. Poor quality of RNA again results in a lower recovery of an already limited amount of input material.

The type of tissue preparation, storage, and fixation have a major impact on analytical success and influence the workup of microdissection studies. First, the time span between onset of ischemia and fixation should be kept as short as possible, as endogenous RNases and ribozymes may be activated or are active for their targets. Next, the fixation of fresh tissues requires several hours, depending on the type of fixative, rate of penetration into the tissue, and size of the tissue sample. This time also influences the molecular quality of the tissue (i.e., due to extensive crosslinking and strand scissions by formaldehyde). On the other hand, persistent formalin fixation for 48 h and more may remarkably change measurement of relative mRNA expression *(5)*. Finally, the embedding procedure may cause an adverse effect on nucleic acids, i.e., high temperature exposure during submersion in melted paraffin.

Snap-freezing of unfixed tissue in liquid nitrogen-cooled isopentane circumvents these problems and results in the best available product for further molecular applications. This advantage, however, is contrasted by a poorer quality in morphology with possible split artifacts and higher costs for tissue storage at –80°C.

Promising trials have involved new strategies for optimal tissue processing, i.e., fixation of small tissue specimens in 70% ethanol followed by embedding in paraffin or low-melting polyester *(6)*. Alternatively, methacarn fixation (60% methanol, 30% chloroform, 10% glacial acetic acid) and paraffin embedding were also shown to yield reasonable quantity and quality of mRNA and protein, significantly superior to that obtained from cross-linking fixatives *(7)*. Low temperature embedding in plastic resin was suggested to obtain a higher degree of integrity of nucleic acids *(8)*, but low-melting polyester also has the disadvantage of drying out and cracking after storage. Finally, there are no data on the integrity of cellular morphology and nucleic acids after long-term storage of tissues processed this way.

In consequence, there is no optimal preparation strategy available. If microdissection is intended, especially with full-length cDNA synthesis, tissue specimens should be frozen airproof at –80°C or in liquid nitrogen. If fixation is desired (because of easier storage and handling), a non-crosslinking fixative should be combined with low-temperature embedding. Only when immunolabeling has to be combined with microdissection should a mild neutral buffered-formalin fixation be considered, accepting a higher degradation rate of nucleic acids. In this case, 4.5% neutral buffered formalin (e.g., Roti-Histofix, Roth, Karlsruhe, Germany) is appropriate. Fixation time of small specimens should not exceed 12 h to limit crosslinking and shearing of nucleic acids.

While the conditions of fresh sample preparation can be controlled, archival tissues are, most often, formalin-fixed and paraffin-embedded (FFPE). Due to easy storage and very good long-term preservation of morphology, such tissue specimens have accumulated over a long time period and represent a valuable resource of tissue. To make these tissues available for molecular approaches, procedures have to be developed to use this type of material. Several reports have adapted existing protocols to the fixation-inherent problems. Due to the crosslinking of all tissue components, proteinase K digestion was shown to be indispensable for releasing nucleic acids from the cells.

## 2. Materials

### 2.1. Sectioning

1. Cryotome for frozen tissues, e.g., CM 3000 (Leica, Bensheim, Germany) with feather microtome blade (PFM, Cologne, Germany).
2. Microtome for FFPE tissues, e.g., Jung SM 2000C (Leica).
3. Tissue Tek® O.C.T.™ Compound (Sakura Finetek Europe, Zoeterwoude, The Netherlands).
4. Parafilm"M"® laboratory film (American National Can™, Chicago, IL).
5. Glass slides, e.g., Superfrost® Plus (Menzel-Gläser, Braunschweig, Germany).
6. Poly-L-lysine and/or 3-aminopropyltriethoxysilane (APES).

### 2.2. Routine Histological Staining

If not mentioned otherwise, reagents are purchased from Sigma Aldrich, in "molecular biology" quality.

1. Hematoxylin, e.g., Mayer's acidic hematoxylin (Division Chroma, Münster, Germany).
2. Xylene.
3. Ethanol, 70%, 90%, 100%.
4. RNase-free $H_2O$ (DEPC-treated).

### 2.3. Immunofluorescence/Immunohistochemistry

#### 2.3.1. Immunofluorescence

1. Primary antibody.
2. Secondary antibody, coupled with fluorophore (e.g., FITC).
3. Tris-buffered saline (TBS): 50 m$M$ Tris-HCl, 150 m$M$ NaCl, pH 7.5.

## 2.3.2. Immunohistochemistry

1. Primary antibody.
2. Secondary "link" antibody, e.g., mouse immunoglobulins 1:40 (Dako Diagnostica, Hamburg, Germany).
3. Enzyme-carrying antibody, e.g., APAAP Mouse Monoclonal 1:50 (Dako).
4. Antibody Diluent Chem Mate™ (Dako).
5. Staining complex, e.g., new fuchsin (Division Chroma), naphthol, levamisole in TBS, pH 8.8 (300 m$M$ Tris-HCl, 750 m$M$ NaCl).
6. TBS : 50 m$M$ Tris-HCl, 150 m$M$ NaCl, pH 7.5
7. Reaction buffer for 50 cells: 52 m $M$ Tris-HCl, pH 8.3, 78 m$M$ KCl, 3.1 m$M$ MgCl$_2$.

## 2.4. RNA Extraction

1. RNA lysis buffer I: 4 $M$ GTC, 25 m$M$ sodium citrate, 0.5% sarcosyl, 0.72% 2-mercaptoethanol, 100 m$M$ Tris-HCl, pH 7.5 (add 2-mercaptoethanol immediately before use).
2. RNA lysis buffer II: 200 µL 1 $M$ GTC, 0.5% sarcosyl, 0.72% 2-mercaptoethanol, 20 m$M$ Tris-HCl, pH 7.5 (add 2-mercaptoethanol immediately before use).
3. RNA lysis buffer III: 10 m$M$ Tris-HCl (pH 8.0), 0.1 m$M$ EDTA (pH 8.0), 2% sodium dodecyl sulfate (pH 7.3).
4. Water saturated phenol, pH 4.3 (light-sensitive, toxic, at 4°C).
5. Chloroform.
6. Isoamylalcohol.
7. Isopropanol.
8. Ethanol 75%.
9. Rnase-free H$_2$O (DEPC treated).
10. Proteinase K (aliquots at –20°C).
11. Glycogen 20 mg/mL (e.g., Roche Diagnostics, Mannheim, Germany).
12. RNase-free DNase I (e.g., Ambion, Austin, TX) (at –20°C).

## 2.5. Reverse Transcription

Purchased from Applied Biosystems, Foster City, CA; stored at –20°C.

1. GeneAmp® 10X PCR Buffer II
2. 25 m$M$ MgCl$_2$.
3. 50 µ$M$ Random hexamers .
4. 20 U/µL RNase-inhibitor.
5. 50 U/µL MMLV-reverse transcriptase.
6. dNTPs (Eurobio, Raunheim, Germany).

## 2.6. Qualitative/Quantitative PCR

Purchased from Applied Biosystems; stored at –20°C.

1. GeneAmp® 10X PCR Gold Buffer II.
2. 25 m$M$ MgCl$_2$.

3. dNTPs.
4. Oligonucleotide primers, probe.
5. AmpliTaq® Gold 5 U/mL.
6. Polymerase chain reaction (PCR) cycler, e.g., GeneAmp™ 2400 PCR cycler (Applied Biosystems).

## 2.7. RNA Preamplification

1. Smart™ PCR cDNA Synthesis Kit (BD Clontech, Palo Alto, CA; stored at –20°C).
2. Advantage® 2-PCR Kit (BD Clontech; stored at –20°C).
3. QIAquick® PCR Purification Kit (Qiagen, Hilden, Germany).

## 3. Methods

The methods described here outline the sectioning, routine staining, immunospecific labeling, and RNA preparation from (1) frozen tissue specimens and (2) FFPE tissue specimens. Depending on the introduced cell amount, specific protocols are provided in both cases for few-cell approaches as well as for higher cell scales. For microdissection the major techniques are (1) laser capture microdissection (LCM), (2) laser microdissection and micromanipulation (LMM), and (3) laser pressure catapulting (LPC). Details are given in **Subheading 3.1.3.**

While special features of each system are mentioned, most procedures can be used with any system. Afterwards, reverse transcription (RT) and PCR are described for the different quantities of cells. Furthermore, a reliable protocol for RNA preamlification is given that increases the initial amount of cDNA from microdissected cells. This allows sufficient material to be obtained for cDNA array and virtual Northern blot hybridization.

### 3.1. Frozen Tissue

#### 3.1.1. Sample Preparation, Sectioning, and Routine Staining

1. Fresh tissue samples are frozen in liquid nitrogen-cooled isopentane and stored at –80°C. To keep them airproof, the specimens are wrapped in parafilmM® and placed in suitable cryo-tubes. Tissue fragments 125 mm$^3$ are prepared in the cryotome at –20°C to avoid thawing of the total specimen (*see* **Note 1**).
2. For slide preparation new disposable knives are used, while all other contact areas are cleaned with 70% ethanol followed by 0.1 *M* NaOH. Cryosections are cut at maximal thickness (5–10 µm), allowing a sufficient precise microscopic recognition, and mounted on glass slides. While nonadhesive glass slides are sufficient for short staining protocols, immunolabeling procedures require slides precoated with poly-L-lysine and/or 3-aminopropyltriethoxysilane (APES) to ensure attachment *(9,10)*. In the case of LPC, sections are mounted onto a 1.35–1.5-mm polyethylene membrane treated with 1% poly-L-lysine or 8% APES solution in acetone. Prior to this procedure, the membrane must be fixed wrinkle-free to the glass slides *(11)*.

**Table 1**
**Routine Staining Procedure for Frozen Tissue Sections**

|                              | Frozen tissue   |
| ---------------------------- | --------------- |
| 100% EtOH                    | optional 30 s   |
| 95% EtOH                     | —               |
| 70% EtOH                     | —               |
| 0.1% Mayer's hematoxylin     | 30–60 s         |
| Rinse in DEPC water          | 2 × 5 s         |
| Eosin (optional)             | 10–30 s         |
| Rinse in DEPC water          | 2 × 5 s         |
| 70% EtOH                     | 15 s            |
| 95% EtOH                     | 15 s            |
| 100% EtOH                    | 15 s            |

3. To prevent RNA degradation and contamination, standard precautions should be followed (*see* **Note 2**).
4. Sectioning should be performed immediately before staining. The routine staining procedure is presented in **Table 1**. After staining, the sections are stored in 100% ethanol to keep the tissue soft for LMM and micromanipulation. In contrast, using membrane and LPC, the ethanol has to evaporate. For LCM in particular, the sections are finally dehydrated in xylene for 2 min twice before air-drying *(12)*. The time between staining and microdissection should be kept as short as possible. We have not seen any disadvantageous effect of routine hematoxylin staining on RNA recovery. Additionally, methyl green, nuclear fast red, and eosin stains are also recommended *(13,14)*.

### 3.1.2. Immunolabeling

1. Precise characterization of defined cell types within complex tissues often requires immunolabeling. Several articles have reported on factors influencing the quality of nucleic acids following this process. Total staining time, number of incubation steps, and staining complex/enzymatic reaction were seen to be crucial for RNA recovery from unfixed and alcohol-fixed tissues *(15)*. RNA is degraded rapidly during exposure to the aqueous phase so that labeling must to be accomplished within a few minutes *(12,16,17)*. Moreover, enzymatic conversion of the staining complex takes time and may affect negatively quality and quantity of RNA *(14,15)*. In consequence, immunofluorescence staining is preferable to immunohistochemical staining. The total incubation time should not exceed 10–15 min in unfixed sections (*see* **Note 3**). This may be achieved by reduction of the antibody incubation steps. Therefore, moderate increases in the antibody working concentrations are often necessary. After acetone or 100% ethanol treatment for 1–5 min, sections are rinsed in Tris-buffered saline (TBS;

pH 7.5) briefly. Depending on the applied antibodies, incubations last about 1–5 min. Optionally, RNase inhibitor (RNasin, 400 U/mL, Promega, Mannheim, Germany) can be added to the antibody solutions.

2. Using micromanipulation, the labeled sections must be kept under ethanol continuously until isolation, as drying leads to marked unspecific background signals.

3. If immunohistochemical staining is essential, the procedure is described as for FFPE tissue in **Subheading 3.2.2.** However, the total staining time should not exceed 20 min for unfixed or ethanol-fixed tissue sections, as RNA is continuously degraded during this time.

### 3.1.3. Microdissection

1. The laser beam either dissects the tissue directly or focally melts a thermoplastic membrane to form a composite with the tissue. In the latter case, the cells become adherent to the film and both can be removed at once. This type of microdissection uses a pulsed low-energy infrared laser and is called laser capture microdissection (LCM; Arcturus Engineering, Mountain View, CA).

2. In the first case, a pulsed ultraviolet laser is coupled to an inverted microscope. After microdissection, the tissue islet or cells can be procured with a sterile needle mounted on a motorized micromanipulator (laser microbeam microdissection and micromanipulation, LMM; PALM Microlaser Technology, Bernried, Germany; *19*).

3. Alternatively, cells can be catapulted by the laser beam into the cap of a reaction tube positioned above. This is called laser pressure catapulting (LPC) (PALM; *20*).

4. A third system also equipped with an ultraviolet laser for microdissection has become available recently. Using an upright microscope, the dissected area falls into a cap that is positioned below (Leica Microsystems, Wetzlar, Germany; *21*).

5. Applying LMM, the stained sections are transferred from 100% ethanol to the microscope table without drying. One drop of 100% ethanol onto the uncovered section improves the histomorphology remarkably. Moreover, the tissue and cells remain flexible and adhere tightly to the needle, which can be lifted easily after rapid evaporation of the alcohol.

6. When LPC is performed from sections mounted on the membrane, any liquid must be removed between membrane and glass slide; otherwise the adhesion forces may prevent the catapulting.

7. For LCM the sections need to dry sufficiently to allow the local fusion with the film. On the other hand, strong adhesion of the section to the glass by prolonged drying may prevent removal of the cells.

### 3.1.4. RNA Preparation

#### 3.1.4.1. Low Cell Number ( 50 Cells)

1. The requirements for microdissection may differ markedly. For some approaches, single cells must be isolated, while other studies require a hundred or even thou-

**Table 2**
**Recommended Treatment for the Investigation of RNA From Microdissected Cells Harvested From Frozen Tissue Sections**

|                          | 50 cells                  | >50 cells                 |
|--------------------------|---------------------------|---------------------------|
| • Resuspension in        | reaction buffer, 10 μL    | 4 *M* GTC-buffer, 200 μL  |
| • Proteinase K digestion; 0.5 μg/μL final conc. | optional; 30 min, 58°C | — |
| • Denaturation           | 7 min, 95°C               | —                         |
| • Extraction             | —                         | phenol/chloroform         |
| • DNase digestion        | optional; 2 U, 30 min, 37°C | optional; 2–10 U, 30 min, 37°C |
| • Denaturation           | 7 min, 95°C               | 7 min, 95°C               |

sands of cells. Usually, nucleic acids have to be extracted for further application. However, isolation of a few cells allows proceeding without an extraction step. The limit is 50 to 100 cells *(22,23)*. Up to this number, cells can be directly transferred to a reaction buffer (see Materials), suitable for necessary digestions, reverse transcription, and PCR.

2. Alternatively, a PCR buffer can be used. In case of LPC, the buffer can be prefilled into the cap of a reaction tube to collect the catapulted cells. In case of the Leica system, the cells drop into the buffer-filled cap. RNase inhibitor can be added (4% v/v).

3. To disrupt the cell membrane of intact cells (e.g., from cytospins), addition of a nonionic detergent, e.g., Igepal CA-630 or Tween-20, can be advantageous. Its concentration should not exceed 1% v/v; otherwise further enzymatic reactions may be inhibited. We usually snap-freeze the samples, followed by thawing three times, and centrifuge the samples at 10,000*g* for 1 min to destroy the cell structure. Afterwards, the samples are kept in liquid nitrogen.

4. When tissues with abundant connective tissue are investigated or proteins are expected to be bound to RNA, proteinase K (≤0.5 μg/μL; 53–60°C) digestion may improve the results. The incubation time should not exceed 30–60 min. For denaturation of the enzyme, 7 min at 95°C are necessary which can be used simultaneously for denaturation before cDNA synthesis. **Table 2** summarizes the preparation steps.

### 3.1.4.2. Higher Cell Number (>50 cells)

1. While fewer than 50 cells can be processed without extraction, this step is necessary for larger numbers of cells. After microdissection of frozen tissue, the cells are resuspended in 200 μL of lysis buffer containing 4 *M* guanidine thiocyanate

(GTC), 25 m*M* sodium citrate, 0.5% sarcosyl, 0.72% 2-mercaptoethanol, and 20 m*M* Tris-HCl, pH 7.5. After incubation for 10 min at room temperature, 20 μL 2 *M* sodium acetate, 220 μL phenol (pH 4.3), and 60 μL chloroform/isoamyl-alcohol (24:1) are added.

2. The samples are vortexed and centrifuged for 15 min at 4°C.
3. The aqueous layer is collected, 1 μL glycogen (10 mg/mL) added and then the samples are precipitated with 200 μL isopropanol.
4. Samples are frozen for 1 h at –20°C and centrifuged for 15 min at 12,000*g*.
5. The pellets are washed with 75% ethanol, air-dried and finally resuspended in 10 μL H$_2$O (**Table 2**).
6. Use of silica-columns for RNA extraction is an interesting alternative, but up to now, the elution requires a considerably higher total volume (at least 30–40 μL). This again has to be reduced by speed vacuum centrifugation. Recently, the first columns with lower binding capacity but also lower elution volume were released (i.e., Qiagen).

### 3.1.5. DNase Digestion

1. Employing RNase-free DNase I digestion (2–10 U, ≤30 min, 37°C) depends on the subsequent analysis and should be limited to those cases when intron-spanning primers are useless, such as presence of pseudogenes or investigation of an intron-free gene.
2. The use of RNA for cDNA libraries or array hybridization requires this kind of digestion as well.
3. A subsequent precipitation to eliminate DNase is optional and depends on the volume of the sample and the planned analysis. Its presence may impair further enzymatic reactions.

### 3.2. FFPE Tissue

### 3.2.1. Tissue Preparation, Sectioning, and Routine Staining

1. FFPE tissue specimens are cooled at 4°C for 1–2 hours for better sectioning. The slides are prepared on a microtome (e.g., Jung SM 2000R, Leica).
2. Contact areas are cleaned with 70% ethanol followed by 0.1 *M* NaOH.
3. The sections (5–10 μm) are mounted on adhesive-coated glass slides (e.g., Superfrost®Plus). They are made at maximal thickness, but still allow precise microscopic cell recognition.
4. For drying, the sections are stored at room temperature for 24 h.
5. In the case of LPC, sections are mounted onto a 1.35–1.5-μm polyethylene membrane treated with 1% poly-L-lysine or 8% APES solution in acetone.
6. The routine staining procedure is presented in **Table 3**. Analogous to frozen tissue, the sections are stored in 100% ethanol after staining to keep the tissue soft for LMM and micromanipulation.
7. When using membrane and LPC or LCM, the ethanol has to evaporate, e.g., by dehydration in xylene for 2 min twice before air drying.

**Table 3**
**Routine Staining Procedure**
**for FFPE Tissue Sections**

|  | FFPE tissue |
| --- | --- |
| Dewax with xylene | $2 \times 5$ min |
| 100% EtOH | 30 s |
| 95% EtOH | 30 s |
| 70% EtOH | 30 s |
| 0.1% Mayer's hematoxylin | 30–60 s |
| Rinse in DEPC water | $2 \times 5$ s |
| Eosin (optional) | 10–30 s |
| Rinse in DEPC water | $2 \times 5$ s |
| 70% EtOH | 30 s |
| 95% EtOH | 30 s |
| 100% EtOH | 30 s |

8. Apart from hematoxylin staining, methyl green, nuclear fast red, and eosin stains are also recommended.

### 3.2.2. Immunolabeling

1. For microdissection of archival FFPE tissue, even prolonged immunohisto-chemical staining could be combined successfully with RNA analysis of a few or single cells.
2. Limit the antibody incubation times to 5 min in combination with the alkaline-phosphatase monoclonal antialkaline phosphatase (APAAP) technique, slightly modified from Cordell et al. *(24)*: After 10 min of deparaffinization in xylene, the slices are immersed in acetone and acetone/Tris-buffered saline (TBS; 1:1) for 10 min each.
3. For antibody dilution, TBS or alternatively the DAKO Chem Mate™ antibody diluent is used.
4. The samples are incubated at room temperature with the primary antibody for 5 min.
5. After short washing in TBS, second "link" antibody and third antibody mouse-APAAP complex (Dako, 1:50) are applied analogously.
6. Alkaline phosphatase substrate reaction is performed at pH 8.8 with new fuchsin (100 µg/mL) and levamisole (400 µg/mL) for 20 to 25 min at room temperature.
7. Afterwards, sections are counterstained with hematoxylin for 45 s, rinsed in water, immersed in 70%, 95%, and 100% ethanol, and stored in 100% ethanol for LMM or dried for LPC/LCM.
8. Thus, for LMM the staining complex has to be resistant to alcohol. Moreover, the complex should not interfere with further reactions; neither DAB nor new fuchsin showed any deleterious effects *(14,25)*.

**Table 4**
**Recommended Treatment for Investigation of RNA From**
**Microdissected Cells Harvested From FFPE Tissue Sections**

|  | 50 cells | >50 cells |
|---|---|---|
| Resuspension in | reaction buffer, 10 µL | 1 *M* GTC buffer, or Tris/EDTA buffer, 200 µL |
| Proteinase K digest.; 0.5 µg/µL final conc. | indispensable; 6–10 h, 58°C | indispensable; 12–16 h, 58°C |
| Denaturation | 7 min, 95°C | — |
| Extraction | — | phenol/chloroform |
| DNase digestion | optional; 2 U, ≤30 min, 37°C | optional; 2–10 U, ≤30 min, 37°C |
| Denaturation | 7 min, 95°C | 7 min, 95°C |

9. High-temperature antigen retrieval treatments should be omitted, as they can adversely affect recovery of nucleic acids.

### 3.2.3. RNA Preparation

#### 3.2.3.1. LOW CELL NUMBER ( 50 CELLS)

1. In analogy to frozen tissue we differentiate a low cell scale (≤50 cells) from higher cell numbers. Up to this amount the isolated cells can be transferred into the 10-µL reaction buffer (**Table 4**).
2. Regardless of the cell number, microdissected cells from FFPE tissue must be digested by proteinase K to release the nucleic acids from the cross-linking network.
3. Therefore, 1 µL with 5 µg/µL proteinase K (final conc: 0.5 µg/µL) is added and incubated for 6 to 10 h at 58°C.
4. Afterwards, the enzyme is denatured for 7 min at 95°C; this can be used as an initial denaturation step for cDNA synthesis.
5. For some applications a DNase digestion may be advantageous (*see* **Subheading 3.1.5.**).

#### 3.2.3.2. HIGHER CELL NUMBER (>50 CELLS)

1. Greater amounts of microdissected cells are resuspended in 200 µL of a lysis buffer containing 10 m*M* Tris-HCl (pH 8.0), 0.1 m*M* EDTA (pH 8.0), and 2% sodium dodecyl sulfate (pH 7.3), and are then digested with proteinase K (0.5 mg/ mL end concentration) at 58°C for 12–16 h (*see* **Note 4**).
2. RNA is then extracted by the aforementioned phenol/chloroform procedure *(26)*.

3. Alternatively, the microdissected cells are suspended in 200 µL 1 *M* GTC, 0.5% sarcosyl, 0.72% 2-mercaptoethanol, and 20 m*M* Tris-HCl, pH 7.5.
4. Adding 0.5 µg/µL proteinase K to the buffer, the samples are digested for 12–16 h (58°C) and phenol/chloroform extraction follows *(27)*.
5. The first technique turned out to be most sensitive and is feasible for lower cell amounts and low copy genes.
6. On the other hand, higher DNA contamination has to be taken into account. The GTC preparation results in lower DNA contamination and is recommended especially when introns are absent and thus intron spanning primers cannot be applied or pseudogenes are present (*see* **Note 5**).

### *3.3. cDNA Synthesis*

1. cDNA synthesis is performed shortly after digestion/extraction or sole denaturation. Especially when processing a small number of cells, microdissection, digestion, and reverse transcription (RT) should be performed directly and cells not stored until RT has been completed.
2. Microdissected cells collected in the reaction buffer, as well as RNA dissolved in $H_2O$, are heated to 70°C for 10 min and then cooled on ice for 5 min.
3. With a preceding digestion step, denaturation of the enzyme simultaneously serves to denaturate the RNA; thus the 70°C step can be omitted. cDNA synthesis from 10 µL $H_2O$-diluted RNA is performed with 4 µL $MgCl_2$ (25 m*M*), 2 µL 10X buffer II (100 m*M* Tris-HCl, pH 8.3, 500 m*M* KCl), 1 µL dNTP (10 m*M* each), 1 µL random hexamers (50 µ*M*), 0.5 µL RNase inhibitor (10 U), and 1 µL MMLV reverse transcriptase in a total volume of 19.5 µL (*see* **Note 6**).
4. Due to the presence of $MgCl_2$ in the cell buffer, only 2 µL $MgCl_2$ are added, resulting in a volume of 17.5 µL for few-cell approaches without RNA extraction. Samples are incubated at 20°C for 10 min followed by 42°C for 60 min. The reaction is stopped by heating to 95°C for 5 min and then cooled to 4°C (*see* **Note 7**).
5. The cDNA from extracted RNA can be applied to several PCRs. In the case of few-cell analyses, samples are split into two identical volumes for further PCR reactions (i.e., for target gene and standard gene analysis). Three or more PCR reactions from one sample reduce the respective recoveries *(22)*.

### *3.4. PCR for Qualitative mRNA Analysis*

1. To distinguish DNA and RNA, intron-spanning primers should be constructed.
2. Selecting an intron with 1000 bp, amplification of DNA can be prevented by suitable PCR conditions. In consequence, short amplification products for cDNA ( 150 bp) spanning a large intron guarantee optimal PCR efficiency ( *see* **Note 8**). The amplicon length is particularly crucial for the investigation of archival FFPE tissue and is ideally limited up to 100 bp.
3. Due to degradation and fragmentation of RNA, increasing length of the PCR product just to 150–300 bp results in a considerably lower recovery or even total loss *(26,28)*.

4. In a final volume of 50 µL, a standard PCR master mix for one sample consists of 5 µL PCR 10X buffer, 4 µL MgCl₂, forward and reverse primer in a final concentration of 300 n$M$, 1 µL dNTP with 10 m$M$ each 0.5 µL AmpliTaq Gold and cDNA. Up to 10 µL of the cDNA may be transferred to the PCR reaction.

5. Because of an inhibitory effect of the RT product on the polymerase activity, not more than 20% of the PCR volume should consist of the cDNA.

6. Typical PCR conditions are: 95°C for 6 min followed by 45 to 50 cycles with 95°C for 20 s and annealing temperature for 30–60 s (*see* **Note 9**).

7. If the mRNA detection fails after first amplification (i.e., low-copy genes from either a few cells or a single cell), the PCR product can be reamplified in a second "nested" PCR *(20)*.

## 3.5. PCR for Quantitative mRNA Analysis

1. Real-time PCR quantification of mRNA molecules has become a reliable analysis, even in combination with few microdissected cells *(29)*. This has been shown for frozen tissue *(30–32)* as well as FFPE tissue *(26,33,34)*.

2. To determine the regulation of gene expression quantitative assays are inevitably needed. Moreover, interaction of mRNA and binding proteins can be assessed by this technique *(35)*.

3. Valid measurement over a large range of initial starting quantities is obtained with known sensitivity of PCR. While absolute quantification by comparing the target to constant dilution series requires a remarkable effort, relative quantification ($\Delta C_T$) is often sufficient to measure gene regulation. Based on **Eq. 1**, the target gene sequence is normalized to an internal reference gene (i.e., almost unregulated "housekeeping gene" mRNA) or to an external standard.

$$\frac{T_0}{R_0} = K \bullet (1 + E)^{(CT,R - CT,T)} \tag{1}$$

$T_0$: initial number of target gene copies; $R_0$: initial number of reference gene/standard copies; $E$: efficiency of amplification; $CT,T$: threshold cycle of target gene; CT,R: threshold cycle of reference gene/standard; K: Constant.

4. In preliminary experiments it must be shown that the amplification efficiency of the target and the reference primer/probe sets—both may vary between 0 and 1—are approximately equal. This has to be determined by calculating the slope of the dilution series. $K$ is assumed to be equal within a definite fluorogenic-labeled primer/probe system and thus does not affect the comparison of the ratios.

5. Apart from the aspects mentioned above (short PCR products), special requirements mainly concern the construction of the primer/probe system. The primers are selected to span a long intron to prevent amplification of DNA by suitable PCR conditions. Additionally, in case of a single probe, it is placed centrally onto the exon-exon transition. In case of two probes, one is positioned centrally onto the exon-exon transition. This assures a complete detachment of the probe from the nucleic acid in case of partial DNA binding.

6. Alternatively, SYBR-Green may be applied for real-time quantification since it is less expensive and, moreover, works independently from any sequence.

7. However, some aspects have to be considered: (a) the amplification of a single PCR product has to be ensured; (b) as primer-dimers may affect the amplification efficiency their generation must be minimized and a sufficient efficiency rate has to be determined by a dilution series; (c) the analysis must be performed and calculated at a temperature level where primer-dimers are already melted.

### 3.6. RNA Preamplification

1. RNA preamplification techniques were introduced to increase the total mRNA amount without affecting the representative expression profile. This allows sufficient RNA to be obtained for several single gene analyses, virtual Northern blotting, or cDNA microarray hybridization even from microdissected cells.

2. For array hybridization, protocols were suggested using either T7-based linear amplification *(36–38)* or PCR-based amplification *(39–41)*. Both of these techniques preserve the expression profile sufficiently.

3. Due to reproducibility, reliability, and efficiency we considered the SMART™ PCR technique to be superior and modified the original protocol by introducing the entire cDNA to PCR amplification *(42)*. The principle of this kind of unspecific amplification is described in **ref. 43**.

4. To generate cDNA profiles representative for the input mRNA, it is crucial to stop the PCR cycling before amplification of any gene reaches its plateau phase. Depending on the initial amount of RNA, timely termination of the PCR has to be evaluated in preliminary experiments; otherwise the expression profile would be inevitably changed.

5. After microdissection the RNA is extracted as described in **Subheading 3.1.4.2.**, resuspended in 10 μL $H_2O$ and DNase digested (1 U, 30 min, 37°C).

6. The extraction is repeated and RNA is finally diluted in 4 μL 1 $H_2O$.

7. Total RNA is reverse-transcribed using the SMART™ PCR cDNA Synthesis Kit (BD Clontech) with slight modifications: 4 μL total RNA, 1 μL CDS Primer (diluted to a concentration of 5 μM) and 1μL SMART II oligonucleotide (diluted to a concentration of 5 μM) are mixed and incubated at 70°C for 8 min.

8. After short spinning, 2 min on ice and 2 min at 42°C, a master mix containing 2 μL 5X buffer, 1 μL DTT (20 mM), 1 mL 1 dNTP (10 mM) and 0.5 μL RNase H⁻ MMLV reverse transcriptase (PowerScript™, BD Clontech) is added and incubated at 42°C for 1 h.

9. Afterwards, cDNA is mixed with 38.5 μL TE buffer (10 mM Tris-HCl, pH 7.6, 1 mM EDTA) and purified by the QIAquick™ PCR Purification Kit (Qiagen).

10. Therefore, 250 μL buffer PB are added to the cDNA to load a column. According to the manufacturer's protocol, the columns are washed once.

11. For elution, 45 μL elution buffer (EB; 10 mM Tris-HCl, pH 8.5) is applied to the center of the column, incubated for 2 min, and centrifuged.

12. To improve the recovery, this step is repeated using the first eluate again.

13. From the eluted cDNA (~44 µL), 2 µL can be separated for further determination of the amplification factor.
14. For PCR-based amplification, the remaining 42 µL cDNA are mixed with the reagents of the Advantage™ 2 PCR Kit (BD Clontech): 5 µL 10X buffer, 1 µL PCR Primer (10 µ*M*), 1 µL dNTP (10 m*M*) and 1 mL Advantage 2 Polymerase Mix. PCR conditions are 95°C for 1 min, followed by the evaluated amount of cycles with 95°C for 15 s, 65°C for 30 s, and 68°C for 3 min.
15. The resulting PCR product is purified using the QIAquick columns as described above. 44 µL elution buffer are applied twice for elution and 2 µL may be separated for determination of the amplification factor.
16. If necessary the final volume is reduced by SpeedVac centrifugation.
17. To determine the factor of preamplification 2 µL nonamplified cDNA, as well as 2 µL amplified PCR product, are introduced to real-time PCR reactions.
18. Applying a primer/probe set for a representative gene the ratio of CT values is a measure for the preamplification efficiency (*see* **Subheading 3.5.**).

## 4. Notes

1. Even under these conditions we could observe a slight deterioration of amplifiability during months of storage especially when analyzing RNA from single or a few cells. Thus, ideally, tissue not older than 3–6 mo is applied in this case.
2. To avoid the potential danger of contamination and false-positive results, tissue preparation, microdissection, extractions, and especially RT/PCR preparations should be carried out in different rooms and separated from the amplification and postamplification settings. To prevent transmission of contaminant every workbench should be equipped with its own set of tools. Disposable gloves should be worn throughout the procedure and removed when leaving the work site. Positive air displacement pipets with sterile filter tips help to prevent liquid and aerosol contamination. Plasticware must be kept sterile, glassware and buffers should be autoclaved and RNase-free solutions are to be used (i.e., DEPC-treated $H_2O$).
3. Applying a mild formalin fixation (short perfusion or ~2 h immersion) allowed us to increase the labeling time and, moreover, to obtain a sufficient RNA recovery. However, the samples have to be treated afterwards similar to FFPE tissue, including proteinase K digestion.
4. RNA extraction from FFPE tissue with oligo-dT-based techniques (supermagnetic beads, columns) resulted in a somewhat lower recovery. This might be due to the high RNA fragmentation, so that only poly-A tails with the linked 3' ends are caught. All other mRNA fragments are lost.
5. However, regardless of the extraction technique, the use of RNA derived from routine FFPE tissue remains problematic for hybridization to cDNA arrays and especially oligonucleotide arrays. Moreover, when employing this RNA for construction of cDNA libraries and for RNA preamplification techniques, the results have to be considered with reservation due to the high amount of degradation and fragmentation.

6.  Random hexamers are preferable, especially when working with FFPE-derived RNA. Due to degradation, all mRNA fragments are reverse-transcribed. In case of oligo-dT priming, only the poly A tail-carrying mRNA fragments are transcribed.

7.  Depending on the requirements, RT-PCR can be performed at once using enzyme mixtures (reverse transcriptase and DNA polymerase). For analysis of a low cell number, the application of recombinant *Thermus thermophilus* enzyme for one-step RT-PCR is not recommended, as both reverse transcriptase and polymerase activities are often lower than those of the respective single enzymes.

8.  To distinguish DNA and RNA when intron-spanning primers cannot be applied, samples have to be split after proteinase digestion/extraction to one +RT (reverse transcriptase added) and one –RT (no enzyme added). The RT master mix should differ only in the enzyme that is replaced in –RT (mock) samples by $H_2O$; all other reagents and incubation steps remain identical. Subsequent quantitative PCR allows determination of the RNA portion in relation to DNA by calculating the ratio of the threshold cycles.

9.  Several internal controls should be incorporated. While many preparation steps may lead to mistakes and negative results, highest sensitivity opens the risk of contamination and false positive results. Suitable negative controls should comprise especially the microdissection procedure and PCR reactions (buffers); positive controls are necessary to assess cDNA synthesis and PCR amplification. The use of dUTP and uracil-*N*-glycosylase digestion for PCR reaction is helpful and should be applied routinely but cannot replace controls.

## References

1.  Kuppers, R., Zhao, M., Hansmann, M. L., and Rajewsky, K. (1993) Tracing B cell development in human germinal centres by molecular analysis of single cells picked from histological sections. *EMBO J.* **12**, 4955–4967.

2.  Moskaluk, C. A. and Kern, S. E. (1997) Microdissection and polymerase chain reaction amplification of genomic DNA from histological tissue sections. *Am. J. Pathol.* **150**, 1547–1552.

3.  Hiller, T., Snell, L., and Watson, P. H. (1996) Microdissection RT-PCR analysis of gene expression in pathologically defined frozen tissue sections. *Biotechniques* **21**, 38–40, 42, 44.

4.  Walch, A., Specht, K., Smida, J., Aubele, M., Zitzelsberger, H., Hofler, H., and Werner, M. (2001) Tissue microdissection techniques in quantitative genome and gene expression analyses. *Histochem. Cell Biol.* **115**, 269–276.

5.  Macabeo-Ong, M., Ginzinger, D. G., Dekker, N., McMillan, A., Regezi, J. A., Wong, D. T., and Jordan, R. C. (2002) Effect of duration of fixation on quantitative reverse transcription polymerase chain reaction analyses. *Mod. Pathol.* **15**, 979–987.

6.  Gillespie, J. W., Best, C. J., Bichsel, V. E., Cole, K. A., Greenhut, S. F., Hewitt, S. M., et al. (2002) Evaluation of non-formalin tissue fixation for molecular profiling. *Am. J. Pathol.* **160**, 449–457.

7. Shibutani, M., Uneyama, C., Miyazaki, K., Toyoda, K., and Hirose, M. (2000) Methacarn fixation: a novel tool for analysis of gene expressions in paraffin-embedded tissue specimens. *Lab. Invest.* **80**, 199–208.
8. Finkelstein, S. D., Dhir, R., Rabinovitz, M., Bischeglia, M., Swalsky, P. A., De Flavia, P., et al. (1999) Cold-temperature plastic resin embedding of liver for DNA- and RNA-based genotyping. *J. Mol. Diagn.* **1**, 17–22.
9. d'Amore, F., Stribley, J. A., Ohno, T., Wu, G., Wickert, R. S., Delabie, J., et al. (1997) Molecular studies on single cells harvested by micromanipulation from archival tissue sections previously stained by immunohistochemistry or nonisotopic in situ hybridization. *Lab. Invest.* **76**, 219–224.
10. Fink, L., Kinfe, T., Seeger, W., Ermert, L., Kummer, W., and Bohle, R.M. (2000) Immunostaining for cell picking and real-time mRNA quantitation. *Am. J. Pathol.* **157**, 1459–1466.
11. Gjerdrum, L.M., Lielpetere, I., Rasmussen, L. M., Bendix, K., and Hamilton-Dutoit, S. (2001) Laser-assisted microdissection of membrane-mounted paraffin sections for polymerase chain reaction analysis: identification of cell populations using immunohistochemistry and in situ hybridization. *J. Mol. Diagn.* **3**, 105–110.
12. Fend, F., Emmert-Buck, M. R., Chuaqui, R., Cole, K., Lee, J., Liotta, L.A., and Raffeld, M. (1999) Immuno-LCM: laser capture microdissection of immuno-stained frozen sections for mRNA analysis. *Am. J. Pathol.* **154**, 61–66.
13. Burton, M. P., Schneider, B. G., Brown, R., Escamilla-Ponce, N., and Gulley, M. L. (1998) Comparison of histologic stains for use in PCR analysis of microdissected, paraffin-embedded tissues. *Biotechniques* **24**, 86–92.
14. Murase, T., Inagaki, H., and Eimoto, T. (2000) Influence of histochemical and immunohistochemical stains on polymerase chain reaction. *Mod. Pathol.* **13**, 147–151.
15. Fink, L., Kinfe, T., Stein, M.M., Ermert, L., Hanze, J., Kummer, W., et al. (2000) Immunostaining and laser-assisted cell picking for mRNA analysis. *Lab. Invest.* **80**, 327–333.
16. Kohda, Y., Murakami, H., Moe, O. W., and Star, R.A. (2000) Analysis of segmental renal gene expression by laser capture microdissection. *Kidney Int.* **57**, 321–331.
17. Murakami, H., Liotta, L., and Star, R.A. (2000) IF-LCM: laser capture microdissection of immunofluorescently defined cells for mRNA analysis. *Kidney Int.* **58**, 1346–1353.
18. Emmert-Buck, M. R., Bonner, R. F., Smith, P. D., Chuaqui, R. F., Zhuang, Z., Goldstein, S. R., et al. (1996) Laser capture microdissection. *Science* **274**, 998–1001.
19. Schutze, K., Becker, I., Becker, K. F., Thalhammer, S., Stark, R., Heckl, W. M., et al. (1997) Cut out or poke in—the key to the world of single genes: laser micromanipulation as a valuable tool on the look-out for the origin of disease. *Genet. Anal.* **14**, 1–8.
20. Schutze, K. and Lahr, G. (1998) Identification of expressed genes by laser-mediated manipulation of single cells. *Nat. Biotechnol.* **16**, 737–742.

21. Kolble, K. (2000) The Leica microdissection system: design and applications. *J. Mol. Med.* **78,** B24–B25.

22. Fink, L., Stahl, U., Ermert, L., Kummer, W., Seeger, W., and Bohle, R. M. (1999) Rat porphobilinogen deaminase gene: a pseudogene-free internal standard for laser-assisted cell picking. *Biotechniques* **26,** 510–516.

23. To, M. D., Done, S. J., Redston, M., and Andrulis, I. L. (1998) Analysis of mRNA from microdissected frozen tissue sections without RNA isolation. *Am. J. Pathol.* **153,** 47–51.

24. Cordell, J. L., Falini, B., Erber, W. N., Ghosh, A. K., Abdulaziz, Z., MacDonald, S., et al. (1984) Immunoenzymatic labeling of monoclonal antibodies using immune complexes of alkaline phosphatase and monoclonal anti-alkaline phosphatase (APAAP complexes). *J. Histochem. Cytochem.* **32,** 219–229.

25. Imamichi, Y., Lahr, G., and Wedlich, D. (2001) Laser-mediated microdissection of paraffin sections from Xenopus embryos allows detection of tissue-specific expressed mRNAs. *Dev. Genes Evol.* **211,** 361–366.

26. Specht, K., Richter, T., Muller, U., Walch, A., Werner, M., and Hofler, H. (2001) Quantitative gene expression analysis in microdissected archival formalin-fixed and paraffin-embedded tumor tissue. *Am. J. Pathol.* **158,** 419–429.

27. Stanta, G. and Schneider, C. (1991) RNA extracted from paraffin-embedded human tissues is amenable to analysis by PCR amplification. *Biotechniques* **11,** 304, 306, 308.

28. Goldsworthy, S. M., Stockton, P. S., Trempus, C. S., Foley, J. F., and Maronpot, R. R. (1999) Effects of fixation on RNA extraction and amplification from laser capture microdissected tissue. *Mol. Carcinog.* **25,** 86–91.

29. Fink, L., Seeger, W., Ermert, L., Hanze, J., Stahl, U., Grimminger, F., et al. (1998) Real-time quantitative RT-PCR after laser-assisted cell picking. *Nat. Med.* **4,** 1329–1333.

30. Bohle, R. M., Hartmann, E., Kinfe, T., Ermert, L., Seeger, W., and Fink, L. (2000) Cell type-specific mRNA quantitation in non-neoplastic tissues after laser-assisted cell. *Pathobiology* **68,** 191–195.

31. Nagasawa, Y., Takenaka, M., Matsuoka, Y., Imai, E., and Hori, M. (2000) Quantitation of mRNA expression in glomeruli using laser-manipulated micro-dissection and laser pressure catapulting. *Kidney Int.* **57,** 717–723.

32. Cohen, C. D., Frach, K., Schlondorff, D., and Kretzler, M. (2002) Quantitative gene expression analysis in renal biopsies: a novel protocol for a high-throughput multicenter application. *Kidney Int.* **61,** 133–140.

33. Cohen, C. D., Grone, H. J., Grone, E. F., Nelson, P. J., Schlondorff, D., and Kretzler, M. (2002) Laser microdissection and gene expression analysis on form-aldehyde-fixed archival tissue. *Kidney Int.* **61,** 125–132.

34. El-Sherif, A. M., Seth, R., Tighe, P. J., and Jenkins, D. (2001) Quantitative analysis of IL-10 and IFN-gamma mRNA levels in normal cervix and human papillomavirus type 16 associated cervical precancer. *J. Pathol.* **195,** 179–185.

35. Steger, K., Fink, L., Klonisch, T., Bohle, R. M., and Bergmann, M. (2002) Protamine-1 and -2 mRNA in round spermatids is associated with RNA-binding proteins. *Histochem. Cell Biol.* **117,** 227–234.
36. Luo, L., Salunga, R. C., Guo, H., Bittner, A., Joy, K. C., Galindo, J. E., et al. (1999) Gene expression profiles of laser-captured adjacent neuronal subtypes. *Nat. Med.* **5,** 117–122.
37. Ohyama, H., Zhang, X., Kohno, Y., Alevizos, I., Posner, M., Wong, D. T., and Todd, R. (2000) Laser capture microdissection-generated target sample for high-density oligonucleotide array hybridization. *Biotechniques* **29,** 530–536.
38. Luzzi, V., Holtschlag, V., and Watson, M. A. (2001) Expression profiling of ductal carcinoma in situ by laser capture microdissection and high-density oligonucleotide arrays. *Am. J. Pathol.* **158,** 2005–2010.
39. Spirin, K. S., Ljubimov, A. V., Castellon, R., Wiedoeft, O., Marano, M., Sheppard, D., et al. (1999) Analysis of gene expression in human bullous keratopathy corneas containing limiting amounts of RNA. *Invest. Ophthalmol. Vis. Sci.* **40,** 3108–3115.
40. Vernon, S. D., Unger, E. R., Rajeevan, M., Dimulescu, I. M., Nisenbaum, R., and Campbell, C. E. (2000) Reproducibility of alternative probe synthesis approaches for gene expression profiling with arrays. *J. Mol. Diagn.* **2,** 124–127.
41. Leethanakul, C., Patel, V., Gillespie, J., Pallente, M., Ensley, J. F., Koontongkaew, S., et al. (2000) Distinct pattern of expression of differentiation and growth-related genes in squamous cell carcinomas of the head and neck revealed by the use of laser capture microdissection and cDNA. *Oncogene* **19,** 3220–3224.
42. Fink, L., Kohlhoff, S., Stein, M. M., Hanze, J., Weissmann, N., Rose, F., et al. (2002) cDNA array hybridization after laser-assisted microdissection from non-neoplastic tissue. *Am. J. Pathol.* **160,** 81–90.
43. Chenchik, A., Zhu, Y. Y., Diachenko, L., Li, R., Hill, J., and Siebert, P. D. (1998) Generation and use of high-quality cDNA from small amounts of total RNA by SMART PCR, in *Gene Cloning and Analysis by RT-PCR*, BioTechniques Books, Westborough, MA, pp. 305–319.

# 15

## Gene Expression Profiling of Primary Tumor Cell Populations Using Laser Capture Microdissection, RNA Transcript Amplification, and GeneChip® Microarrays

**Veronica I. Luzzi, Victoria Holtschlag, and Mark A. Watson**

### Summary

Gene expression profiling from microdissected cell populations is a powerful approach to explore molecular processes involved in development and solid tumor biology. In this chapter, we detail robust and validated methods for tissue preparation and isolation of high-quality RNA from microdissected cell populations. A protocol is also provided for linear transcript amplification using as little as 10 ng of total RNA to produce labeled cRNA targets for hybridization to GeneChip® high-density oligonucleotide microarrays. Particular emphasis is placed on troubleshooting each technical step in the protocol and measures of quality assurance for both RNA isolation and resulting microarray data.

**Key Words:** Laser capture microdissection; transcript amplification; oligonucleotide microarray; gene expression profiling.

## 1. Introduction

Gene expression profiling is a powerful tool to survey the transcriptional landscape of human cancer (1–3). However, the cellular heterogeneity of many solid tumors such as breast and prostate adenocarcinoma may obscure significant patterns of gene transcription present in subpopulations of cells when a homogenized tumor tissue is considered in its entirety. At the very least, this approach precludes a comparative analysis of expression profiles from independent tumor cell populations present at different stages of malignant progression in the same specimen. Laser capture microdissection (LCM) can be used to isolate discrete cell populations from heterogeneous tissue sections. Downstream applications of this technology have included DNA and protein

From: *Methods in Molecular Biology, vol. 293: Laser Capture Microdissection: Methods and Protocols*
Edited by: G. I. Murray and S. Curran © Humana Press Inc., Totowa, NJ

analysis, and more recently RNA expression analysis using microarray technology *(4–8)*. Gene expression profiling may be performed using spotted cDNA microarrays *(4,5)* or Affymetrix high-density oligonucleotide GeneChip® microarrays *(6–8)*, each presenting distinct challenges in sample preparation and data interpretation for microdissected cellular specimens. In this chapter, we will review optimized methods for tissue preparation, RNA isolation, and biotinylated cRNA target synthesis for hybridization to GeneChip microarrays. Using examples from an ongoing study involving microdissected cell populations obtained from human breast tumor tissue, guidelines for data interpretation and technical troubleshooting will be provided.

## 2. Materials

1. Cryostat.
2. Finest/Premium Superfrost slides (Fisher cat. no. 12-544-7).
3. Disposable cytology slide mailers (Fisher cat. no. 23-034804 ).
4. 100% Ethanol.
5. Mayer's hematoxylin (Sigma cat. no. MHS-16).
6. Automation buffer (Biomeda cat. no. M30).
7. Alcoholic eosin (Sigma cat. no. HT110-1-128).
8. Mixed xylenes (Sigma cat. no. X2377).
9. 0.5-mL and 1.5-mL nuclease-free microcentrifuge tubes.
10. RNA isolation reagents, e.g., Micro RNA Isolation system (Stratagene cat. no. 200344) or Picopure RNA isolation system (Arcturus cat. no.KIT0202), or individual reagents including:
    - Denaturation buffer (4 $M$ guanidine isothiocyanate/20 m$M$ sodium citrate/ 0.5% sarcosyl).
    - 14.4 $M$ 2-Mercaptoethanol.
    - 2 $M$ Sodium acetate, pH 4.0.
    - 24:1 chloroform:isoamyl alcohol.
    - Acid phenol, pH 5.3–5.7.
    - Isopropanol.
    - 2 mg/mL glycogen solution (Ambion cat. no. 9510).
11. Nuclease-free (non-DEPC-treated) water (Ambion cat. no. 9930).
12. Bioanalyzer 2100 and RNA 6000 Nano LabChip (Agilent).
13. Programmable thermal cycler or regulated heat block.
14. Linear amplification reagents, e.g., Riboamp OA system (Arcturus cat. no. KIT0206), or MessageAmp system (Ambion cat. no. 1750), or individual enzymes and reagents including:
    - HPLC-purified T7T$_{24}$ oligonucleotide 5'-GGCCAGTGAATTGTAATACG-ACTCACTATAGGGAGGCGGT(24)-3'.
    - Superscript II reverse transcriptase (Invitrogen cat. no. 18064-014).
    - 10 m$M$ dNTP solution (Invitrogen cat. no. 18427-013).

- 100 m*M* DTT.
- 5X second-strand buffer (Invitrogen cat. no. 10812-014).
- DNA ligase (Invitrogen cat. no. 18052-019).
- *E. Coli* DNA polymerase (Invitrogen cat. no. 18010-025).
- RNase H (Invitrogen cat. no. 18021-014).
- T4 DNA polymerase (Invitrogen cat. no. 18005-025).
- Linear acrylamide (Ambion cat. no. 9520).
- Random hexamers (Invitrogen cat. no. 48190-011).
- Megascript T7 reagent system (Ambion cat. no. 1334).
- RNeasy mini clean-up kit (Qiagen cat. no. 74104).

15. BioArray HighYield biotin transcript labeling system (Affymetrix cat. no. 900182).
16. Affymetrix GeneChip microarray system and GeneChip microarrays.

## 3. Methods

The following methods describe (1) tissue preparation and staining to optimize RNA retrieval; (2) total RNA extraction, purification, and analysis; (3) preparation of labeled cRNA targets for Affymetrix GeneChip microarray analysis; and; (4) interpretation and quality assessment of microarray data.

### 3.1. Tissue Preparation and Staining

The isolation of high-quality RNA is critical to the success of all expression profiling experiments. When properly prepared, frozen tissue yields the highest quality RNA, although suitable RNA also may be obtained from tissues properly fixed in ethanol and embedded in low melting point paraffin *(9)*. The protocol in this chapter will focus only on fresh-frozen tissue. Because RNA is extremely labile in the context of fresh frozen tissue, careful attention to tissue processing and staining is paramount to the successful isolation of intact RNA. Tissue microdissection is performed using the PixCell II Laser Capture Microdissection instrument (Arcturus) and tissue preparation protocols described below have been optimized for this film transfer system. Use of other microdissection instruments may require alternate tissue preparation methods.

1. Tissues should be properly prepared and frozen (*see* **Note 1**). Mount the frozen tissue block on the cryostat and cut rough sections until a parallel face is achieved. The rough cuts of tissue may be saved and used to isolate total RNA as a positive control for proper tissue collection and preservation (*see* **Note 2**).
2. Cut a 6–8-μm section and transfer it immediately to the center of a chilled, untreated glass slide. **Immediately** place the mounted tissue section in a cytology slide mailer containing 70% ethanol at 4°C (*see* **Note 3**). The slide should be maintained in 70% ethanol for at least 1 min prior to staining or storage (*see* **Note 4**).
3. After cutting a sufficient number of slides for microdissection, cut one additional slide. This slide should be stained with conventional hematoxylin and eosin and cover-slipped so that it may be used as a reference "roadmap" for dissection.

**Table 1**
**Frozen Tissue Section Staining Protocol**
**for LCM and RNA Retrieval**

| Solution | Time/exposure |
|----------|---------------|
| 70% ethanol no. 1 | 60 s |
| Deionized water | 5 dips |
| Mayer's hematoxylin | 10 dips |
| Deionized water | 10 dips |
| 1X automation buffer | 10 dips |
| Deionized water | 10 dips |
| 70% ethanol no. 2 | 60 s |
| 95% ethanol no. 1 | 60 s |
| Eosin-Y | 15 s |
| 95% ethanol no. 2 | 10 dips |
| 95% ethanol no. 3 | 10 dips |
| 100% ethanol no. 1 | 10 dips |
| 100% ethanol no. 2 | 60 s |
| 100% ethanol no. 3 | 60 s |
| Xylene no. 1 | 10 dips |
| Xylene no. 2 | 3 min |
| Xylene no. 3 | 3 min |

4. Staining solutions may be made in disposable cytology slide mailers, which allow convenient and small amounts of staining solutions (~30 mL) to be utilized for staining two to four slides at one time. Change the staining solutions frequently, as they become depleted. Proper selection of staining reagents is also critical to the isolation of high-quality RNA from microdissected tissue sections (*see* **Note 5**). Preparation and use of staining reagents should be conducted under a fume hood. In particular, xylene is a volatile liver toxin. Depending on laboratory space constraints, it may be useful to establish a portable staining hood near either the cryostat and/or LCM instrument *(10)*.

5. Sequentially dip slides in the solutions as noted in **Table 1**. The slide should be dipped in and out of the solution to allow solutions to completely cover the tissue section. Do not allow the slide to dry between solutions. After staining, allow slides to air-dry under a fume hood for 5 min before dissecting (*see* **Note 4**).

## *3.2. RNA Isolation and Analysis*

RNA purification may be performed using a number of commercially available reagent systems (*see* **Note 6**). The RNA purification described in this chapter uses an organic extraction method with commercial reagents (Stratagene MicroRNA isolation system) followed by assessment via microcapillary gel electrophoresis.

## 3.2.1. RNA Isolation

1. Prepare lysis buffer by adding 7.2 µL of 14.4 *M* 2-mercaptoethanol (supplied with kit) to 1 mL of denaturation solution (supplied with kit; 4 *M* guanidine isothiocyanate/20 m*M* sodium citrate/0.5% sarcosyl). Mix well.

2. Tissue microdissection is performed using the PixCell II LCM instrument (Arcturus) and CapSure™ LCM transfer films. When all target cells within the transfer film area have been captured, remove the cap and place it on a 0.5-mL microcentrifuge tube containing 100 µL of lysis buffer. Invert the tube so that the lysis buffer covers the cap and incubate at room temperature for at least 10 min to allow tissue digestion off the cap.

3. While tissue captured on the first transfer film is being lysed, place a new film over an additional area of the section or on a new slide. Repeat laser capture of the desired cell populations.

4. Briefly spin the first cap and tube to collect the lysis buffer and lysed tissue. Remove the first cap and replace it with a new cap of dissected tissue. In this manner, multiple dissections of the same target cell population may be pooled into a single volume of lysis buffer.

5. After dissecting all areas of interest, scrape the remaining tissue from the slide into a tube containing 100 µL of lysis buffer to serve as a positive control for RNA isolation (*see* **Note 2**).

6. Transfer the tissue lysate from the 0.5-mL tube to a 1.5-mL tube. Add 2 µL of nuclease-free water (*see* **Note 7**) and 10 µL of 2 *M* sodium acetate (pH 4.0). Vortex briefly.

7. Add 100 µL of water-saturated acid phenol and vortex briefly.

8. Add 20 µL of chloroform:isoamyl alcohol (24:1) and vortex vigorously for 1 min.

9. Place on wet ice for 15 min.

10. Centrifuge at 12,000*g* for 5 min at room temperature to separate the aqueous and organic phases (*see* **Note 7**).

11. Transfer the upper aqueous layer (~100 µL) to a new 1.5-mL tube.

12. Add 2 µL of glycogen (20 µg) as a carrier and mix well.

13. Add 100 µL of cold isopropanol and mix well.

14. Place in a –20°C freezer for at least 30 min. Samples may be stored indefinitely in isopropanol.

15. Centrifuge at 16,000–20,000*g* for 30 min at 4°C to precipitate RNA.

16. Carefully wash RNA pellet with 400 µL of chilled 70% ethanol.

17. Centrifuge at 16,000–20,000*g* for 5 min at 4°C.

18. Remove as much supernatant as possible without disturbing the pellet.

19. Allow the pellet to air-dry for 5–10 min. Be certain that no ethanol is remaining in the tube (although a small amount of water may be present). Do not allow the pellet to overdry as the precipitated RNA pellet may become difficult to resuspend in water.

20. Resuspend the RNA pellet in 10 µL of nuclease-free water. Do not resuspend the pellet in DEPC-treated water, Tris-EDTA, or RNA stabilization buffers, as these may inhibit downstream enzymatic reactions. Store the RNA at –80°C.

21. Note that treatment with DNase is not necessary for subsequent GeneChip microarray analysis and, in fact, may result in significant degradation or total loss of RNA.

### 3.2.2. Assessment of Total RNA Quality

Generation of high-quality gene expression microarray data is dependent on high-quality input RNA. Therefore, whenever possible, RNA isolated from microdissected tissue should be directly assessed by UV spectroscopy and/or electrophoresis. In this chapter, we demonstrate qualitative RNA analysis using the Agilent 2100 Bioanalyzer system with RNA Nano Chip microcapillary gel electrophoresis. However, a number of alternative assessments of RNA quality are available (*see* **Note 8**).

1. Mix 1 μL of isolated RNA (at least 10 ng) and 1 μL of nuclease-free water. Heat the sample and load 1 μL onto the RNA 6000 Nano Chip as directed by the manufacturer.
2. **Table 2** summarizes representative yields of RNA obtained from LCM. Depending on the tissue and cell type dissected, 1000 laser pulses with a 30-μm beam should yield 20–60 ng of total RNA. Estimation of RNA yield by electrophoresis and ribosomal band intensities may be as much as twofold discrepant with calculated yields based on other methods (*see* **Note 8**).
3. **Figure 1** demonstrates representative Nano Chip traces from isolated RNA. Ideally, the intensity ratio of 28S to 18S ribosomal RNA bands should be 2.0. This is seldom the case. Ribosomal 28S:18S band ratios may be as low as 1.0, and RNA may still yield satisfactory microarray data (**Table 2**). Of greater importance is the presence or absence of a low molecular weight "smear" below the 18S band. The presence of such a smear suggests severe RNA degradation. Also, the complete absence of a 28S ribosomal band (**Fig. 1C,D**) is indicative of degraded RNA that will not perform well for microarray analysis.
4. If too few cells have been captured to analyze the microdissected RNA directly, RNA isolated from the remaining dissected tissue section (*see* **Note 2**) should be assessed as described above (**Fig. 1E,F**).
5. While this method is a convenient and sensitive approach to quantify and qualitatively assess RNA, it can not provide data regarding RNA purity. If a sufficient quantity of RNA has been isolated, a UV spectrophotometer (e.g., NanoDrop ND-1000 spectrometer, NanoDrop Technologies) may be used to measure the absorbance ratio at 260 nm vs 280 nm. When suspended in nuclease-free water, RNA samples with a 260/280 ratio of less than 1.6 are likely contaminated with trace organics or protein, even if their electrophoresis profile appears acceptable. This is particularly problematic when RNA is isolated by organic extraction rather than spin-column-based-protocols. Trace organics or protein contaminants can inhibit downstream enzyme reactions. Therefore, if measured 260/280 ratios are significantly lower than 1.6, RNA should be reprecipitated with ammonium acetate and ethanol.

**Table 2**
**Representative Yield and Quality of Cellular RNA Isolated From LCM,
Resulting Yield of Biotin-Labeled cRNA Synthesized From 20 ng of Each Total
Cellular RNA, and Corresponding Quality of Genechip® Microarray Data
Derived From Each cRNA Hybridization**

| # Shots | RNA yield (ng) | 28S/18S | cRNA yield (μg) | SF | %P |
|---------|---------------|---------|-----------------|------|-----|
| 1097 | 45 | 1 | 20 | 5.9 | 27 |
| **1100** | **40** | **0.2** | **18** | **5.7** | **23** |
| 1428 | 55 | 1.2 | 20 | 3.3 | 36 |
| **1500** | **110** | **0.8** | **24** | **2.3** | **36** |
| 1552 | 120 | 1 | 23 | 2.3 | 36 |
| 1700 | 30 | 0.7 | 18 | 12.5 | 14 |
| 1700 | 70 | 0.9 | 18 | 2.4 | 31 |
| 2357 | 40 | 0.8 | 18 | 2.8 | 38 |
| 2454 | 74 | 0.3 | 18 | 13.3 | 15 |
| 2500 | 130 | 1.3 | 18 | 2.2 | 33 |
| 2750 | 140 | 1 | 30 | 3.3 | 34 |
| 2800 | 190 | 1.2 | 20 | 7.0 | 27 |
| 3000 | 180 | 1.1 | 40 | 2.7 | 31 |
| 3100 | 130 | 1.2 | 40 | 1.3 | 38 |
| 3245 | 190 | 1.6 | 18 | 3.1 | 35 |
| 3245 | 190 | 1.6 | 18 | 21.5 | 19 |
| 3300 | 100 | 1.3 | 40 | 1.6 | 39 |
| **3400** | **110** | **0.9** | **24** | **2.4** | **34** |
| 3467 | 40 | 1.2 | 40 | 7.2 | 17 |
| **3700** | **150** | **1.5** | **40** | **2.1** | **33** |
| 4212 | 100 | 0.9 | 18 | 6.9 | 29 |
| 4323 | 130 | 0.9 | 21 | 6.9 | 24 |
| 5425 | 220 | 1 | 30 | 4.3 | 30 |

RNA samples in **bold** are also presented in **Fig. 1**. The number of 30-μm laser shots performed for each dissection is indicated; 28S/18S, ratio of ribosomal RNA band intensities determined by Agilent RNA 6000 Nano Chip; SF, scale factor calculated by Affymetrix Microarray Analysis Suite (MAS) 5.0 software setting target intensity scaling at 150; %P, percentage of transcripts called detected (P) by the MAS5.0 software. All targets were hybridized to Affymetrix U95A human GeneChip microarrays, representing approx 12,600 gene transcripts.

### 3.3. Preparation of Labeled Targets for GeneChip Microarray Analysis

Starting with 10–100 ng of total RNA (*see* **Note 9**), transcript amplification and biotinylated cRNA target preparation involves two successive rounds of cDNA synthesis and in vitro transcription. This is in contrast to a single round of cDNA synthesis and in vitro transcription that is routinely utilized for pre-

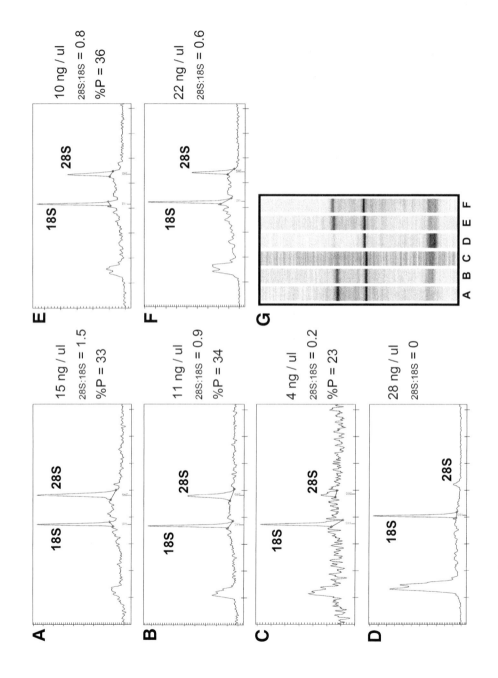

E    10 ng / ul    28S:18S = 0.8    %P = 36

28S    18S

F    22 ng / ul    28S:18S = 0.6

28S    18S

G

A B C D E F

A    15 ng / ul    28S:18S = 1.5    %P = 33

28S    18S

B    11 ng / ul    28S:18S = 0.9    %P = 34

28S    18S

C    4 ng / ul    28S:18S = 0.2    %P = 23

28S    18S

D    28 ng / ul    28S:18S = 0

28S    18S

194

paring targets for GeneChip hybridization starting with 5 µg of total RNA (standard Affymetrix protocol). In this chapter, we describe our initially reported amplification protocol *(6)* that relies on standard Gubler-Hoffman double-stranded cDNA synthesis *(11)* followed by T7 RNA polymerase mediated in vitro transcription as originally described by VanGelder et al. *(12)*. More recently, commercial reagent systems have become available that are equally reliable and easy to use. For example, we have utilized the RiboAmp reagent system (Arcturus) to generate amplified targets *(13)*, although several other acceptable approaches are available (*see* **Note 10**).

Throughout the protocol, the following points should be remembered:

1. Practice care in preventing RNase contamination. Generally, utilizing RNase-free reagents and disposables on a clean laboratory bench surface will suffice. DEPC treatment, high-temperature baking, and RNase cleansing agents are usually not necessary.
2. Whenever possible, target amplification of all experimental samples should be performed in parallel. Utilize "master mixes" for all reagents to avoid pipetting small volumes and to ensure uniformity of reagent concentrations across all samples. For $N$ number of samples, make sufficient master mix for $N + 1$ reactions.
3. Make certain that all components of master mixes are mixed thoroughly, after addition of each component. Unless specifically instructed, do not vortex samples. Mix by repeatedly pipetting reagents in the tube, or gently mix the tube

---

Fig. 1. (*opposite page*) Representative RNA isolations from microdissected tissue sections. Cellular RNA was evaluated using the Agilent Bioanalyzer and RNA 6000 Nano Chips. Electrophoresis profiles (**A–F**) and pseudo-gel image (**G**, Lanes **A–F**) are displayed for six RNAs of differing quality. Electrophoresis profiles show fluorescent intensity of nucleic acid (*y*-axis) as a function of migration time (*x*-axis). Peaks corresponding to 28S and 18S ribosomal RNA bands are indicated. Next to each profile is displayed the calculated concentration of RNA, the calculated 28S:18S ribosomal RNA peak ratio, and the resulting percentage of transcripts scored as "detected" (P) on the GeneChip microarray to which biotinylated cRNA derived from the RNA was hybridized (*see also* **Table 2**). Note that RNA with 28S:18S ratios of 1.5 (**A**) to 0.9 (**B**) yield comparable microarray data. RNA with 28S:18S ratios of 0.2 or lower (**C**) yield considerably poorer microarray results. In many cases, large quantities of RNA can be isolated from microdissected tissue, but this RNA may be too degraded for use in microarray experiments (**D**). Also note a lack of complete correspondence between 28S:18S ratios (a measure of RNA "quality") and resulting microarray data, as many samples with 28S:18S < 1.0 yield high-quality data (**B,E**). As described in the text (**Note 2**), analysis of RNA derived from tissue remaining after microdissection (**F**) can be used as a surrogate marker for the quality of RNA derived from the actual microdissected tissue (**E**).

by hand. After mixing, briefly spin reaction tubes to collect the reaction at the bottom of the tube.

4. Reactions should be performed in a regulated heat block. A thermal cycler is ideal, particularly for 16°C reactions. For prolonged incubations at 37°C or 42°C, it is important to occasionally (i.e., every 30–60 min) mix the reaction components and then briefly spin the tube to collect any condensate. Prolonged incubations in a thermal cycler with a heated lid may increase the reaction temperature above 37°C and therefore, should be avoided.

### 3.3.1. cDNA Synthesis and Purification (Round 1)

1. Add the following components in a 0.5-mL reaction tube:
   - 11 µL 10–100 ng RNA in nuclease-free water.
   - 1 µL 100 µM T7T$_{24}$ primer.

   Mix the reagents, incubate at 70°C for 10 min, spin briefly, and place on ice.

2. Add the following components to the reaction:
   - 4 µL 5X Superscript II reaction buffer.
   - 2 µL 0.1 M DTT.
   - 1 µL 10 mM dNTP mix.

   Mix reagents and incubate at 42°C for 2 min. Then add
   - 1 µL Superscript II reverse transcriptase (100 U/ mL).

   Incubate the final 20-µL reaction at 42°C for 1 h.

3. Briefly spin reactions and place on ice.

4. Add the following components to the first strand reaction for second strand synthesis:
   - 91 µL Nuclease free water.
   - 30 µL 5X Second strand buffer.
   - 3 µL 10 mM dNTPs.
   - 1 µL 10 U/µL DNA ligase.
   - 4 µL 10 U/µL *E.coli* DNA polymerase I.
   - 1 µL 2 U/µL RNase H.

   Gently tap the tube to mix; spin briefly and incubate the 150-µL reaction at 16°C for 2 h.

5. Add 2 µL (10 U) T4 DNA polymerase to the reaction and incubate for 5 min at 16°C.

6. Add 10 µL 0.5 M EDTA to stop the reaction. At this point the double-stranded cDNA may be stored indefinitely at –20°C.

7. Transfer the cDNA reaction to a 1.5-mL tube. Add 162 µL of phenol:chloroform:isoamyl alcohol solution to the second strand cDNA reaction mixture, vortex for 30 s, and then spin for 5 min at room temperature and 12,000g to extract the cDNA.

8. Remove the aqueous (upper phase) into a clean 1.5-mL tube. Add 1 µL of 20 µg/µL glycogen and mix thoroughly. Add 0.5 vol (~80 µL) of 7.5 M ammonium acetate and mix thoroughly. Add 2.5 vol (~400 µL) of 100% ethanol and mix thoroughly to precipitate the extracted cDNA.

9. Centrifuge immediately at 16,000g for 20 min at room temperature (*see* **Note 11**)

10. Wash the pellet thoroughly with 70% ethanol. Spin at 16,000g for 5 min. Carefully pour off the 70% ethanol without displacing the pellet, although a pellet may not always be visible. Wash the pellet again, spin, and pour off the second wash. Carefully aspirate any remaining ethanol with a pipet, again being careful not to displace any visible pellet.
11. Allow the cDNA pellet to dry and dissolve it in 8 μL of nuclease free water (*see* **Note 12**).

### 3.3.2. In Vitro Transcription and Purification (Round 1)

1. To synthesize complementary RNA (cRNA) from the double stranded cDNA template, the Ambion Megascript in vitro transcription kit (T7 enzyme) is utilized. Following the recommended reaction conditions, add reaction components as follows:
   - 8 μL cDNA from **step 11** above.
   - 2 μL 10X reaction buffer.
   - 2 μL ATP solution.
   - 2 μL CTP solution.
   - 2 μL GTP solution.
   - 2 μL UTP solution.
   - 2 μL T7 enzyme Mix.

   Add the reaction components to the 1.5-mL tube containing the cDNA from **step 11** above. Mix thoroughly and then transfer the entire reaction to a 0.5-mL tube. Incubate at 37°C for 6 h. Gently mix the reaction by hand every hour and centrifuge briefly to collect reaction tube condensate. After in vitro transcription, the reaction may be frozen at −70°C until purification.
2. cRNA is purified using Qiagen's RNeasy or Arcturus' PicoPure RNA purification columns. For Qiagen RNeasy columns, the manufacturer's recommended protocol is utilized. The cRNA is then eluted twice, with 40 μL of nuclease-free water.
3. Add 1 μL of 20 μg/μL glycogen, 0.5 vol (40 μL) of 7.5 *M* ammonium acetate, and 2.5 vol (200 μL) of 100% ethanol to the eluted 80 μL of cRNA. Mix thoroughly and spin immediately.
4. Centrifuge at 16,000g for 30 min.
5. Wash the pellet, which may not be visible, thoroughly with 70% ethanol (*see* **Note 11**).
6. Allow the cRNA pellet to dry and resuspend in 11 μL of nuclease-free water (*see* **Note 12**).
7. Depending on the initial amount of starting material, 1 μL of the synthesized cRNA may be quantified by UV absorbance (*see* **Note 8**) or qualitatively assessed by microcapillary gel electrophoresis (*see* **Fig. 2**).

### 3.3.3. cDNA Synthesis and Purification (Round 2)

1. Add the following components in a 0.5-mL reaction tube:
   - 11 μL cRNA in nuclease-free water.
   - 1 μL 1 μg/μL random hexamers.

   Mix the reagents, incubate at 70°C for 10 min, spin briefly, and place on ice.

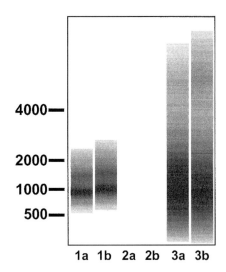

Fig. 2. Electrophoresis analysis of amplified cRNA. One microliter of cRNA was analyzed on an RNA 6000 Nano Chip after one round (Lanes 2a,b) and two rounds (Lanes 3a,b) of amplification. The cRNA was synthesized starting with 10 ng of total RNA. For comparison, 1 µL of cRNA synthesized from 10 µg of total RNA and one round of amplification (the standard Affymetrix target synthesis protocol) is also displayed (Lanes 1a,b). Note that starting with 10 ng of RNA, no cRNA is detectable after a single round of amplification. The size distribution of final amplified product is similar using standard target synthesis (1a,b) and the protocol in this chapter (3a,b). Alternative methods for generating cRNA (*see* **ref. 13**) will result in shorter amplified target.

2. Add the following components to the reaction:
   - 4 µL 5X Superscript II reaction buffer.
   - 2 µL 0.1 $M$ DTT.
   - 1 µL 10 mM dNTP mix.

   Mix reagents and incubate at 42°C for 2 min. Then add 1 µL Superscript II reverse transcriptase (100 U/µL). Incubate the final 20 µL reaction at 42°C for 1 h.
3. Briefly spin reactions and place on ice.
4. Add 1 µL (2 U/µL) RNase H to the reaction and incubate for 20 min at 37°C. Terminate the reaction by heating it to 95°C for 5 min, followed by a brief spin and incubation on ice.
5. To prime second-strand synthesis add 1 µL (100 µM) T7T$_{24}$ oligonucleotide primer to the reaction and heat at 70°C for 10 min. Briefly spin the reaction and hold on ice.
6. Add the following second strand synthesis reaction reagents:
   - 91 µL Nuclease-free water.
   - 30 µL 5X second-strand reaction buffer.

- 3 µL 10 m*M* dNTP.
- 4 µL 10 U/ µL *E. coli* DNA polymerase.

Incubate the 150-µL reaction at 16°C for 2 h.

7. Add 2 µL (10 U/µL) T4 DNA polymerase and incubate for 10 min at 16ºC.
8. Add 10 µL 0.5 *M* EDTA to terminate the reaction. The double-stranded cDNA may be stored at –20°C until purification.
9. The cDNA is purified exactly as described in **Subheading 3.3.1., steps 7–11** except that purified cDNA should be resuspended in 22 µL of nuclease-free water in preparation for biotin-abeled in vitro transcription.

### 3.3.4. Biotin-Labeled In Vitro Transcription and Purification (Round 2)

1. The Enzo BioArray™ HighYield™ in vitro transcription kit is used to generate biotin-labeled cRNA for hybridization to Affymetrix GeneChip microarrays. The recommended protocol is utilized by mixing the following reaction components:
   - 22 µL cDNA template.
   - 4 µL 10X HY reaction buffer.
   - 4 µL 10X Biotin label ribonucleotides.
   - 4 µL 10X DTT.
   - 4 µL 10X RNase inhibitor mix.
   - 2 µL 20X T7 RNA polymerase.

   Add the reaction components to the 1.5-mL tube containing the cDNA from **step 9** above. Mix thoroughly and then transfer the entire reaction to a 0.5-mL tube. The reaction is carried out for 6 h at 37°C, mixing the reaction components every hour and spinning briefly to collect the reaction condensate.
2. After 6 h, the reaction may be frozen at –70°C until purification.
3. The biotin labeled cRNA is purified using Qiagen's RNeasy or Arcturus' PicoPure RNA purification columns. For Qiagen RNeasy columns, the manufacturer's recommended protocol is utilized. The cRNA is eluted with 40 µL of nuclease-free water and the eluate is passed over the column a second time to increase the yield.
4. Quantify 1 µL of the purified, labeled cRNA by UV absorbance (*see* **Note 8**). **Table 2** provides representative data of cRNA yields, starting with 20 ng of total RNA from microdissected human breast epithelium. Using this protocol, approx 20 µg of biotin-labeled cRNA should be generated for every 10 ng of input RNA. This corresponds to approx 40,000-fold amplification. Significantly lower yields than this suggest amplification failure (*see* **Note 13**). The cRNA may also be qualitatively assessed by gel electrophoresis (*see* **Fig. 2**). The amplified cRNA should have a similar molecular weight distribution compared to the standard Affymetrix cRNA synthesis protocol starting with 5 µg of RNA. A very low- or very high-molecular-weight cRNA distribution suggests anomalous amplification and will generally result in poor hybridization signal if applied to the microarray.
5. Fragment 20 µg of the labeled cRNA using the standard protocol supplied by Affymetrix.

### 3.4. Target Hybridization and Data Quality Assessment

In this final section, we describe the critical parameters used in assessing the technical success and data quality of GeneChip microarray data generated from microdissected tissue specimens.

1. 15 μg of fragmented, biotin-labeled cRNA is prepared in a 300-μL hybridization cocktail following the standard Affymetrix GeneChip protocol.
2. For initial experiments, the cocktail should be hybridized to a test array to assess the target quality before hybridization to larger, whole-genome arrays. Once the investigator is comfortable with the relationship between cRNA yield and data quality, hybridization to a test array may be omitted.
3. Following standard hybridization and washing of the test array, the array image should be processed using Affymetrix Microarray Analysis Suite 5.0 (MAS5.0). The following parameters may then be assessed:
   a. *GAPDH 3'/ GAPDH 5' probe set signal ratios*: Traditionally, the ratio of signal intensities from control probe pair sets directed to the 3' and 5' ends of the glyceraldehyde phosphate dehydrogenase (GAPDH) transcript (human, mouse, or rat) is a measure of full-length reverse transcription and in vitro transcription. Using the standard Affymetrix protocol and 5 μg of total RNA, the signal ratio of the 3':5' probe sets is usually 1–3:1. The additional round of amplification utilized in this protocol will often result in 3':5' GAPDH probe set signal ratios as high as 5–10:1. A modest increase in the GAPDH ratio is *not* an indicator of failed target synthesis. The 3':5' signal ratio of actin control probe pair sets (e.g., AFFX-HSAC07) may be even higher (20:1), even from targets that yield good hybridization data.
   b. *Detection* p-*values*: Depending on the species origin of the tissue (human, mouse, rat), the detection *p*-values for several other endogenous transcripts represented on the Test3 GeneChip microarray should be examined. The detection *p*-values for abundant "housekeeping" transcripts such as GAPDH and actin should be much less than 0.05. Less abundant transcripts such as human and murine transferrin receptors, mouse pyruvate carboxylase, human STAT1, and rat hexokinase should also have significant ($p < 0.05$) detection calls. Detection of only ribosomal RNAs suggests target synthesis failure. Detection of only high abundant transcripts (e.g., actin and/or GAPDH) suggests suboptimal target synthesis.
4. If the quality of the test array hybridization is deemed acceptable, the hybridization cocktail should be hybridized to the genome array of the appropriate species. After hybridization and washing using standard protocols, analyze the microarray images with MAS5.0 software and examine the following parameters to assess the technical success of the experiment:
   a. *Percentage of genes scored*: Depending on the tissue of origin and the microarray utilized, the percentage of genes scored detected (P) may range from 25 to 40% (*see* **Table 2**). This is slightly lower than that obtained using five micrograms of starting RNA and the standard Affymetrix synthesis pro-

tocol. Arrays with fewer than 25% of genes scored P are usually the result of poor target synthesis (*see* **Note 13**), although valid data have been obtained even when as few as 5% of all transcripts are scored as detected. The most reliable (accurate) results will be obtained when microarrays with similar %P scores are compared.

b. *Scaling factor*: The scaling factor for each microarray will depend on the target intensity value chosen in the Microarray Analysis Suite software. Generally, the scaling factor is inversely related to the percentage of genes scored P. Scaling factors for amplified targets may be three to five times higher than targets made using the standard synthesis protocol (*see* **Table 2**). When comparing microarray data, it is more important that the scaling factors for all arrays be comparable.

c. *Replicate data*: When independent dissections of the same cell population are performed, RNA independently purified, and cRNA targets independently prepared, only 4–5% of all transcripts represented on the array demonstrate greater than a twofold change in signal intensity (**Fig. 3C**). When the same RNA is amplified in duplicate, less than 1% of represented transcripts demonstrate a greater than twofold change (**ref. *13***; **Fig. 3B**). To produce lowest variability between samples, all target amplifications for a given experiment should be performed in parallel and data from unamplified and amplified samples should not be directly compared (**Fig. 3D**).

## 4. Notes

1. Proper preservation of the tissue specimen to be dissected is the most critical factor in obtaining high-quality RNA for amplification. Modest delays associated with obtaining and freezing human tissue specimens does not qualitatively affect the isolation of total cellular RNA (*10*). However, if the tissue is already frozen, avoid thawing the tissue during the embedding process. Tissue can also be processed by precipitating fixatives such as ethanol, methanol/acetic acid, and acetone followed by paraffin wax embedding (*14,15*). Although these strategies preserve the histology, RNA degradation is significantly greater when compared to snap-frozen material (*10*).

2. Positive controls can be generated throughout the procedure. First, cut one single frozen tissue section and immediately lyse the tissue in the appropriate RNA extraction buffer. RNA isolated from this specimen will serve as a positive control for tissue preservation and RNA isolation techniques. Second, after completing tissue dissection, scrape the remainder of the stained tissue off the slide and into 100 µL of RNA extraction buffer. RNA isolated from this specimen will serve as a positive control for tissue staining, dehydration, and RNA stability during the time of dissection.

3. Mounting of sections to slides is critical for efficient LCM transfer. The section must be placed completely flat on the slide without folds or wrinkles. The slide must then be immediately immersed in 70% ethanol fixative. Delays in placing the slide in the fixative (e.g., when trying to mount more than one section per

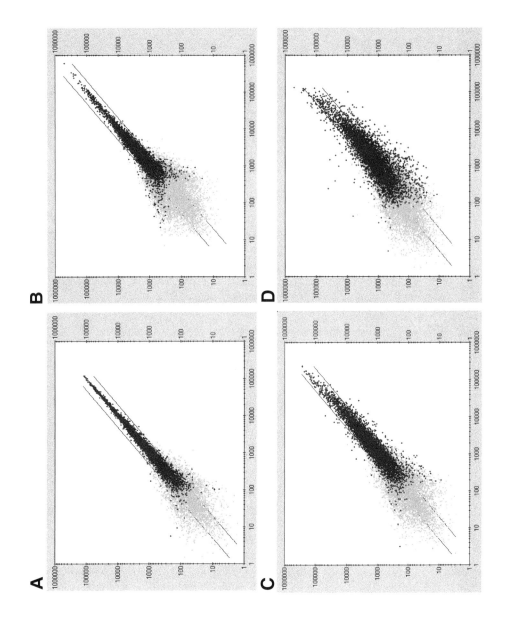

slide) will result in air-drying, which both adversely affects histological detail and prevents efficient transfer from the slide to the LCM transfer film.

4. Slides may be stored in 70% ethanol for several hours at 4°C without affecting histology, tissue transfer, or RNA quality. For optimal results, however, cutting, staining, and dissecting should be performed all within a single day. After staining and dehydration, it is critical that slides remain dry. Rehydration in ambient laboratory humidity will reactivate tissue RNases and prevent efficient transfer of tissue from the slide to the transfer film. If more than a single slide is stained at one time, each slide should be held in a vacuum desiccator until ready for dissection. Do not store stained, desiccated slides in a vacuum desiccator for more than 4–6 h, as significant RNA degradation *in situ* will begin to occur. Although storage of frozen section slides at –80°C is common practice in some laboratories, prolonged storage may result in desiccation or ice crystal formation, which destroys histological detail and impedes tissue transfer from the slide to the LCM film.

5. RNA recovery is very sensitive to the histological stains used *(10)*. Traditional Harris' hematoxylin and bluing reagent causes severe chemical degradation of RNA. This protocol uses Mayer's formulation of hematoxylin and an alternative developing reagent that results in much better preservation of cellular RNA *(14)*, although the histological detail achieved is not as good. Staining with methyl green, eosin-Y, toluidine blue, and Nissl stain *(16)* have also yielded high-quality RNA from microdissected tissues.

6. To avoid the use of organic extraction, several "spin-column"-based microRNA isolation kits are commercially available. The Arcturus PicoPure RNA isolation

---

Fig. 3. (*opposite page*) Scatter plots demonstrating GeneChip microarray data reproducibility. Each point on the plot corresponds to one of approx 12,600 transcripts represented on the Affymetrix U95A microarray. Expression level of each transcript is plotted on a logarithmic scale. Dark points represent transcripts scored as "detected" (P) on one or both microarrays; gray points represent transcripts scored as "not detected" (A) in both samples. Diagonal lines represent the limits of twofold change in expression between the two samples. (**A**) Expression values from two duplicate cRNA target preparations, starting with 10 µg of total RNA and using a single round of transcript amplification (standard Affymetrix protocol); (**B**) Expression values from two duplicate cRNA target preparations, starting with 10 ng of the same total RNA as in (**A**) and using two rounds of amplification as described in this chapter; (**C**) Expression values from the same target population of nonmalignant breast epithelial cells, isolated from two independent microdissections of serial tissue sections. RNAs from each LCM sample were independently isolated and 20 ng of each RNA was used in two rounds of amplification as described in this chapter; (**D**) Expression values of target generated from 10 µg of RNA and the standard Affymetrix protocol (*x*-axis) vs target generated from 10 ng of the same RNA and two rounds of amplification as described in this chapter (*y*-axis).

system has been used by the authors successfully and is a reasonable alternative to the method described in this chapter.

7. Occasionally, there will be no phase separation after centrifugation of the organic extraction. Because dehydrated, microdissected tissue has no water content, the high salt concentration of the undiluted lysis buffer may prevent the formation of an aqueous phase. If no phase separation occurs, add 1–2 µL of nuclease-free water. Vortex the sample, and re-spin.

8. To quantify RNA, the authors utilize a fiberoptic spectrophotometer (NanoDrop ND-1000, NanoDrop Technologies). Small amounts of RNA may also be quantified by dye-binding fluorescence assays (e.g., RiboGreen assay, Molecular Probes, Eugene, OR). Qualitative assessment of RNA is also important. If access to specialized equipment such as an Agilent Bioanalyzer is not available, a small amount of RNA may be analyzed by agarose gel electrophoresis followed by staining with sensitive fluorescent dyes (e.g., SybrGold dye, Molecular Probes). Although less convenient, this approach has been successfully utilized to qualitatively assess nanogram quantities of total RNA from microdissected tissue *(9)*.

9. The authors routinely use 20 ng of total RNA for target synthesis, with resulting yields of 20–40 µg of labeled cRNA (*see* **Table 2**). If RNA is of very high quality, as little as 5 ng of total RNA may be used to create 15–20 µg of cRNA, sufficient for hybridization to one set of GeneChip whole-genome arrays. Conversely, as much as 100 ng of total RNA may be used if the starting material is of questionable quality or higher yields of target are needed.

10. The protocol described in this chapter has been successfully used by the authors *(6)*. Commercially available reagent systems (e.g., RiboAmp, Arcturus; MessageAmp, Ambion) have also been tested and work well, providing the additional convenience (and cost) of a standardized kit format *(13)*. Minor modifications to this protocol and additional validation data are also available as a Technical Note from Affymetrix (www.affymetrix.com).

11. Ammonium acetate precipitation is a popular method for nucleic acid precipitation. However, many downstream enzymes (e.g., T7 RNA polymerase) are inhibited by this salt. Therefore, it is crucial that no salt is carried over during the precipitation step. Centrifugation should be performed at room temperature to avoid salt precipitation and all precipitated pellets should be thoroughly washed with 70% ethanol.

12. Nucleic acid pellets should be air-dried for a sufficient length of time to allow for complete ethanol evaporation. However, overdrying pellets may make them difficult to resuspend into subsequent reaction buffers. After washing with 70% ethanol and allowing pellets to air-dry, a small amount of liquid may remain in the tube.

13. When yields of cRNA target are significantly below 20 µg per 10 ng of input RNA and/or target hybridizations result in a low percentage of genes being scored as detected (P) by Affymetrix Microarray Analysis software, the following troubleshooting algorithm should be performed:

   a. *Control RNA works; experimental samples do not*: Ten nanograms of control RNA, consisting of a high-quality RNA source, diluted to 10 ng/µL, should be initially utilized with all experimental samples for target amplification and GeneChip test array analysis. Successful amplification of the control RNA, but only a few or none of the experimental RNAs suggests problems with RNA isolation. Qualitatively reassess the RNA by any of the methods discussed in **Subheading 3.2.2.** If sufficient RNA is not available, assess the RNA isolated from the remaining tissue section after performing LCM (*see* **Note 2**).

b. *Neither control nor experimental samples work*: The most common reagent failure resulting in failed target syntheses of control and experimental samples is the T7T$_{24}$ oligonucleotide primer. Any degradation of the promoter primer can destroy the T7 RNA polymerase site and result in cDNA templates with nonfunctional transcription promoters. The primer should be HPLC- or gel-purified to ensure that only full-length primer is used. Fresh primer should be utilized and primer stocks should be aliquoted to avoid repeated freeze-thaw cycles. The second most common source of systematic error is carryover of ammonium acetate salts from precipitation into subsequent enzyme reactions. Be certain to centrifuge samples at room temperature and perform thorough washings with 70% ethanol prior to drying the nucleic acid pellets. Conversely, because pelleted nucleic acid is often invisible, extra care should be exercised to avoid washing away pellets.

c. *RNA from LCM tissue is degraded or undetectable*: RNA isolated from tissue remaining after LCM should be evaluated as a positive control. If the quality of this RNA is acceptable, this suggests that an insufficient amount of tissue was dissected or that RNA was lost or degraded during the isolation process.

d. *RNA from control tissue section is degraded*: If RNA isolated from tissue remaining after LCM is degraded, consider that the tissue was not properly preserved, the staining procedure was not performed correctly, or RNA isolation was not performed correctly. Cut a single fresh-frozen section from the tissue block and place it immediately into lysis buffer. If the resulting RNA is also degraded, this suggests a problem with tissue preservation or RNA isolation procedure. Consider using freshly isolated tissue or performing a control RNA isolation using tissue culture cells. If RNA from the tissue block is of high quality, but post-LCM tissue RNA is degraded, the staining protocol should be reevaluated. Make certain that compatible stains are used. Do not allow slides to stand more than 1–3 h after staining and dehydration.

## Acknowledgments

Tissue processing and laser capture microdissection were performed with the assistance of the Siteman Cancer Center Tissue Procurement Core. GeneChip hybridization and processing were performed in the Siteman Cancer Center Multiplexed Gene Analysis Core. The authors wish to thank Kate Hamilton for technical assistance with microarray processing. This work was supported in part by the Alvin J. Siteman Cancer Center and grant number 016-99 from the Mary Kay Ash Charitable Foundation (M.A.W.).

## References

1. van 't Veer, L. J., Dai, H., van de Vijver, M. J., He, Y. D., Hart, A. A., Mao, M., et al. (2002) Gene expression profiling predicts clinical outcome of breast cancer. *Nature* **415**, 530–536.
2. Sorlie, T., Perou, C. M., Tibshirani, R., Aas, T., Geisler, S., Johnsen, H., et al. (2001) Gene expression patterns of breast carcinomas distinguish tumor subclasses with clinical implications. *Proc. Natl. Acad. Sci. USA* **98**, 10,869–10,874.
3. Ramaswamy, S., Ross, K. N., Lander, E. S., and Golub, T. R. (2003) A molecular signature of metastasis in primary solid tumors. *Nat. Genet.* **33**, 49–54.
4. Miura, K., Bowman, E. D., Simon, R., Peng, A. C., Robles, A.I., Jones, R.T., et al. (2002) Laser capture microdissection and microarray expression analysis of lung adenocarcinoma reveals tobacco smoking- and prognosis-related molecular profiles. *Cancer Res.* **62**, 3244–3250.
5. Ma, X. J, Salunga, R., Tuggle, J. T., Gaudet, J., Enright, E., McQuary, P., et al. (2003) Gene expression profiles of human breast cancer progression. *Proc. Natl. Acad. Sci. USA* **100**, 5974–5979.
6. Luzzi, V., Holtschlag, V., and Watson, M.A. (2001) Expression profiling of ductal carcinoma *in situ* by laser capture microdissection and high-density oligonucleotide arrays. *Am. J. Pathol.* **158**, 2005–2010.
7. Alevizos, I., Mahadevappa, M., Zhang, X., Ohyama, H., Kohno, Y., Posner, M., et al. (2001) Oral cancer in vivo gene expression profiling assisted by laser capture microdissection and microarray analysis. *Oncogene.* **20**, 6196–6204.
8. Chen, Z., Fan, Z., McNeal, J. E., Nolley, R., Caldwell, M. C., Mahadevappa, M., et al. (2003) Hepsin and maspin are inversely expressed in laser capture microdissectioned prostate cancer. *J. Urol.* **169**, 1316–1319.
9. Kabbarah, O., Pinto, K., Mutch, D. G., and Goodfellow, P. J. (2003) Expression profiling of mouse endometrial cancers microdissected from ethanol-fixed, paraffin-embedded tissues. *Am. J. Pathol.* **162**, 755–762.
10. Huang, L. E., Luzzi, V., Ehrig, T., Holtschlag, V., and Watson, M. A. (2002) Optimized tissue processing and staining for laser capture microdissection and nucleic acid retrieval. *Methods Enzymol.* **356**, 49–62.
11. Gubler, U. and Hoffman, B. J. (1983) A simple and very efficient method for generating cDNA libraries. *Gene* **25**, 263–269.
12. Van Gelder, R.N., von Zastrow, M. E., Yool, A., Dement, W. C., Barchas, J. D., and Eberwine, J. H. (1990) Amplified RNA synthesized from limited quantities of heterogeneous cDNA. *Proc. Natl. Acad. Sci. USA* **87**, 1663–1667.
13. Luzzi, V., Mahadevappa, M., Raja, R., Warrington, J.A., and Watson, M.A. (2003) Accurate and reproducible gene expression profiles from laser capture microdissection, transcript amplification, and high density oligonucleotide microarray analysis. *J. Mol. Diagn.* **5**, 9–14.
14. Goldsworthy, S. M., Stockton, P. S., Trempus, C. S., Foley, J. F., and Maronpot, R. R. (1999) Effects of fixation on RNA extraction and amplification from laser capture microdissected tissue. *Mol. Carcinog.* **25**, 86–91.

15. Gillespie, J. W., Best, C. J., Bichsel, V. E., Cole, K. A., Greenhut, S. F., Hewitt, S. M., et al. (2002) Evaluation of non-formalin tissue fixation for molecular profiling studies. *Am. J. Pathol.* **160,** 449–457.
16. Betsuyaku, T., Griffin, G. L., Watson, M. A., and Senior, R. M. (2001) Laser capture microdissection and real-time reverse transcriptase/polymerase chain reaction of bronchiolar epithelium after bleomycin. *Am. J. Respir. Cell. Mol. Biol.* **25,** 278–284.

# 16

# Quantification of Gene Expression in Mouse and Human Renal Proximal Tubules

## Jun-ya Kaimori, Masaru Takenaka, and Kousaku Okubo

### Summary

The kidney consists of many functional modules called nephrons. Each nephron has a tubular structure made up of several structurally and functionally distinct segments. The analysis of individual segments requires the use of microdissection techniques. We describe protocols that have been used to successfully isolate messenger RNA from proximal tubules of both freshly prepared and archival samples using laser capture microdissection and laser-manipulated microdissection.

**Key Words:** Proximal tubule; laser capture microdissection; real-time PCR.

## 1. Introduction

### 1.1. Kidney Structure

The human kidney consists of approx 1 million functional units called nephrons. Each nephron has a convoluted tubular structure made up of five morphologically and functionally distinct segments. Various manipulative techniques have been developed for the selective isolation of these segments *(1)*. Laser microdissection techniques *(2)* enable isolation of specific cell types from renal biopsy specimens that are routinely generated for laboratory diagnosis *(3,4)*. In this chapter, we summarize the protocols that have been successfully employed in recovering messenger RNA from proximal tubules of both freshly prepared and archival samples *(5)* using laser capture microdissection (LCM) *(3)* and laser-manipulated microdissection (LMM) *(4)*.

### 1.2. Tissue Preservation

It is easy to isolate RNA from kidney tissues compared with other tissues that are rich in RNase (pancreas), fibrous proteins (muscular tissue), or mucus

From: *Methods in Molecular Biology, vol. 293: Laser Capture Microdissection: Methods and Protocols*
Edited by: G. I. Murray and S. Curran © Humana Press Inc., Totowa, NJ

secretions (airways, alimentary tracts). On the other hand, it has a complex histological structure. In general, microscopic structure and RNA are both preserved either by ethanol, formalin, or RNA-preserving solutions such as RNAlater. Ethanol dehydrates the tissue and stops the action of nuclease and protease without any deleterious effect on RNA even after prolonged incubation.

The fixative action of formalin is via inactivation of degradation enzymes by methylol addition or methylene bridge formation within or across protein molecules. It similarly affects bases in nucleic acids and makes them a poor template for molecular biological applications (*6*). Moreover, prolonged incubation of tissues in formalin results in extensive protein crosslinking that subsequently makes tissue resistant to solubilization. RNAlater is a widely used reagent for tissue RNA preservation. Its action seems to be precipitation of proteins by high ion concentration similar to the action of ammonium sulfate. It preserves RNA in renal specimens but tissue morphology is not well maintained. Accordingly, for laser microdissection of renal tissue, ethanol is the preferred fixative, especially as it can be perfused through the tissue. For biopsy specimens we treat tissue blocks briefly with RNAlater and then store them in ethanol. We find that prolonged digestion with proteinase K is necessary to obtain a satisfactory yield of RNA from archived tissue samples (*6*).

### *1.3. Messenger RNA Content*

The average epithelial cell contains 0.4 pg of mRNA (200,000–300,000 molecules). In a differentiated cell, about 20,000 genes are expressed. There is wide variation in expression of individual genes: the most abundant mRNA amounts to 10% of total mRNA and the least abundant occurs at only a few copies per cell. As an empirical rule, the fraction of transcripts from one gene is approximately equal to $0.1/r$, where r stands for activity rank of the gene in a cell (*7*). For example, mRNA from the hundredth most abundant gene exists at 200 copies ($0.1/100 \times 200{,}000$) in a cell. Assuming the sensitivity of the following RNA detection assay as 100 copies and the yield of viable RNA at 10%, it is necessary to collect 1000 cells for detection of all active transcribed genes, 50 cells for the 1000 most abundant genes, and 5 cells for the 100 most common genes. In mice, the most abundant mRNA is for the androgen-regulated protein (NM_010594 in NCBI RefSeq) in proximal tubules (*8*) and α-B crystallin (NM_001885) in collecting ducts (*9*).

## 2. Materials
### *2.1. RNA Isolation From Mouse Proximal Tubules*
1. Phosphate-buffered saline (PBS), pH 7.4.
2. 99.5% Ethanol (RNA grade).
3. 30% Sucrose/PBS (RNA grade).

4. Tissue-Tek optimum cutting temperature (OCT) compound (Sakura Company, CA, USA).
5. Glass slides.
6. 0.1% poly-L-lysine solution (Sigma Diagnostics, St. Louis, MO).
7. 1.35-μm polyethylene foils (Laser Pressure Catapulting (LPC) membrane; P.A.L.M. Bernried, Germany).
8. Cryostat (JUNG CM3000, Leica, Germany)
9. Carrazzi's hematoxylin solution (RNA grade) (Wako Pure Chemical Industries, Osaka, Japan).
10. Laser capture microdissection (LCM) system (Arcturus Engineering Inc. Mountain View, CA)
11. LCM transfer film (CapSure TF-100; Arcturus Engineering).
12. Superscript preamplification System (Invitrogen, Carlsbad, CA).
13. Microcon YM-10 (Millipore, Bedford, MA).
14. The TSA-1 TaqMan probe 5'-CTGTGGCCAGTTTCATGCCAGGAGAAAGA-3' (accession no. U47737, 3373 bp-3401 bp) with FAM as reporter and TAMRA as quencher, the TSA-1 forward primer sequence 5'-GATGTGCTTCTCAT-GTACCGATCAG-3' (3330 bp-3354 bp), and its reverse primer sequence 5'-CAGCGGCAGATAACGTGATACAG-3' (3408 bp-3430 bp).
15. ABI Prism 7700 Sequence Detection System (Applied Biosystems, Foster City, CA).
16. TaqMan Universal PCR Master Mix (Applied Biosystems).
17. TaqMan ribosomal RNA Control Reagents (Applied Biosystems).

## 2.2. RNA Isolation From Human Proximal Tubules

1. RNAlater (Ambion Inc., Austin, TX).
2. 99.5% ethanol (RNA grade).
3. Xylene.
4. Wax (Paraplast+56°C, Oxford Lab Ware).
5. Microtome: Reichert-Jung, Hn 40 (Cambridge Instrument, Germany).
6. 1.35-μm polyethylene foils (Laser Pressure Catapulting, LPC membrane; PALM).
7. 76 × 26 mm glass cover slips (Deckglaser, Germany).
8. Laser-manipulated microdissection (LMM) system (PALM system) (Carl Zeiss, Oberkochem, Germany).
9. Laser capture microdissection (LCM) system (Arcturus Engineering).
10. Proteinase K solution: 200 m$M$ Tris-HCl pH 7.5, 200 m$M$ NaCl, 1.5 m$M$ MgCl$_2$, 2% SDS, 500 μg/mL proteinase K, 50 μg/mL tRNA, RNA grade.
11. TRIzol reagent (Invitrogen).
12. Superscript preamplification system (Invitrogen).
13. Microcon YM-10 (Millipore).
14. The TaqMan probe for human osteopontin (OPN) sequence was 5'-CGA-AGTTTTCACTCCAGTTGTCCCCACA-3' (accession no. JO4765, 495 bp-522 bp) with FAM as reporter and TAMRA as quencher. The forward primer sequence of human OPN was 5'-TCACTGATTTTCCCACGGACC-3' (464 bp-484 bp) and the reverse was 5'-CCTCGGCCATCATATGTGTCTA-3' (524 bp-545 bp).

**Mouse kidney tissue**                **Human kidney biopsy tissue**

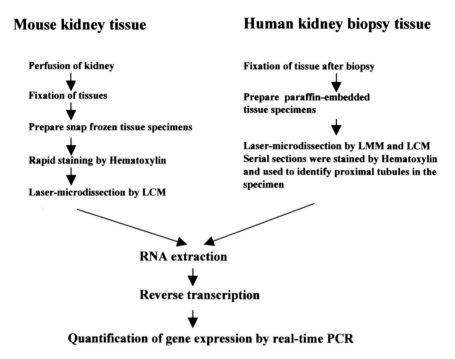

Fig. 1. Procedure for laser microdissection and real-time PCR. Mouse renal tissue and human renal biopsy tissue are processed differently. Mouse tissue is cut into frozen sections, while human renal biopsy specimens are embedded in wax. A combination of LMM and LCM is recommended to specifically collect proximal tubular epithelial cells.

15. TaqMan Human GAPDH Control Reagents (Applied Biosystems).
16. ABI Prism 7700 Sequence Detection System (Applied Biosystems).

## 3. Methods

The methods described below are RNA isolation from (1) mouse proximal tubules from histological tissue specimens and (2) human proximal tubules from biopsy specimens, followed by quantification of specific gene expression. The procedure is shown schematically in **Fig. 1**. Careful preparation of histological specimens is important for successful quantification.

### *3.1. RNA Isolation From Mouse Proximal Tubules*

#### *3.1.1. Mouse Kidney Tissue Processing and Laser Microdissection by LCM*

1. Remove mice kidneys after perfusion with PBS and then with 99.5% ethanol (*see* **Note 1**).
2. Fix the removed kidneys for 4 h in 99.5% ethanol in Rnase-free glass (*see* **Note 2**).

3. Dehydrate with 30% sucrose/PBS overnight.
4. Embed the samples in OCT compound and freeze them on a small stainless board pre-chilled by liquid nitrogen (–196°C).
5. Store them in a deep freezer (–80°C) for later use.
6. Cut the frozen embedded sample with a cryostat into sections of 5-μm thickness.
7. Mount sections onto polyethylene foils on a glass slide (*see* **Note 3**).
8. Stain the membrane-mounted specimens rapidly with Carrazzi's hematoxylin solution for 10 s (*see* **Note 4**).
9. Wash with DEPC-treated water for 10 s.
10. Dehydrate with 99.5% ethanol for 5 min (*see* **Note 4**).
11. Dry at room temperature for 30 min.
12. Mount a glass slide with specimens onto an LM200 Image Archiving Workstation.
13. Cover sections with a transfer film (CapSure TF-100).
14. Microdissect the proximal tubules (PTs) with the laser beam set at 7.5 μm diameter and 50–75 mV.
15. Approximately 50 PTs were collected for the following analysis.
    **Figures 2A** and **2B** show examples of tubule isolation following the above protocol. Without laying polyethylene foils over glass slides (**step 7**), tubules resist microdissection.

## 3.1.2. RNA Extraction

1. In a 0.6-mL reaction tube, soak the collected PTs on a transfer film with 200 μL TRIzol solution that includes 1 μg of tRNA (*see* **Note 5**) for 5 min at room temperature.
2. Vortex for 10 s.
3. Add 40 μL chloroform into reaction mixture.
4. Vortex for 15 s and leave the sample at room temperature for 5 min.
5. Centrifuge at 10,000 rpm for 15 min at 4°C.
6. Transfer the upper phase into another 0.6 mL tube.
7. Add 200 μL isopropanol and 20 μg glycogen and leave at room temperature for 10 min.
8. Centrifuge at 12,000 rpm for 15 min at 4°C.
9. Wash the pellet with 500 μL of 99.5% ethanol.
10. Dry the pellet and resuspend with 10 μL distilled water.

## 3.1.3. cDNA Synthesis

1. Add the following into the RNA solution: 0.5 μL 50 ng/μL. random hexamers; 1.5 μL DEPC-treated water.
2. Incubate at 70°C for 10 min.
3. Add the following into the reaction solution according to the manufacturer's instructions: 2 μL 10X PCR buffer; 2 μL 25 m$M$ MgCl$_2$ solution; 2 μL 1 μL 10 m$M$ dNTP mix; and 0.1 $M$ DTT.
4. Incubate for 5 min at 25°C.
5. Add 1 μL SuperScript II reverse transcriptase (200 U/μL) into reaction mixture.

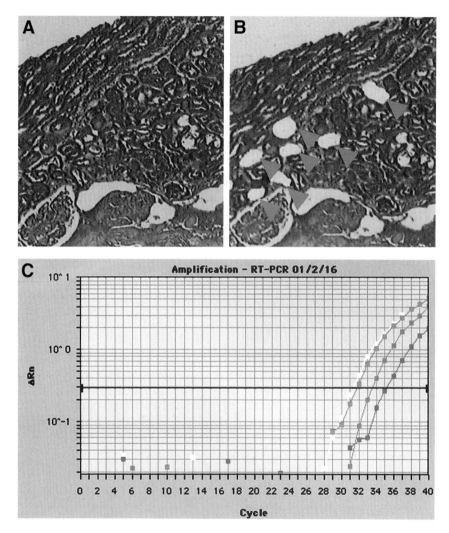

Fig. 2. Isolation of mouse proximal tubules by laser microdissection using LCM and detection of gene expression in microdissected tissue specimens. Figures show samples before (**A**) and after (**B**) laser microdissection of proximal tubules from 5-μm-thick mouse renal tissue section. Arrows represent the original location of isolated proximal tubules. (**C**) Mouse TSA-1 gene expression was quantified in RNA from 50 proximal tubule segments isolated by LCM. The real-time PCR data are shown. Different colors in the amplification curve represent different experiments using different mice.

6. Incubate for 10 min at 25°C.
7. Incubate for 50 min at 42°C.
8. Incubate for 15 min at 70°C.
9. Chill on ice for 5–10min.
10. Add 200 μL distilled water into the reaction solution.
11. Transfer diluted reaction solution into Microcon YM-10 (*see* **Note 6**).
12. Centrifuge at 10,000 rpm for 15 min at 4ΥC (approx 40 μL of reaction mixture remained in YM-10).
13. Use 10 μL of the solution as a template for real-time PCR.

### 3.1.4. Quantification of mRNA by Real-Time PCR

1. Assemble the following reaction mixture in each tube for real-time PCR (*see* **Note 7**):
   - 25 μL TaqMan Universal PCR Master Mix (2X).
   - 5 μL 10 μ*M* Mouse TSA-1 forward primer (*see* **Note 8**).
   - 5 μL 10 μ*M* Mouse TSA-1 reverse primer.
   - 5 μL Mouse TSA-1 TaqMan probe.
   - 10 μL Sample.
   - 50 μL Total.
2. Perform real-time PCR using the following cycle profiles: Initial step 50°C for 10 min; 40 cycles 95°C for 2 min, then 95°C 15 s (denaturation) 60°C for 1 min (annealing/extension); final extension 70°C 10 for 10 min.
   TaqMan ribosomal RNA Control reagents can be employed for standardization. In the example shown in **Fig. 2C**, the increased expression of TSA-1 mRNA in disease model PT was confirmed (*see* **Note 9**).

## 3.2. RNA Isolation From Human Proximal Tubules

### 3.2.1. Tissue Processing of Renal Biopsy Specimens

1. Immerse human renal biopsy specimens in RNAlater reagent for 5 s (*see* **Note 10**) as soon as they are obtained.
2. Fix in 99.5% ethanol overnight at 4°C.
3. Incubate in xylene for 1 h at room temperature and repeat this step three times.
4. Immerse in 100% wax for 1 h at 60°C and repeat this step three times.
5. Embed tissue samples in 100% wax.
6. Store in a deep freezer (–80°C).
7. Cut 10-μm sections using a microtome, taking appropriate precautions to ensure no contamination by ribonuclease.
8. Mount biopsy sections onto polyethylene foils as described above on 76 × 26 glass cover slips for LMM.
9. Dry biopsy sections on cover slips overnight at 37°C.
10. Dewax tissue sections by incubating in xylene for for three 5 min periods.
11. Dry specimens for 30 min at room temperature.
12. Stain several serial sections from each sample with freshly prepared Carrazzi's hematoxylin solution for 10 s.

### 3.2.2. Selective Tissue Collection by Combination of Laser-Mediated Microdissection and Laser Capture Microdissection Methods

1. Place the section mounted on a glass slide onto the laser-manipulated microdis-section (LMM) system.
2. Identify as many proximal tubules as possible (*see* **Note 11**) by confirming their brush borders in the neighboring serial sections stained by hematoxylin (*see* **Note 12**).
3. Cut carefully inside the basement membrane of the proximal tubules with the laser beam to isolate the proximal tubular cells specifically.
4. Collect the samples by using LCM as described above (*see* **Subheading 3.1.1.**). Sections of proximal tubules are immediately processed for the following RNA extraction procedures.

Examples of human renal biopsy sections before and after dissection are shown in **Fig. 3A,B**. With the combination of LMM and LCM methods, only targeted cells were successfully isolated with virtually no contamination with other cells.

### 3.2.3. RNA Extraction From Collected PT Sections and Quantification of Osteopontin mRNA Expression by Real-Time PCR

1. Proximal tubules collected by the combined procedure of LMM and LCM were treated with 20 µL of proteinase K solution including 1 µg tRNA for 15 min at 45°C.
2. Add 200 µL TRIzol reagent into the solution.
3. Extract RNA from TRIzol solution according to the manufacturer's instructions (*see* **Subheading 3.1.2.**).
4. Synthesize first strand cDNA (*see* **Subheading 3.1.2.**).
5. Perform real-time PCR of human OPN (**Fig. 3C**) and glyceraldehyde-3-phos-phate dehydrogenase (GAPDH) using the ABI Prism 7700 Sequence Detection System (*see* **Subheading 3.1.2.**) (*see* **Note 14**).
6. Quantify the expression of human OPN/ GAPDH mRNA .

## 4. Notes

1. Perfusion with ethanol is the best fixative. Do not use a formalin-based fixative at any step.
2. The glassware used in fixation, dehydration, washing, and staining should be heat-sterilized for 3 h at 180°C to prevent contamination of RNase.
3. The thin polyethylene foil is attached onto a slide glass with nail polish. This foil allows us to isolate samples efficiently by the LCM method. The slide glasses are rinsed by a 0.1% poly-L-lysine solution.
4. Carrazzi's hematoxylin solution and 99.5% ethanol should be freshly prepared for each experiment. This is necessary to minimize degradation of RNA.
5. We usually add tRNA into the solution as a carrier that prevents loss of RNA by adhesion to surfaces of the tips and tubes.

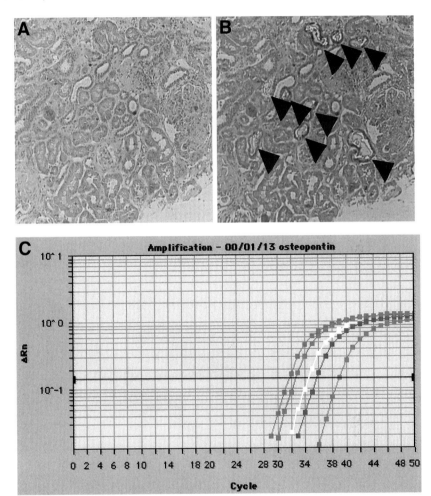

Fig. 3. Isolation of human proximal tubules by laser microdissection using LMM/ LCM and detection of gene expression in microdissected tissue specimens. Figures show proximal tubules before (**A**) and after (**B**) laser-microdissection from 10-μm-thick human renal biopsy tissue sections. Arrows represent proximal tubules isolated from the specimen. (**C**) Human OPN gene expression was detected in RNA from 10 segments isolated by LMM/LCM. Each curve of a different color represents amplification kinetics of a sample from a different patient.

6. The excess random primer for reverse transcriptase reaction should be removed by column filter YM-10 before quantitative PCR.

7. The PCR experiment and RNA isolation procedure should be performed in separate laboratories in order to prevent contamination.

8. It is important to use a good primer set for real-time PCR. Preliminary PCR should be performed with several candidate primers selected by using primer design software, e.g. Primer Express. The set of primers that gives most efficient PCR should be chosen.

9. The quantitation of mRNA expression of TSA-1 was performed with standardization achieved by using rRNA representation. TaqMan ribosomal RNA control reagents (Perkin Elmer Applied Biosystems, Foster City, CA) were used as internal controls for mRNA expression *(8)*.

10. Immersing the biopsy specimen in RNAlater reagent for more than 10 s usually results in disintegration of the tissue.

11. The number of cells required for mRNA detection depends on the target gene abundance *(see* **Subheading 1.3.**). 50 pulses of 10-μm-thick PT cells from a well-preserved section seems to be adequate for detection of any mRNA expressed.

12. We identified proximal tubules by confirming their brush borders in sections stained with hematoxylin. The mRNA in mouse kidney fixed by perfusion remains intact after the staining procedure but in human biopsy specimens usually the RNA shows evidence of degradation.

13. The human GAPDH TaqMan probe and the forward and reverse primers were obtained from TaqMan Human GAPDH control reagents. To prepare positive controls for the quantification of mRNA, human GAPDH and OPN cDNA were cloned by means of PCR. The relative amounts of OPN and GAPDH mRNA in each sample should be calculated in terms of their own standard lines by using real-time PCR *(5)*.

## References

1. Moriyama, A., Murohy, H. R., Martin, B. M., and Garcia, P. A. (1990) Detection of specific mRNA in single nephron segments by use of polymerase chain reaction. *Am. J. Physiol.* **258,** F1470–F1474.

2. Cornea, A. M. (2002) Comparison of current equipment, in *Methods in Enzymology* (Conn, P. M., ed.), Academic Press, San Diego, CA 356, pp. 3–12.

3. Emmert-Buck, M. R., Bonner, R. F., Smith, P. D., Chuaqui, R. F., Zhuang, Z., Goldstein, S. R., et al. (1996) Laser capture microdissection. *Science* **274,** 998–1001.

4. Schutze, K and Lahr, G. (1998) Identification of expressed genes by laser-mediated manipulation of single cells. *Nat. Biotechnol.* **16,** 737–742.

5. Kaimori, J. Y., Takenaka, M., Nagasawa, Y., Nakajima, H., Izumi, M., Akagi, Y., et al. (2002) Quantitative analyses of osteopontin mRNA expression in human proximal tubules isolated from renal biopsy tissue sections of minimal change nephritic syndrome and IgA. Glomerulonephropathy patients. *Am. J. Kidney Dis.* **39,** 9948–9957.

6. Masuda, N., Ohnishi, T., Kawamoto, S., Monden, M., and Okubo, K. (1999) Analysis of chemical modification of RNA from formalin-fixed samples and optimization of molecular biology applications for such samples. *Nucleic Acids Res.* **27,** 4436–4443.
7. Okubo, K. and Hishiki, T. (2003) Knowledge discovery from the human transcriptome, in *Introduction to Bioinformatics* (Krawetz, S. A., ed.) Humana Press, Totowa, NJ, pp. 693–710.
8. Nakajima, H., Takenaka, M., Kaimori, J. Y., Nagasawa, Y., Kosugi, A., Kawamoto, S., et al. (2002) Gene expression profile of renal proximal tubules regulated by proteinuria. *Kidney Int.* **61,** 1577–1587.
9. Takenaka, M., Imai, I., Nagasawa, Y., Matsuoka, Y., Moriyama, T., Kaneko, T., et al. (2000) Gene expression profile of the collecting duct in the mouse inner medulla. *Kidney Int.* **57,** 19–24.

# 17

## Laser Capture Microdissection for Analysis of Macrophage Gene Expression From Atherosclerotic Lesions

### Eugene Trogan and Edward A. Fisher

### Summary

Macrophage foam cells are critical mediators in atherosclerosis plaque development. A better understanding of the in vivo transcript profile of foam cells during the formation and progression of lesions may lead to novel therapeutic interventions. Toward this goal, we demonstrate for the first time that foam cell-specific RNA can be purified from atherosclerotic arteries, a tissue of mixed cellular composition. Foam cells from apolipoprotein (apo) $E^{-/-}$ mice were isolated by laser capture microdissection (LCM); RNA was extracted and used for molecular analysis by real-time quantitative polymerase chain reaction. Compared to whole tissue, a significant enrichment of foam cell-specific RNA transcripts was achieved. Furthermore, to test the ability to quantify differences in gene expression in response to an inflammatory stimulus, apoE$^{-/-}$ mice were injected with lipopolysaccharide, after which the transcriptional induction of the inflammatory mediators, VCAM, ICAM, and MCP-1, was observed in lesional macrophage foam cell RNA. These approaches will facilitate the study of macrophage gene expression under various conditions of plaque formation, regression, and response to genetic and environmental perturbations.

**Key Words:** Apolipoprotein; atherosclerosis; gene expression.

## 1. Introduction

Macrophage foam cells are critical in the development of atherosclerosis *(1–3)*. Therefore, a better understanding of the gene expression changes in foam cells during disease progression and regression has become an important goal in order to develop potential therapies and interventions *(4–6)*. However, the study of macrophage foam cell gene expression in arterial lesions is hampered by the cellular heterogeneity of arterial tissue, which, besides macrophages, also contains lymphocytes, smooth muscle cells, endothelial cells, and adven-

From: *Methods in Molecular Biology, vol. 293: Laser Capture Microdissection: Methods and Protocols*
Edited by: G. I. Murray and S. Curran © Humana Press Inc., Totowa, NJ

titial fibroblasts. To overcome these technical obstacles, we describe here a method for the use of laser capture microdissection (LCM) *(7,8)* to selectively procure macrophage foam cells from arterial lesions (identified by the macrophage-specific marker, CD68/Macrosialin *(9,10)*. RNA extracted from the laser captured foam cell material was used to quantify, by real-time quantitative reverse transcriptase polymerase chain reaction (RT-PCR), the degree of enrichment and to measure the induction of specific transcripts in response to an inflammatory stimulus. These methods make possible the quantitative analysis of gene expression in macrophage foam cells and add a powerful dimension to the study of atherosclerosis.

## 2. Materials

### 2.1. Animals and Tissue Processing

1. Apolipoprotein $E^{-/-}$ mice (The Jackson Laboratory, Bar Harbor, ME).
2. Isoflurane (Baxter, Deerfield, IL).
3. Phosphate-buffered saline (PBS).
4. OCT cryoembedding medium (VWR, San Diego, CA).
5. ColorFrost Plus slides (Fisher Scientific, Pittsburgh, PA).

### 2.2. CD68 Immunodetection of Macrophages for LCM

1. Normal goat serum (Vector Laboratories, Burlingame, CA).
2. Anti-CD68/macrosialin antibody (Serotec, Kidlington, UK).
3. SUPERaseIn (Ambion, Austin, TX).
4. Biotinylated rabbit-anti-rat IgG mouse-adsorbed secondary antibody (Vector Laboratories).
5. Vectastain ABC alkaline phosphatase enzymatic detection antibody (Vector Laboratories).
6. Vector Red substrate (Vector Laboratories).
7. Harris modified hematoxylin (Fisher Scientific, Pittsburgh, PA).

### 2.3. LCM Thermoplastic Caps (Arcturus Engineering, Mountain View, CA)

1. MicroRNA Isolation Kit (Stratagene, La Jolla, CA).
2. DNA-free (Ambion).
3. GlycoBlue (Ambion).
4. Ribogreen RNA quantification kit (Molecular Probes, Eugene, OR).

### 2.4. Real-Time Quantitative RT-PCR

1. ABI Prism 7700 Sequence Detection System (Applied Biosystems, Foster City, CA).
2. Gene-specific Taqman primers and probes (*see* **Table 1**).
3. 1X first-strand buffer: 50 m$M$ Tris-HCl, pH 8.3, 75 m$M$ KCl, 3 m$M$ MgCl$_2$ (Invitrogen Life Technologies, Carlsbad, CA).
4. 100 m$M$ DTT (Invitrogen Life Technologies).

**Table 1**
**Sequences of Real-Time Quantitative RT-PCR Primers and Probes for Measured Gene Transcripts**

| Gene/Primer | Sequence (5' → 3')[a] | Position | Amplicon Length (bp) |
|---|---|---|---|
| *Cyclophilin A (NM_008907)* | | | |
| Fwd | GGCCGATGACGAGCCC | 74–89 | (64) |
| Probe | TGGGCCGCGTCTCCTTGA | 91–109 | |
| Rev | TGTCTTTGGAACTTTGTCTGCAA | 115–148 | |
| *CD68/Macrosialin (X68273)* | | | |
| Fwd | TTGGGAACTACACACGTGGGC | 472–490 | (67) |
| Probe | CGGCTCCCAGCCTTGTGT TCAGC | 494–516 | |
| Rev | CGGATTTGAATTTGGGCTTG | 519–538 | |
| *α-actin (NM_007392)* | | | |
| Fwd | AACGCCTTCCGCTGCCC | 814-829 | (66) |
| Probe | AGACTCTCTTCCAGCCATCTTTCATTGGGA | 832–861 | |
| Rev | CGATGCCCGCTGACTCC | 863–879 | |
| *VCAM-1 (NM_011693)* | | | |
| Fwd | CCCCAAGGATCCAGAGATTCA | 402–422 | (63) |
| Probe | TTCAGTGGCCCCCTGGAGGTTG | 424–445 | |
| Rev | ACTTGACCGTGACCGGCTT | 448–466 | |
| Fwd | ATCTCAGGCGCGCAAGGG | 600–616 | (66) |
| Probe | TGGCATTGTTCTCTAATGTCTCCGAGGC | 617–645 | |
| Rev | CGAAAGTCCGGAGGCTCC | 648–665 | |
| *MCP-1 (M19681)* | | | |
| Fwd | TTCCTCCACCACCATGCAG | 535–553 | (64) |
| Probe | CCCTGTCATGCTTCTGGGCCTGC | 556–578 | |
| Rev | CCAGCCGGCAACTGTGA | 582–598 | |

[a]The probes are labeled at the 5' and 3' positions with 6-carboxyfluorescein (FAM) reporter and a 6-carboxy-tetramethyl-rhodamine (TAMRA) quencher, respectively. The positions of the primers and probes are annotated according to the sequences derived from GenBank (accesion numbers given in parentheses).

223

5. 10 m$M$ dNTP mix (Invitrogen Life Technologies).
6. SuperScript II reverse transcriptase enzyme (Invitrogen Life Technologies).
7. Taq DNA polymerase (Invitrogen Life Technologies).
8. RnaseOut RNase inhibitor (Invitrogen Life Technologies).
9. 1 µg/mL Acetylated BSA (Promega, Madison, WI).
10. 50X 5-carboxy-$X$-rhodamine (ROX) internal reference dye (Invitrogen Life Technologies).
11. MicroAmp optical tubes (Applied Biosystems).
12. MicroAmp optical caps (Applied Biosystems).
13. Nusieve 3:1 agarose (BioWhittaker Molecular Applications, Rockland, ME).

## 3. Methods

### 3.1. Animals and Tissue Processing

All reagents were maintained under RNase-free sterile conditions. Apolipoprotein E$^{-/-}$ mice, a standard model of human atherosclerosis, were fed a standard chow diet for 20 wk. Mice were sacrificed by exsanguination (under general anesthesia with isoflurane) by intravascular perfusion with phosphate-buffered saline (PBS). The thorax was opened and a 21-gage cannula inserted to the left ventricle. The right atrium was incised to allow efflux of blood. Perfusion was at physiological pressure (100 mmHg) with PBS. The heart and aorta were removed *en bloc*. The heart was transected at the lower poles of the atria and perpendicular to them. This plane is parallel to that of the aortic valve. The upper half of the heart (containing the atria) was placed in an embedding mold filled with OCT cryoembedding medium, which was allowed to infiltrate the cavity of the heart. The heart was positioned flat with the cut surface on the bottom of the embedding mold. The embedding mold was filled with OCT and frozen on dry ice. The specimens are stored at –80°C.

Tissue blocks were sectioned (6 µm thickness) and mounted on positively charged Color Frost Plus slides. To demonstrate that gene expression changes can be detected by LCM and real-time RT-PCR, two additional groups of mice were administered either an intraperitoneal injection of the bacterial endotoxin lipopolysaccharide (LPS) (100 µg) or vehicle only and, ~4 h after treatment, the proximal aortas were processed for each mouse as described above.

### 3.2. CD68 Immunodetection of Macrophages for LCM

Standard immunohistochemical staining protocols usually require prolonged incubation periods in aqueous media, which results in significant degradation of RNA. To overcome this limitation, a modified rapid immunostaining protocol was developed for macrophage-specific CD68/macrosialin that does not significantly affect RNA yields (<12% reduction of total RNA in immunostained vs nonimmunostained whole tissue sections) *(11)*.

1. Allow frozen sections to air-dry for 1 min and then fix in 70% ethanol for 15 s followed by cold acetone for 5 min.
2. Incubate slides with anti-mouse CD68/macrosialin IgG (1:10 dilution) in 4% goat blocking serum (made in PBS) supplemented with SuperaseIn RNase inhibitor (0.4 U/μL) for 1 min at room temperature (RT).
3. Rinse slides in PBS 3 times for 5 min each.
4. Incubate slides with biotinylated rabbit-anti-rat IgG mouse-adsorbed secondary antibody (20 μg/mL final) in 4% goat blocking serum supplemented with SuperaseIn RNase inhibitor (0.4 U/μL) for 1 min at RT.
5. Rinse slides in PBS three times for 5 min each.
6. Incubate slides with Vectastain ABC alkaline phosphatase enzymatic detection reagent (prepared at least 30 min in advance according to manufacturer's instructions) for 1 min at RT (*see* **Note 1**).
7. Rinse slides in 100 m*M* Tris-HCl buffer, pH. 7.5, for 1 min at RT.
8. Apply Vector Red chromogenic substrate until adequate intensity of staining develops (~5–10 min). Immerse slides in RNase-free water to stop reaction.
9. Sections are counterstained with Harris modified hematoxylin for approx 5 s.
10. Sections are subsequently dehydrated in graded ethanols (95% 2X, 5 min each, 100% 3X, 5 min each) and cleared in xylene (3X, 5 min each). Allow slides to air-dry under a vacuum hood for 30 min (*see* **Note 2**).

### 3.3. LCM and RNA Extraction

Laser capture was performed under direct microscopic visualization on the CD68-positively stained areas by melting of selected regions onto a thermoplastic film mounted on optically transparent LCM caps (**Fig. 1**). The PixCell II LCM System was set to the following parameters: 15 μm laser spot size, 40 mW power, 3.0 ms duration.

1. Incubate the thermoplastic LCM cap containing the microdissected cells inverted in 200 μL guanidinium isothiocyanate (GITC) buffer and 1.6 μL β-mercaptoethanol for approx 10 min at 37°C. Examine the transfer film under the microscope to ensure complete cell lysis.
2. Pulse spin. Transfer the denaturing solution to a new 1.5-mL tube. Add 20 μL (0.1X vol) of 2 *M* sodium acetate, pH 4.0.
3. Add 220 μL of water-saturated phenol.
4. Add 60 μL of chloroform-isoamyl alcohol.
5. Vortex tube vigorously and put on ice for 15 min.
6. Centrifuge in a microfuge at maximum speed (~15,000 rpm) for 30 min at 4°C.
7. Transfer the upper aqueous layer to a fresh tube. Add 1–2 μL GlycoBlue carrier (10 μg/ μL).
8. Add 200 μL cold isopropanol. Place tube at –80°C for at least 30 min (or, if necessary, leave overnight at –80°C and continue the following morning).
9. Centrifuge for 30 min at maximum speed at 4°C.

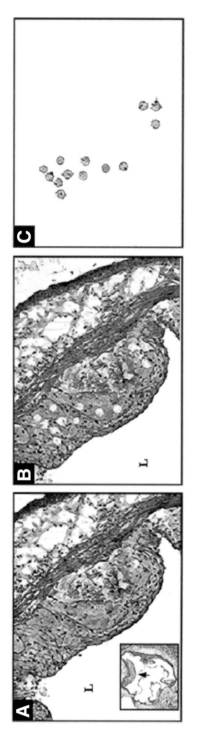

Fig. 1. LCM of macrophage foam cells from atherosclerotic lesions of apolipoprotein E$^{-/-}$ mice. Selection of cells for laser capture was guided by immunohistochemical detection of macrophage specific marker CD68/macrosialin (red staining). (**A**) A proximal aortic lesion is shown prior to laser capture (original magnification ×200; inset: ×30). The CD68-positively stained cells are targeted for laser capture using a 15-μm laser diameter. Arrow in inset denotes a representative lesion from which CD68-positive cells were isolated by LCM. (**B**) After the thermoplastic film is removed, the indentations left removal of the captured material are seen in the remaining heterogeneous tissue. (**C**) The homogeneity of the captured material is confirmed under microscopic visualization prior to processing for RNA extraction (adapted from **ref. 13**).

10. Carefully remove all of the supernatant without disturbing the RNA pellet.
11. Rinse pellet with 400 µL cold ethanol. Centrifuge for 5 min at 4°C.
12. Remove all the supernatant and allow the pellet to air-dry on ice to remove any residual ethanol. Pellet can be subsequently stored at –80°C.
13. To eliminate potential genomic DNA contamination, RNA samples are treated with DNase (4 U, DNA-free; Ambion) in the presence of 20 U RNase inhibitor (RNaseOut, Invitrogen Life Technologies) at 37°C for 1 h.
14. Inactivate the DNase using DNase inactivation reagent per manufacturer's instructions.
15. RNA concentration measured by a Ribogreen sensitive nucleic-acid dye binding assay according to manufacturer's instructions (*see* **Note 3**).
16. Store LCM RNA sample at –80°C.

## 3.4. Analysis of Macrophage Foam Cell Gene Expression by Real-Time Quantitative RT-PCR

Real-time quantitative RT-PCR is a very sensitive method that allows for measurements of low-abundance transcripts and, unlike Northern or RNase protection assays, requires only a very small amount of total RNA (typically 100 pg–1 ng) (for a comprehensive review, *see* **ref. *12***) We have previously used real-time quantitative RT-PCR to demonstrate the following: (1) the selective enrichment from arterial tissue of lesional macrophage RNA by LCM (**Fig. 2C,D**) and (2) the induction of inflammatory genes in lesional macrophage cells upon stimulation with an inflammatory stimulus (described in **ref. *13***; *see* **Note 4**).

1. RT-PCR and subsequent PCR are both carried out in a single sealed optical tube using gene-specific primers and fluorogenic probes.
2. Prepare a master mix containing the following for each reaction: 1X first-strand buffer (50 m$M$ Tris-HCl, pH 8.3, 75 m$M$ KCl, 3 m$M$ MgCl$_2$), 5 m$M$ DTT, 0.3 m$M$ dNTP mix, 20 U SuperScript II reverse transcriptase enzyme, 2.5 U Taq DNA polymerase, 40 U RNaseOut RNase inhibitor, 0.625 µg acetylated BSA, and 1X 5-carboxy-*X*-rhodamine internal reference dye (Invitrogen Life Technologies) in optical tubes.
3. Add 50 n$M$ of the forward primer and reverse primer, and 100 n$M$ of the probe to the master mix for each reaction.
4. To 5 µL of samples (100 pg) and appropriate standard RNA (10 ng–1 pg serial dilutions), add separately 20 µL of the master mix.
5. Set the Sequence Detection System 7700 to the following cycling conditions: RT-PCR stage (95°C, 10 s; 45°C, 50 min; 95°C, 2 min) immediately followed by 40 cycles of PCR amplification (denaturation: 95°C, 15 s; annealing/extension: 60°C, 1 min). The reaction products are separated on a 2% Nusieve 3:1 agarose gel (**Fig. 2B**) to verify the appropriate size of the amplicons (~63–67 bp) (*see* **Note 5**).

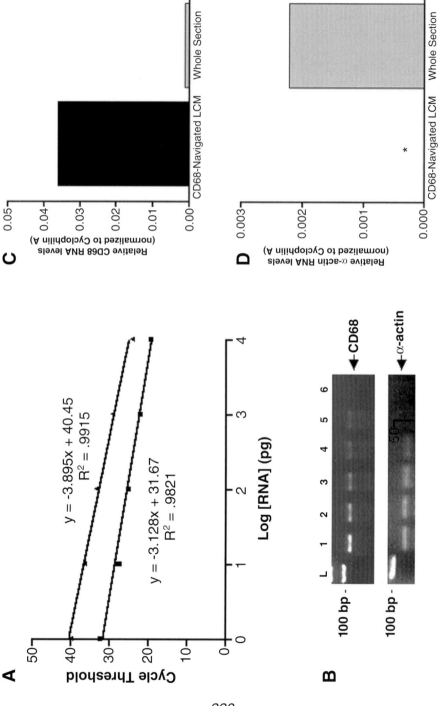

## 4. Notes

1. We have observed that addition of the SuperaseIn RNase inhibitor to the alkaline phophatase enzyme completely inhibits the reaction and does not produce any visible staining. Hence, the RNase inhibitor should not be added at this step.

2. It is important that the alcohols and the xylenes be prepared fresh and not be reused since insufficient dehydration will dramatically affect the ability to perform successful LCM.

3. RNA extraction of the laser captured material yielded 3.5 ng from 30 6-μm-thick sections, equivalent to volume of $1.35 \times 10^6$ μm$^3$ of CD68$^+$ immunostaining.

4. To test whether RNA derived by LCM of macrophage foam cells can be utilized to quantitatively assess the transcriptional regulation of target genes implicated in atherosclerosis, laser captured lesional macrophage RNA from proximal aortas of LPS and control stimulated apolipoprotein E$^{-/-}$ mice were used to measure the levels of the following genes: vascular cell adhesion molecule-1 (VCAM-1), intercellular adhesion molecule-1 (ICAM-1), and monocyte chemoattractant-1 (MCP-1) (*see* **Table 2**).

5. The probe, labeled at the 5' and 3' ends with 6-carboxyfluorescein (6-FAM) reporter and tetramethyl-6-carboxyrhodamine (TAMRA) quencher, respectively, is hydrolyzed by the 5' exonuclease activity of Taq DNA polymerase, causing an increase in fluorescent signal that is measured in "real-time" after each cycle of PCR amplification. Standard curves were constructed by plotting $\log_{10}$ RNA starting quantity vs cycle threshold (**Fig. 2A**). On the basis of appropriate serially diluted standard RNA, the amount of input standard RNA yielding the same amount of PCR product measured from an unknown sample was calculated.

---

Fig. 2. (*opposite page*) LCM selectively enriches lesional macrophage RNA, as assessed by mRNA transcript levels of cell-specific markers (adapted from **ref. *13***). (**A**) Representative standard curves from real-time quantitative RT-PCR of CD68 (▲) and β-actin (■). The plot of cycle threshold versus $\log_{10}$ of input starting concentration shows a linear relationship for 5 orders of magnitude of starting RNA. (**B**) Electrophoretic analysis of the quantitative RT-PCR products. The products for CD68 (top) and β-actin (bottom) verify the specific amplification of appropriately sized amplicons (67 and 66 bp, respectively). The semiquantitative relationship (i.e., graded intensity of ethidium bromide stained bands) for the serial diluted standards (10 ng–1 pg; lanes 1–5) corresponds to that measured in **A**. (**C**) The levels of macrophage specific CD68 were markedly increased in the LCM sample (range: 0.035–0.037) compared to the whole-section RNA sample (range: 0.00096–0.0014). The 33.6-fold increase in CD68 levels in the LCM procured sample attests to the enrichment of macrophage cell derived RNA transcripts. (**D**) The β-actin levels (mean [range]: 0.0022 [0.0019–0.0025]; normalized to cyclophilin A) in LCM derived foam cells are compared to those in whole-section RNA. While the levels of the control reference gene, cyclophilin A, were comparable between the two sample groups, there was no detectable amplification for β-actin in the LCM foam cell derived RNA (the * indicates where the data bar would have been) (adapted from **ref. *13***).

**Table 2**
**Induction of Pro-Atherogenic mRNAs by LPS**

| mRNA | Control LPS | Ratio of means transcript | (LPS:Control) |
|------|-------------|---------------------------|---------------|
| VCAM | 2.1 (1.6–2.1) | 22.1 (11.9–32.3) | 11.9 |
| ICAM | 0.3 (0.2–0.4) | 9.8 (4.1–15.4) | 32.5 |
| MCP-1 | 0.1 (0.126–0.128) | 3.9 (2.4–5.5) | 31.0 |

Real-time quantitative RT-PCR measurements (normalized to the control gene cyclophilin A) of RNA derived from LCM-isolated lesional macrophages from control and LPS-treated apoliprotein E$^{-/-}$ animals are given as a mean (range) calculated from two animals per treatment condition.

## Acknowledgments

This work was supported by NIH grant HL-61814 (EAF) and by a Sigma Xi grant-in-aid (ET). ET is supported by an NIH Training Grant (HL-07824) in Molecular and Cellular Cardiology.

## References

1. Smith, J. D., Trogan, E., Ginsberg, M., Grigaux, C., Tian, J., and Miyata, M. (1995) Decreased atherosclerosis in mice deficient in both macrophage colony- stimulating factor (op) and apolipoprotein E. *Proc. Natl. Acad. Sci. USA* **92,** 8264–8268.
2. Qiao, J. H., Tripathi, J., Mishra, N. K., Cai, Y., Tripathi, S., Wang, X. P., et al. (1997) Role of macrophage colony-stimulating factor in atherosclerosis: studies of osteopetrotic mice. *Am. J. Pathol.* **150,** 1687–1699.
3. de Villiers, W. J., Smith, J. D., Miyata, M., Dansky, H. M., Darley, E., and Gordon, S. (1998) Macrophage phenotype in mice deficient in both macrophage-colony- stimulating factor (op) and apolipoprotein E. *Arterioscler Thromb. Vasc. Biol.* **18,** 631–640.
4. Chong, P.H. and Bachenheimer, B. S. (2000) Current, new and future treatments in dyslipidaemia and atherosclerosis. *Drugs* **60,** 55–93.
5. Brewer, H. B., Jr. (2000) The lipid-laden foam cell: an elusive target for therapeutic intervention. *J. Clin. Invest.* **105,** 703–705.
6. Plutzky, J. (1999) Atherosclerotic plaque rupture: emerging insights and opportunities. *Am. J. Cardiol.* **84,** 15J–20J.
7. Emmert-Buck, M. R., Bonner, R. F., Smith, P. D., Chuaqui, R. F., Zhuang, Z., Goldstein, S. R., et al. (1996) Laser capture microdissection. *Science* **274,** 998–1001.
8. Bonner, R. F., Emmert-Buck, M., Cole, K., Pohida, T., Chuaqui, R., Goldstein, S., and Liotta, L. A. (1997) Laser capture microdissection: molecular analysis of tissue. *Science* **278,** 1481, 1483.

9. Ramprasad, M. P., Fischer, W., Witztum, J. L., Sambrano, G. R., Quehenberger, O., and Steinberg, D. (1995) The 94- to 97-kDa mouse macrophage membrane protein that recognizes oxidized low density lipoprotein and phosphatidylserine-rich liposomes is identical to macrosialin, the mouse homologue of human CD68. *Proc. Natl. Acad. Sci. USA* **92,** 9580–9584.

10. Ramprasad, M. P., Terpstra, V., Kondratenko, N., Quehenberger, O., and Steinberg, D. (1996) Cell surface expression of mouse macrosialin and human CD68 and their role as macrophage receptors for oxidized low density lipoprotein. *Proc. Natl. Acad. Sci. USA* **93,** 14,833–14,838.

11. Fend, F., Emmert-Buck, M. R., Chuaqui, R., Cole, K., Lee, J., Liotta, L. A., and Raffeld, M. (1999) Immuno-LCM: laser capture microdissection of immuno-stained frozen sections for mRNA analysis. *Am. J. Pathol.* **154,** 61–66.

12. Bustin, S. A. (2000) Absolute quantification of mRNA using real-time reverse transcription polymerase chain reaction assays. *J. Mol. Endocrinol.* **25,** 169–193.

13. Trogan, E., Choudhury, R. P., Dansky, H. M., Rong, J. X., Breslow, J. L., and Fisher, E. A. (2002) Laser capture microdissection analysis of gene expression in macrophages from atherosclerotic lesions of apolipoprotein E-deficient mice. *Proc. Natl. Acad. Sci. USA* **99,** 2234–2239.

# 18

# Analysis of Pituitary Cells by Laser Capture Microdissection

Ricardo V. Lloyd, Xiang Qian, Long Jin, Katharina Ruebel, Jill Bayliss, Shuya Zhang, and Ikuo Kobayashi

## Summary

The anterior pituitary gland consists of a heterogeneous population of various cell types. To study a single cell type with a homogeneous cell population, one can perform laser capture microdissection (LCM). Because different pituitary cells have unique immunophenotypic profiles, it is possible to perform immunohistochemical staining before LCM (immuno-LCM) for the collection of a phenotypically homogeneous cell population. These techniques were developed and applied to dissociated anterior pituitary cells and cultured pituitary cells. When combined with reverse transcriptase polymerase chain reaction, it is possible to analyze gene expression in as few as one to 10 pituitary cells. We have used the immuno-LCM technique to prepare homogeneous populations of folliculostellate cells. These cells were analyzed for expression of peptides and receptors. Anterior pituitary hormones were not expressed by these cells. These results show the utility of immuno-LCM for cellular and molecular studies of gene expression.

**Key Words:** Pituitary; folliculostellate cells; immunohistochemistry; LCM, $GH_3$ cells, TtT/GF cells; RT-PCR.

## 1. Introduction

The collection of homogeneous populations of pituitary cells for biologic and molecular analyses has been a difficult challenge for some time. The anterior pituitary represents a complex heterogeneous mixture of cells with six secretory hormone cell types, including prolactin (PRL), growth hormone (GH), adrenocorticotroph (ACTH), thyrotroph (TSH), follicle-stimulating hormone (FSH), and luteinizing hormone (LH), as well as other cells such as the folliculostellate (FS) cells, endothelial cells, fibroblasts, and other stromal cells. The recent development of laser capture microdissection (LCM) has provided a rapid and efficient method to capture pure cell population for molecular and other studies *(1–9)*.

From: *Methods in Molecular Biology, vol. 293: Laser Capture Microdissection: Methods and Protocols*
Edited by: G. I. Murray and S. Curran © Humana Press Inc., Totowa, NJ

Analysis of pituitary cells with LCM involves two separate procedures. The first is immunostaining of specific cell types using antibodies unique to certain cell types such as specific hormones produced by these cells. This is followed by laser capture microdissection (immuno-LCM) *(3,4)*. Immuno-LCM has also been used to obtain pure populations of FS cells by immunostaining for S-100 protein with subsequent LCM and molecular analyses of the cells using reverse transcriptase polymerase chain reaction (RT-PCR). Analyses of FS cells *(10–19)* along with a mouse FS cell line (TtT/GF) by RT-PCR for gene expression has been used to analyze expression of specific mRNA in these cells.

## 2. Materials

1. Anterior pituitaries from 60- to 90-d-old female Wistar-Furth rats.
2. $GH_3$ cells maintained in culture.
3. Folliculostellate (TtT/GF) cells maintained in culture.
4. Dulbecco's modified Eagle medium (DMEM) with 15% horse serum, 2.5% fetal calf serum, and 1% antibiotics.
5. Transforming growth factor (TGF) β-1 growth factor.
6. TRIzol.
7. Glycogen.
8. Chloroform.
9. Isopropanol.
10. Ethanol.
11. Diethyl pyrocarbonate.
12. cDNA synthesis kit (Stratagene, La Jolla, CA).
13. Agarose gel.
14. Ethidium bromide.
15. Nylon membrane.
16. Standard saline citrate (SSC).
17. X-ray film (XOMAT AR, Kodak, Rochester, NY).
18. Taq polymerase.
19. PCR buffer.
20. PCR primers.

## 3. Methods

The methods described below outline (1) preparation of cells; (2) immunostaining; (3) LCM with the PixCell system; (4) extraction of DNA; and (5) RNA RT-PCR and sample analysis.

### 3.1. Tissue Preparation

1. Normal rat pituitary cells are dissociated with 0.25% trypsin. The dispersed rat pituitary cells are attached to uncoated glass slide using between $1 \times 10^3$ to $1 \times 10^4$ cells per slide prepared by cytocentrifugation (*see* **Notes 1–3**).
2. Cells are fixed in 100% ethanol for 5 min, air-dried for 5 min, and then used for immunohistochemistry and LCM (**Fig. 1**).

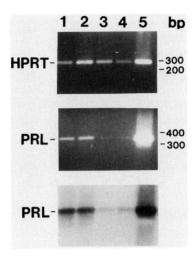

Fig. 1. RT-PCR analysis of GH₃ cells after fixation in ethanol (lanes 1 and 2) or paraformaldehyde (lanes 3 and 4). Lane 5 represents a positive pituitary tissue control. Top figure—HPRT is the housekeeping gene; the middle figure represents the RT-PCR reaction product run on an ethidium bromide stain gel; and the lower part of the figure shows the Southern hybridization with an internal probe for prolactin. These results show that fixation in ethanol is superior to fixation in paraformaldehyde for preservation of RNA quality.

3. Immunohistochemistry is done with specific antibodies to pituitary hormones obtained from the National Pituitary Agency, Baltimore, MD. Negative controls for immunostaining consist of substituting normal rabbit serum for the primary antibody, which should result in no staining of the cells.
4. The cells are lightly counterstained with hematoxylin and placed in 3% glycerol in RNase-free water for 20 min to facilitate cell detachment during LCM.
5. The slides are dehydrated with 95% and 100% ethanol.
6. This is followed by incubation in xylene for 10 min and air-drying at room temperature prior to LCM.
7. During immunostaining, 400 U/mL of RNasin is added to all solutions to decrease RNase contamination.
8. Incubate slides with the primary antibody from 2 to 20 min followed by avidin biotin peroxidase reaction for 20 min and diaminobenzidine for 1 to 5 min (*see* **Note 4**).
9. LCM is done with a PixCell II laser capture microdissection system (Arcturus Engineering, Inc., Mountain View, CA).
10. The immunopositive cells are captured directly onto a thermoplastic polymer film-coated cap by a one-step transfer method (*see* **Note 5**). An infrared laser with 60 mw of laser power and a laser beam with a 7.5- to 15-μm diameter are used to capture the cells of interest.

11. The laser melts the ethylene vinyl acetate from a plastic film directly onto the targeted cells, embedding the captured cells.
12. After transfer to the cap, the samples are used for RNA analysis.

### 3.2. Operating the PixCell System

1. Turn on the power strip located behind the computer monitor. Also, turn on power for the monitor.
2. On computer screen, click mouse on "shortcut to Arc 100."
3. When instrument serial number is given, click "continue."
4. Select name or enter new name and click "acquire data."
5. Highlight study or enter new study and enter "select."
6. Enter slide number, spot size, cap lot, and thickness (usually 10 μm, but this may vary depending on slides cut), and click "continue."
7. Enter laser power of approx 45 and pulse of approx 55 (to start), length needed (see procedures for RNA and DNA extractions for this information).
8. Place slide on microscope, and, using TV monitor, find an appropriate starting area that will be easy to find after finishing the microdissection.
9. Click on "before" to obtain a picture of the area selected.
10. Insert a row of Arcturus caps with transfer film into slot on the right side of the microscope.
11. If the laser power chosen is below 60, place the optic beam-adjust piece without the filter in the indentation on the end of the placement "arm." If laser power chosen is above 60, place the beam-adjust piece with the filter on the arm.
12. Using the placement arm, pick up a cap, move the arm all the way over to the left (this will put it directly above the slide), and gently release it so that the cap slowly drops onto the slide.
13. Pick up the white cord with the red button on the end and press the button to get a laser pulse. A dark circle will appear around the area if it "melted." (However, just because an area melts does not necessarily mean that it transferred to the cap.)
14. Use the joystick located to the bottom left of the microscope to move the slide around and get pulses in different areas.
15. After finishing the microdissection, use the arm to pick up the cap very gently (so as not to pick up any other tissue), move it over to the right, and place it on a sterile 0.5-mL microcentrifuge tube.
16. Find the original starting area and clock on "after" to get an image of the completed microdissection. Then look at the slide under a microscope to make sure that most of the areas transferred.
17. Click "done" and then "exit" or "continue."

### 3.3. DNA Extraction From Collected Cells

1. Stain slides with hematoxylin and eosin.
2. Aliquot 100 μL 0.05 *M* Trizma buffer, pH 8.3, and 4 μL proteinase K (10 mg/mL into sterile 0.5-mL Eppendorf microfuge tubes in 0.05 *M* Trizma buffer.

3. Use LCM to capture the desired number of cells from the slide.
4. Place the film cap containing the captured cells on the top of the microfuge tube and use the black cap tool to snap the cap in. (It is *critical* to place the bottom of the film cap just inside the lip of the microfuge tube, leaving about a 1-cm gap. Using the black cap tool ensures the correct spacing needed to prevent leakage during incubation.)
5. Invert the tubes and incubate in a 55°C water bath for 48 h.
6. Remove tubes from the water bath and spin down in microcentrifuge. Discard the film cap and transfer the solution to a new sterile 0.5-mL tube.
7. Boil samples for 8 min at 95°C in a hot block and place on ice. Store DNA at 4°C until ready to use.

### 3.4. RNA Extraction From Collected Cells

1. Obtain samples in cap with transfer film using LCM.
2. Aliquot 200 μL TRIzol reagent into a *sterile* 0.5-mL microcentrifuge tube, then place cap with film on top, and invert the tube (*see* **Note 6**).
3. Leave samples with TRIzol inverted at room temperature for more than 1 h.
4. Take off cap, add 1 μL glycogen and 40 μL chloroform to each tube, and shake vigorously for 15 s.
5. Incubate samples for 3 min at room temperature.
6. Using Eppendorf centrifuge in cold room, spin for 15 min at 2500$g$.
7. Transfer aqueous phase to a new 0.5-mL tube, add 10 μL isopropanol, and vortex.
8. Incubate at room temperature for 10 min.
9. Place in a –70°C freezer for 1 h or –20°C for 3 h. Take out and allow to thaw.
10. Centrifuge for 30 min at 2500$g$ in Eppendorf centrifuge.
11. Very carefully discard the supernatant, as you will probably not see a pellet.
12. Add 200 μL 75% ethanol and vortex.
13. Centrifuge at 2000$g$ for 5 minutes in Eppendorf centrifuge.
14. Carefully pour off the supernatant, invert the tube, and air-dry for approx 5 min.
15. Resuspend the pellet in 10 μL of diethyl pyrocarbonate (DEPC) water and use this directly for the RT reaction.

### 3.5. RT-PCR Analysis

1. The first-strand cDNA is prepared from the total RNA by using a first strand synthesis kit from Stratagene, LaJolla, CA. The RT reaction was performed in a final volume of 50 μL with 10 μL of total RNA from the LCM transfer cells and 300 ng of oligoDT primers at 37°C for 90 min.
2. Heat at 90°C for 5 min and immediately place on ice.
3. Omission of the reverse transcriptase enzyme during the RT reaction is used as a negative control.
4. PCR amplification was performed with 40 cycles for the LCM samples. The annealing temperatures ranged from 55 to 60°C.
5. After the final cycle, the elongation step is extended by 10 min at 72°C.

6. Housekeeping gene (HPRT) is amplified from the same RT products and used as an internal control. Analysis of other pituitary cells including GH, prolactin, and POMC is done to determine the homogeneity of the cell population (*see* **Note 7**).
7. Southern hybridization is performed with a 20-μL aliquot of the PCR product. The PCR product is analyzed by electrophoresis on a 2% agarose gel with ethidium bromide staining.
8. Titration studies with different amounts of cDNA is performed to verify that each amplification is in the linear range.
9. The PCR products are transferred to nylon membrane filters, and southern hybridization is performed with 33P label internal probes at 42°C for 18 h.
10. After washing in 6X SSC with 0.1% SDS at 23°C for 20 min and at 42°C for 20 min, autoradiography is performed with Kodak XOMAT AR film (**Fig. 2**).

## 4. Notes

1. To prevent difficulties with capturing dissociated cells from the slides, it is important to use uncharged slides. Use of charged slides will prevent easy capture of the cells by LCM.
2. The use of $GH_3$ pituitary cell line and normal rat pituitary cells show that 1 to 10 cells were sufficient to examine gene expression for specific hormones. Up to 100 cells could be captured in 30 min for analysis of multiple gene expression *(4)*.
3. The specificity of the immuno-LCM experiments can be verified by immunostaining for specific hormones such as prolactin and ACTH and then performing RT-PCR to show that amplified prolactin cDNA was obtained only from prolactin stained cells and not from ACTH-stained cells, whereas POMC cDNA was obtained only from ACTH-stained cells but not from prolactin cells. When liver cells were dissociated and analyzed by immuno-LCM, these were negative for all hormones.
4. It is important to use the antibody for as short a time as possible in immunostaining. Titrate the antibody to obtain optimal staining in 2 to 5 min if possible. Use of RNasin is probably helpful to prevent RNase contamination.
5. FS cells are S-100 positive, so immuno-LCM was performed using S-100-antibody for immunostaining as the first step. The FS cells do not contain secretory granules, while the hormone-producing cells in the anterior pituitary glands have secretory granules. The FS cells produce several peptides and growth factors that have paracrine regulatory functions in the anterior pituitary *(13–19)*.
6. TRIzol for RNA extraction or other buffers should be added to the collected cells as soon as possible to avoid degradation of nucleic acids, proteins, and so on.
7. FS cells as well as the FS cell line (TtT/GF), when analyzed for gene expression by RT-PCR, are shown to express messenger RNAs for glial fibrillary acidic protein, transforming growth factor beta, transforming growth factor beta receptors, interleukin-6, leptin, leptin receptor, pituitary adenyl cyclase activating polypeptide (PACAP), and PACAP receptors. These cells do not express the messenger RNAs for prolactin, growth hormone, or POMC, indicating the specificity of the immuno-LCM procedure *(9)*.

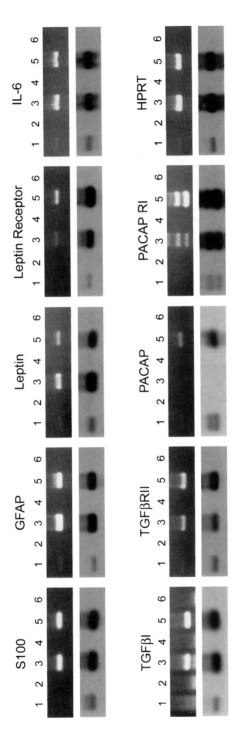

Fig. 2. RT-PCR analysis of folliculostellate cells. Approximately 400 S-100-positive cells were collected by immuno-LCM and analyzed by RT-PCR. (A) Analysis of various mRNA in folliculostellate cells. Lane 1—Rat pituitary FS cell. Lane 3—TtT/GF cells. Lane 5—Normal rat pituitary without RT. The top panel is the ethidium bromide stained gel (*9*), and the bottom panel is a Southern hybridization with the internal probes. Analysis of PRL, GH, and POMC expression in the bottom panel shows that only the normal pituitary control expressed these hormone messenger RNAs. Reproduced with permission of the Endocrine Society.

## References

1. Emmert-Buck, M. R., Bonner, R. F., Smith, P. D., Chuaqui, R. F., Zhuang, Z., Goldstein, S. R., et al. (1996) Laser capture microdissection. *Science* **274,** 998–1001.
2. Bonner, R. F., Emmert-Buck, M., Cole, K., Pohida, T., Chuaqui, R., Goldstein, S., and Liotta, L. A. (1997) Laser capture microdissection: molecular analysis of tissue. *Science* **278,** 1481,1483.
3. Fend, F., Emmert-Buck, M. R., Chuaqui, R., Cole, K., Lee, J., Liotta, L. A., and Raffeld, M. (1999) Immuno-LCM: laser capture microdissection of immuno-stained frozen sections for mRNA analysis. *Am. J. Pathol.* **154,** 61–66.
4. Jin, L., Thompson, C. A., Qian, X., Kuecker, S. J., Kulig, E., and Lloyd, R. V. (1999) Analysis of anterior pituitary hormone mRNA expression in immuno-phenotypically characterized single cells after laser capture microdissection. *Lab. Invest.* **79,** 511–512.
5. Kuecker, S., Jin, L., Kulig, E., Ondraogo, G., Roche, P., and Lloyd, R. (1999) Analysis of PRL, PRL-R, TGFβ1 and TGFβ-RII gene expression in normal and neoplastic breast tissues after laser capture microdissection. *Appl. Immuno-histochem. Mol. Morphol.* **7,** 193–200.
6. Goldsworthy, S. M., Stockton, P. S., Trempus, C. S., Foley, J. F., and Maronpot, R. R. (1999) Effects of fixation on RNA extraction and amplification from laser capture microdissected tissue. *Mol. Carcinog.* **25,** 86–91.
7. Simone, N. L., Bonner, R. F., Gillespie, J. W., Emmert-Buck, M. R., Liotta, L. A. (1998) Laser-capture microdissection: opening the microscopic frontier to molecular analysis. *Trends Genet.* **14,** 272–276.
8. Suarez-Quian, C. A., Goldstein, S. R., Pohida, T., Smith, P. D., Peterson, J. I., Wellner, E., et al. (1999) Laser capture microdissection of single cells from complex tissues. *Biotechniques* **26,** 328–335.
9. Jin, L., Tsumanuma, I., Ruebel, K. H., Bayliss, J. M., and Lloyd, R. V. (2001) Analysis of homogeneous populations of anterior pituitary folliculostellate cells by laser capture microdissection and reverse transcription-polymerase chain reaction. *Endocrinology* **142,** 1703–1709.
10. Cocchia, D. and Miani, N. (1980) Immunocytochemical localization of the brain-specific S-100 protein in the pituitary gland of adult rat. *J. Neurocytol.* **9,** 771–782.
11. Hofler, H., Walter, G. F., and Denk, H. (1984) Immunohistochemistry of folliculo-stellate cells in normal human adenohypophyses and in pituitary adenomas. *Acta Neuropathol. (Berl.)* **65,** 35–40.
12. Girod, C., Trouillas, J., and Dubois, M. P. (1985) Immunocytochemical localization of S-100 protein in stellate cells (folliculo-stellate cells) of the anterior lobe of the normal human pituitary. *Cell Tissue Res.* **241,** 505–511.
13. Lloyd, R. V. and Mailloux, J. (1988) Analysis of S-100 protein positive folliculo-stellate cells in rat pituitary tissues. *Am. J. Pathol.* **133,** 338–346.
14. Baes, M., Allaerts, W., and Denef, C. (1987) Evidence for functional communication between folliculo-stellate cells and hormone-secreting cells in perifused anterior pituitary cell aggregates. *Endocrinology* **120,** 685–691.

15. Allaerts, W., Carmeliet, P., and Denef, C. (1990) New perspectives in the function of pituitary folliculo-stellate cells. *Mol. Cell Endocrinol.* **71,** 73–81.
16. Marin, F., Stefaneanu, L., and Kovacs, K. (1991) Folliculo-stellate cells of the pituitary. *Endocr. Pathol.* **2,** 180–192.
17. Ferrara, N. and Henzel, W. J. (1989) Pituitary follicular cells secrete a novel heparin-binding growth factor specific for vascular endothelial cells. *Biochem Biophys. Res. Commun.* **161,** 851–858.
18. Ferrara, N., Schweigerer, L., Neufeld, G., Mitchell, R., and Gospodarowicz, D. (1987) Pituitary follicular cells produce basic fibroblast growth factor. *Proc. Natl. Acad. Sci. USA* **84,** 5773–5777.
19. Vankelecom, H., Matthys, P., Van Damme, J., Heremans, H., Billiau, A., and Denef, C. (1993) Immunocytochemical evidence that S-100-positive cells of the mouse anterior pituitary contain interleukin-6 immunoreactivity. *J Histochem Cytochem* **41,** 151–156.

# IV

## MICRODISSECTION TECHNIQUES AND APPLICATIONS IN PROTEOMICS

# 19

# Laser Capture Microdissection and Colorectal Cancer Proteomics

## Laura C. Lawrie and Stephanie Curran

### Summary

The ability to define protein profiles of normal and diseased cells is important in understanding cell function. Laser capture microdissection permits the isolation of specific cell types for subsequent molecular analysis. In this study we have established conditions for obtaining proteomic information from laser capture microdissected colorectal cancer cells. Laser capture microdissection was performed on toluidine blue-stained frozen sections of colorectal cancer. Proteins were solubilized from microdissected cells and the solubilized proteins were separated by two-dimensional gel electrophoresis: protein spots were characterized by peptide mass mapping using matrix assisted laser desorption ionization-time of flight mass spectrometry. Proteins isolated from laser capture microdissected tissue retained their expected electrophoretic mobility and peptide mass mapping was also unaffected. The ability to study the protein expression profile of specific cell types will allow for the identification of novel disease markers and therapeutic targets and also provide for the enhanced understanding of pathogenetic mechanisms.

**Key Words:** Colorectal cancer; 2-D gel electrophoresis; laser capture microdissection; mass spectrometry; proteomics.

## 1. Introduction

Interest in proteomics has increased steadily with the realization that characterization of DNA and RNA alone will not be sufficient to elucidate mechanisms of disease and to identify new drug targets and diagnostic markers (*1*). Proteomics includes the identification and characterization of proteins and the determination of post-translational modifications (*1,2*). One of the primary applications of proteomics is to identify differentially expressed proteins by comparing the protein expression patterns between normal and diseased cells, such as tumor cells. Proteins that show enhanced expression in the diseased

From: *Methods in Molecular Biology, vol. 293: Laser Capture Microdissection: Methods and Protocols*
Edited by: G. I. Murray and S. Curran © Humana Press Inc., Totowa, NJ

sample are potential diagnostic or prognostic markers or therapeutic targets. Proteomics encompasses a wide range of technologies, of which two-dimensional (2D) gel electrophoresis and mass spectrometry are currently the most widely used.

Molecular analysis of tumors requires the isolation of specific populations of cells since the presence of an admixture of cell types within a sample remains a major obstacle to meaningful biological analysis *(3–5)*. Laser capture microdissection (LCM) permits the rapid and reliable procurement of a specific type of cell from a tissue section, in one step, under direct microscopic visualization *(6–9)*. As part of our program of colorectal cancer research, we have developed a method to isolate colorectal cancer cells using LCM and combined this with proteomics analysis *(10)*. We have shown that proteins solubilized from microdissected colorectal cancer cells can be separated by 2-D gel electrophoresis and that individual proteins can be identified by mass spectrometry *(10)*. These findings provide the basis for the identification of novel markers of prognosis in colorectal cancer and enhanced understanding of the biology of colorectal cancer *(11)*.

## 2. Materials

1. Toluidine blue.
2. Ethanol.
3. Xylene.
4. Coomassie Plus Protein Assay Reagent (Pierce).
5. Thiourea, (Amersham Biosciences).
6. Chaps [3-(3-cholamidopropyl)dimethylammonio-1-propanesulfonate] (Amersham Biosciences).
7. Mega 10 (*N*-decanoyl-*N*-methylglucamine, Sigma).
8. OBG (1-*O*-Octyl-β-D-glucopyranoside, Aldrich).
9. Triton X-100 (polyoxyethylene-*p*-isooctylphenol, Amersham Biosciences).
10. Tris [Tris(hydroxymethyl)aminomethane] (Amersham Biosciences).
11. Dithiothreitol (DTT, Amersham Biosciences).
12. IPG 3-10 NL (immobilized pH gradient buffer, Amersham Biosciences).
13. β-Mercaptoethanol (Amersham Biosciences).
14. Tributylphosphine (Amersham Biosciences).
15. Immobiline Drystrips pI 3-10 NL (Amersham Biosciences).
16. Dry strip cover fluid (Amersham Biosciences).
17. Hydrochloric acid.
18. SDS (dodecyl sulfate, sodium salt).
19. Glycerol.
20. Iodoacetamide (Amersham Biosciences, UK).
21. NuPAGE, 7 cm, 4–12%, 1 well, Bis-Tris gel (Invitrogen).
22. Low melting point agarose (Amersham Biosciences).
23. Colloidal blue staining kit ( Invitrogen).

24. Methanol.
25. Acetic acid.
26. Porcine trypsin, sequencing grade (Promega).
27. α-cyano-4-hydroxycinnamic acid (Aldrich).
28. Water, HPLC grade.
29. Preparation of Lysis Buffer

    The protocol for the preparation of the protein lysis buffer was originally obtained from the National Cancer Institute protocols for the 2-D gel and mass spectrometric analysis of prostate tissue. We have used the same lysis buffer to successfully solubilize samples of colorectal tissue.

    To prepare 50 mL of lysis buffer:

    - Add 21 g urea to 35 mL HPLC-grade water in a 100-mL tube.
    - Vortex until the urea has dissolved.
    - Add sequentially:
        a. 7.6 g of thiourea.
        b. 2 g of CHAPS.
        c. 0.5 g of Mega 10.
        d. 0.5 g of OBG.
        e. 250 µL of Triton X-100.
        f. 0.25 g of Tris.
        g. 0.4 g of DTT.
        h. 500 µL of IPG buffer pI 3-10NL.
        i. 500 µL of β-mercaptoethanol.
        j. 10 µL of tributylphosphine.
        k. Trace of bromophenol blue.
    - Make up volume to 50 mL with distilled water.
    - Vortex until all components have dissolved.
    - Store in 1-mL aliquots at –20°C.
30. Preparation of Equilibration Buffer
    - Mix together 18 g urea and 10 mL of 0.5 M Tris-HCl, pH 6.9.
    - Vortex until dissolved.
    - Add 10 mL of 20% SDS and 200 mg of DTT.
    - Invert gently to mix.
    - Add 15 mL of glycerol.
    - Vortex until thoroughly mixed.
    - Add a trace of bromophenol blue as an indicator.

## 3. Methods

The methods described below outline (1) laser capture microdissection, (2) the solubilization of samples, (3) preparation of samples for 2-D gel electrophoresis, (4) separation and visualization of proteins by 2-D gel electrophoresis, (5) preparation of samples for mass spectrometry, and (6) identification of proteins by database searching.

## 3.1. Laser Capture Microdissection

1. Frozen sections (10 μm in thickness) of either colorectal cancer or normal colorectal mucosa were cut on a cryostat (Leica, UK).
2. Four LCM sections were mounted onto clean uncoated glass microscope slides, air-dried for 5 s, and then fixed at room temperature in 70% ethanol for 1 min (*see* **Note 1**). For comparison some sections were not subject to LCM and were placed directly in cold microfuge tubes (*see* **step 7** below and **Subheading 3.2.**).
3. The fixed sections were then stained with toluidine blue (*see* **Note 2**). Staining with toluidine blue was carried out by covering the sections with 0.25% toluidine blue (pH 4.5) for 5 s at room temperature and washing in 100% ethanol to remove excess dye; sections were then washed again in 100% ethanol and dehydrated in xylene.
4. Following complete evaporation of the xylene (*see* **Note 3**) the sections were microdissected using a PixCell II laser capture microdissection system (Arcturus Engineering, CA). The laser capture system was equipped with PixCell II image archiving software for Windows 95 (Arcturus Engineering).
5. The laser was operated with the following parameters: spot diameter 15 μm, pulse duration 50 ms and power 50 mW (*see* **Note 4**). Microdissected cells were obtained from 8–10 sections of each sample and a separate "cap" was used to capture cells from each section. Approximately 2500 laser pulses were used per cap (*see* **Note 5**).
6. Following microdissection the thermoplastic film containing the microdissected cells was separated from the rest of the cap, and all the films containing cells from one sample placed in a single 1.5-mL microfuge tube and 125 μL of protein lysis solution added (*see* **Note 6** and **Subheading 3.2.**). These samples were then subject to 2-D gel electrophoresis (*see* **Note 7** and **Subheading 3.2.**).
7. To determine the effects of histological processing and subsequent LCM on protein recovery, some frozen sections (whole tissue samples; *see* **Subheading 3.2.2.**) were placed directly into microcentrifuge tubes followed by the addition of protein lysis solution without the sections being subjected to either toluidine blue staining or LCM.

## 3.2. Solubilization of Tissue Samples

1. Use a Bradford protein assay kit (Pierce Coomassie Plus Protein Assay Reagent kit) to assess the amount of protein in each sample, with a view to adding around 500 μg of protein to the gel.
2. For "whole" colon tissue samples we have found that solubilizing 30 10-μm frozen sections of normal tissue and 30 10-μm frozen sections of tumor tissue in 350 μL and 500 μL of lysis buffer respectively produced a final protein loading on the gel of approx 500 μg for each type of sample.

   After adding lysis buffer to the samples:
3. Vortex samples for 1 min.

4. Centrifuge samples at 14,000*g* for 2 min.
5. Crush remaining solid material with a pestle to aid solubilization.
6. Vortex for a further 1 min.
7. Centrifuge at 14,000*g* for a further 2 min.

### 3.3. Preparation of Samples for 2-D Gel Electrophoresis

1. Add 125 µL of normal sample (in duplicate) to IPGPhor strip holders.
2. Add 125 µL of tumor sample (in duplicate) to IPGPhor strip holders.
3. Place IPG strips pI 3-10NL, gel side down, on top of sample.
4. Slowly add 800 µL dry strip cover fluid to each strip holder to prevent strips from drying out.
5. Allow strips to absorb protein solution overnight.
6. Remove strips from strip holders.
7. Clean strip holders.
8. Cover electrodes of strip holders with slightly damp electrode strips (approx 4 mm x 4 mm), to help soak up excess salt during first dimension run.
9. Return rehydrated strips to strip holders, gel side down.
10. Add 800 µL dry strip cover fluid to each strip holder.

### 3.4. Isoelectric Focusing

An IPGphor system (Amersham Biosciences) was used to separate the proteins in the first dimension according to their isoelectric point (pI), under the following conditions:

1. 30 min at 20 V.
2. 90 min at 200 V.
3. 90 min gradient to 3500 V.
4. 35 h at 3500 V.

On completion of focusing the strips were equilibrated in 5 mL of equilibration buffer for 30 min. Strips were equilibrated for a further 30 min in the above equilibration buffer, but this time DTT was replaced with 500 mg iodoacetamide.

### 3.5. SDS-PAGE Separation

1. Proteins were separated in the second dimension according to their molecular weight on a 7-cm NuPAGE 4–12%, 1 well, Bis-Tris gel (Invitrogen).
2. The IPG strips were attached to the second dimension gel with a 4% low melting point agarose solution (Amersham Biosciences, UK). Normal and tumor samples from the same patient were run in the same gel tank to eliminate any differences in protein mobility that could be caused by variation in the gel electrophoresis.
3. Gels were run at a constant 120 V until the bromophenol dye front reached the end of the gel.

## 3.6. Visualization of Proteins

1. Proteins were visualized using a colloidal blue staining kit (Invitrogen). Gels were initially fixed in a solution containing methanol (50% v/v), acetic acid (10% v/v) for 30 min, then transferred to a staining solution containing methanol (20% v/v), stainer A (20% v/v), stainer B (5% v/v) for overnight staining to visualize the proteins.
2. Gels were destained using HPLC-grade water with microwave heating (5 × 1 min on full power, replacing the water each time) to facilitate the destaining process.

## 3.7. Detection of Differential Protein Expression

1. Destained gels were immediately photographed to produce a black-and-white image.
2. Gel photographs were scanned on a flatbed scanner to produce a computer image, which was then enlarged and printed onto sheets of acetate. Overlaying the normal and tumor acetate gel pictures allowed proteins that were differentially expressed to be rapidly detected by visual inspection.
3. Differentially expressed protein spots were cut from the individual gels in preparation for identification by mass spectrometry.

## 3.8. Identification of Proteins From Gels

1. Individual proteins were identified by peptide mass mapping *(12–14)*.
2. Protein spots were cut from the gel, washed to remove Coomassie stain, reduced with DTT, and alkylated with iodoactetamide, then digested with trypsin (*see* **Note 8**).
3. The resultant tryptic peptides were extracted from the gel pieces under full automation (Pro-Gest Robot, Genomic Solutions).
4. The tryptic fragments were desalted using micro-porous tips (Millipore), and deposited onto a sample plate along with a matrix chemical (α-cyano-4-hydroxycinnamic acid) under full automation (Pro-MS, Genomic Solutions).
5. The masses of the tryptic fragments were then determined by matrix-assisted laser desorption ionization time-of-flight mass spectrometry (MALDI-TOF MS) (*see* **Note 9**).

## 3.9. Protein Identification Through Database Searching

1. To identify the original protein, the masses of the tryptic peptides were entered into a protein database-searching program. Database-searching programs attempt to match the experimentally obtained masses of tryptic peptides with the theoretically calculated masses of tryptic peptides derived from all proteins within a database.
2. We used the MS-Fit database-searching program (http://prospector. ucsf.edu/ucsfhtml4.0/msfit.htm) to identify differentially expressed proteins.
3. The results from a database search consist of a list of protein "hits," ranked according to a statistical scoring system that takes into account how many pep-

tides have been matched to the protein in the database and the accuracy of these matches. The database search was restricted to search only for human proteins; no restriction was placed on either the molecular weight or the isoelectric point of the protein.

4. To be confident that the correct protein was identified, a clear difference in statistical score between the proteins ranked first and second in the results list should be obtained. It is also important to check that the major tryptic peptides from the mass spectrum are assigned to the matching peptides in the database.

## 4. Notes

1. It is important to use uncoated microscope slides, as any coating of the slides with adhesive, e.g., poly-L-lysine or aminopropylethoxy silane-coated slides, will prevent the successful transfer of microdissected cells.
2. A rapid single-step histological staining method was selected to minimize or avoid any loss of or alterations to cellular proteins during the staining procedure.
3. It is crucial to the success of laser capture microdissection that the stained sections are completely dehydrated prior to attempting microdissection. Incompletely dehydrated sections will not be successfully microdissected. We have found that even a trace of moisture will inhibit successful transfer of microdissected cells.
4. The optimum setting for the laser, i.e., the settings that ensure transfer (pick-up) of selected cells, has to be determined experimentally for each type of tissue and in our experience often varies on a day-to-day to basis. The values that we have quoted are the ones that we have generally found to be useful. Occasionally we have found that it is not possible to microdissect an individual section even when the section has been completely dehydrated, as the cells remain adherent to the slide. The reasons for this are obscure and we are aware that other users have also made the same observation.
5. The number of laser pulses required to obtain sufficient numbers of cells for proteomics is considerably higher than the number of cells required for most nucleic acid-based molecular analyses. There is no protein equivalent of the polymerase chain reaction to permit target amplification.
6. This modification of the manufacturer's recommended protocol for processing the caps ensured that small volumes of protein lysis solution could be used to obtain a sufficiently concentrated protein solution suitable for subsequent 2D gel electrophoresis.
7. The solubilized LCM samples were analyzed directly by 2D gel electrophoresis with the entire volume of protein lysis solution applied to a single IPG strip.
8. Trypsin cleaves proteins (at peptide bonds) after arginine and lysine residues. This action produced a set of tryptic fragments unique to each protein.
9. In MALDI-TOF MS the tryptic fragments are mixed with α-cyano-4-hydroxycinnamic acid and deposited on a sample plate. An ultraviolet laser is fired at the mixture, which causes the peptides to be desorbed from the sample plate and ionized. A high-voltage gradient is then applied, which causes the ion-

ized peptides to be accelerated into a flight tube. The ions separate in the flight tube according to their mass-to-charge ratio, with lighter ions arriving at the detector before heavier ions of the same charge. The time the ions take to reach the detector (time of flight), compared with standard proteins of known mass, is used to calculate the masses of the tryptic peptides.

## Acknowledgments

LCL was the Jean V. Baxter Fellow of the Scottish Hospital Endowments Research Trust. The Aberdeen Colorectal Cancer Research Initiative is supported by a grant from The University of Aberdeen Development Trust.

## References

1. Lawrie, L. C., Fothergill, J. E., and Murray G.I. (2001) Spot the differences: Proteomics in cancer research. *Lancet Oncol.* **2,** 270–277.
2. Lawrie, L. C. and Murray, G. I. (2002) The proteomics of colorectal cancer. *Applied Genom. Proteom.* **1,** 169–181.
3. Zou, T. T., Selaru, F. M., Xu, Y., Shustova, V., Yin, J., Mori, Y., et al. (2002) Application of cDNA microarrays to generate a molecular taxonomy capable of distinguishing between colon cancer and normal colon. *Oncogene* **21,** 4855–4862.
4. Sugiyama, Y., Sugiyama, K., Hirai, Y., Akiyama, F., and Hasumi, K. (2002) Microdissection is essential for gene expression profiling of clinically resected cancer tissues. *Am. J. Clin. Pathol.* **117,** 109–116.
5. Notterman, D. A., Alon, U., Sierk, A. J., and Levine, A. J. (2001) Transcriptional gene expression profiles of colorectal adenoma, adenocarcinoma and normal tissue examined by oligonucleotide arrays. *Cancer Res.* **61,** 3124–3130.
6. Curran, S., McKay, J. A., McLeod, H. L., and Murray, G. I. (2000) Laser capture microscopy. *J. Clin. Pathol. Mol. Pathol.* **53,** 64–68.
7. Curran, S. and Murray G. I. (2002) Tissue microdissection and its applications in pathology. *Curr. Diagnos. Pathol.* **8,** 183–192.
8. Dundas, S. R., Curran, S., and Murray, G. I. (2002) Laser capture microscopy: application to urological cancer research. *UroOncology* **2,** 33–35.
9. Craven, R. A., Totty, N., Harnden, P., Selby, P. J., and Banks, R. E. (2002) Laser capture microdissection and two-dimensional polyacrylamide gel electrophoresis: evaluation of tissue preparation and sample limitations. *Am. J. Pathol.* **160,** 815–822.
10. Lawrie, L., Curran, S., McLeod, H. L., Fothergill, J. E., and Murray, G. I. (2001) Application of laser capture microdissection and proteomics in colon cancer. *J. Clin. Pathol: Mol. Pathol.* **54,** 253–258.
11. McLeod, H. L. and Murray, G. I. (1999) Tumour markers of prognosis in colorectal cancer. *Brit. J. Cancer* **79,** 191–203.
12. Wilm, M., Shevchenko, A., Houthaeve, T., Breit, S., Schweigerer, L., Fotsis, T., and Mann, M. (1996) Femtomole sequencing of proteins from polyacrylamide gels by nano-electrospray mass spectrometry. *Nature* **379,** 466–469.

13. Shevchenko, A., Wilm, M., Vorm, O., and Mann, M. (1996) Mass spectrometric sequencing of proteins from silver stained polyacrylamide gels. *Anal. Chem.* **68,** 850–858.
14. Ritchie, H., Lawrie, L. C., Crombie, P. W., Mosesson, M. W. and Booth, N. A. (2000) Cross-linking of plasminogen activator 2 and $\alpha$2-antiplasmin to fibrino(gen). *J. Biol. Chem.* **275,** 24,915–24,920.

# 20

## Proteomic Analysis of Human Bladder Tissue Using SELDI® Approach Following Microdissection Techniques

**Rene C. Krieg, Nadine T. Gaisa, Cloud P. Paweletz, and Ruth Knuechel**

### Summary

Lysing of a complete biopsy sample results in a mixture of desired and undesired proteins, reflecting the originating cell types. Therefore microdissecting tissue material is mandatory prior to sample lysis and all downstream applications of protein analysis (proteomics). The two most important dissecting methods for bladder tissue specimens are manual microdissection and laser microdissection. Sample transfer can further be separated into manual laser pressure catapulting (LPC) and laser capture microdissection (LCM). One of the possible downstream applications of protein analysis is surface-enhanced laser desorption ionization (SELDI) time-of-flight mass spectrometry. The small quantities of tissue obtained by microdissection are sufficient for use in the SELDI technique.

**Key Words:** Proteomics; SELDI; mass spectrometry; lysis; microdissection.

### 1. Introduction

Following completion of the human genome project (*1,2*) the study of proteins is assuming much greater importance. The proteome is by definition (*3*) the totality of all proteins expressed in a cell; its analysis is called proteomics. Progress in sophisticated, sensitive, and high-throughput analytical methods is impressive, but should not distract from the importance of primary sample preparation. The inherent cellular heterogeneity of tissues is a major problem when dealing with sample preparation for proteomics (*4*). A tumor biopsy, for example, will contain—in addition to the diseased tissue—surrounding normal tissue including epithelium, endothelium, connective tissue, and infiltrating lymphocytes. Microdissection is the method of choice for addressing this problem. Regarding the complex tissue architecture, it is necessary to use consecu-

From: *Methods in Molecular Biology, vol. 293: Laser Capture Microdissection: Methods and Protocols*
Edited by: G. I. Murray and S. Curran © Humana Press Inc., Totowa, NJ

tive sections with every 10th section hematoxylin and eosin (H&E) stained for diagnostic purposes. From two consecutive H&E-stained sections the tissue architecture of the nine unstained sections in between can be estimated without generating unacceptable artifacts. Microdissection itself is a well-known tool from genomics and has been adapted slightly for proteomics. For large tissue areas manual microdissection using syringe needles for dissecting and transferring tissue is still the method of choice, while observing under a low-magnifying stereo microscope. As soon as the areas of interest become smaller, a higher-magnifying microscope is required to observe the manipulations; therefore, laser-assisted microdissection then becomes the method of choice. The two most important laser-based microdissection methods are laser pressure catapulting (LPC) *(5,6)* and laser capture microdissection (LCM) *(7,8)*. Since LCM allows for direct procurement isolation of the cells of interest, LCM seems to have become the preferred tool for proteomics.

SELDI® mass spectrometry has established itself as a sensitive, high-throughput method for protein analysis. As the obtained protein patterns are highly reproducible, this method is nowadays one of the leading technologies in screening samples for the presence of a specific protein, e.g., a tumor marker. Finding such a specific protein will enable enhancement of current diagnosis and may even lead to new therapeutic approaches.

## 2. Materials

1. Snap-frozen normal urothelium and pTaG1 papillary bladder tumor biopsies.
2. Storage equipment: dry ice, –80°C freezer.
3. KILLIK frozen section medium (Bio-optica, Milano, Italy).
4. Cryostat.
5. Glass slides (Engelbrecht, Edermuende, Germany).
6. PEN-polyethylene-membrane, 1.35 mm thick (P.A.L.M. Microlaser Technologies, Bernried, Germany).
7. Adhesive tape (Tesa AG, Hamburg, Germany).
8. 50-mL polypropylene centrifuge tubes.
9. Ethanol 100%, 90%, 70%.
10. Ultrapure water ("Millipore water").
11. Mayer's hematoxylin solution (Sigma Diagnostics, St. Louis, MO).
12. $NaHCO_3$.
13. Xylene.
14. Complete™ Protease inhibitors cocktail tablets (Roche Diagnostics).
15. Nonidet NP40 10% aqueous solution (Roche Diagnostics).
16. Dulbeco's Phosphate-Buffered Saline (PBS, GIBCO™ Invitrogen Corporation, Grand Island, NY).
17. Stereo microscope 40X.
18. Microlancets (Microlance3, Becton Dickinson).
19. Arcturus® PixCell II (Arcturus, Mountain View, CA, USA).

20. CapSure™ LCM caps (Arcturus).
21. Safe-Lock-Tubes 2,0 mL (Eppendorf AG, Hamburg, Germany).
22. PALM® MicroBeam (P.A.L.M.).
23. PCR tubes PCR-02D-C 0.2 mL (Axygen Scientific, Inc., Union City, CA).
24. AdhesiveCaps (P.A.L.M.).
25. Proteinchip Biology System II (Ciphergen, Fremont, CA).
26. SAX2 Proteinchip Array (Ciphergen).
27. WCX2 Proteinchip Array (Ciphergen).
28. IMAC3 Proteinchip Array (Ciphergen).
29. Bioprocessor 96-Well (Ciphergen).
30. Bioprocessor accessory, 96-well disposable cartridge and gasket (Ciphergen).
31. 200-µL 8-channel pipettor (e.g., Eppendorf, Eppendorf AG, Hamburg, Germany).
32. 10 m$M$ HCl.
33. Autoclaved ultrapure water ("Millipore water").
34. 10 m$M$ ammonium acetate.
35. 50 m$M$ nickel sulfate.
36. 0.1% Triton X100 in PBS.
37. Plastic lid of a pipet tip tray.
38. Vacuum device.
39. Trifluoroacetic acid.
40. Acetonitril (Sigma, HPLC grade).
41. Cinnamic acid (Sigma).

## 3. Methods

The subsequent descriptions are divided into (1) sample acquisition, (2) sample preparation and treatment, (3) performing microdissection, (4) sample lysis, and (5) SELDI analysis.

### 3.1. Sample Acquisition

All biopsies are taken within an ethically approved prospective study on photodynamic diagnosis with 5-aminolevulinic acid *(9,10)* for bladder cancer. The study involves consequent molecular analysis of biopsy material. Visibly manifest tumors, suspicious areas, and normal urothelium were collected endoscopically with biopsy forceps and snap-frozen in liquid nitrogen immediately thereafter. For subsequent sample treatment the tissue was stored at –80°C (long-term) or on dry ice (short-term).

### 3.2. Sample Preparation and Treatment

#### 3.2.1. Sectioning

The frozen tissue pieces were embedded with KILLIK frozen section medium and cut into 5-µm sections with a routine cryostat, using the recommended D-shaped knife.

### 3.2.2. Tissue Preparation

For (1) manual microdissection, (2) laser capture microdissection (LCM) with Arcturus PixCell II, and (3) laser microdissection with PALM using software function "auto laser pressure catapulting (autoLPC)," the frozen sections can be applied directly onto common glass slides using routine handling procedures.

If using PALM software function "LPC," the slides must be prepared in a special way (*see* **Note 1**):

1. Trim the PEN-membrane to cover 75% of the clear area of a slide with a sterile scalpel.
2. Immerse the slides into 100% ethanol for 2–3 min to clean and degrease the slides.
3. Apply the PEN-membrane onto the wet slides.
4. Fix the film by gluing with adhesive tape at the edges.
5. Apply the frozen section onto the film.

### 3.2.3. Staining

Every tissue was sectioned in a consecutive sequence. The first section was H&E-stained, embedded with a mounting medium, and furnished with a cover slip using routine protocols *(11,12)* to ensure diagnosis by a pathologist. The regions of interest (ROIs) were then marked under microscopic surveillance using a permanent marker with a fine tip. These slides were used as a template to enhance orientation on the slides for actual microdissection.

The slides for actual microdissection were fitted with the consecutive sections and stained according to this protocol (*see* **Note 2**):

1. 70% Ethanol (30 s).
2. Millipore water (10 s).
3. Mayer's hematoxylin solution (30 s).
4. Millipore water (10 s).
5. Aqueous solution of $NaHCO_3$ (0.5%; 10 s).
6. 70% Ethanol (30 s).

To all of the solutions above the Complete protease inhibitor was added into every other solution with the concentration according to the manufacturer's recommendation (*see* **Note 3**).

The prepared slides are now ready for performing manual microdissection and microdissection with PALM. Until microdissection the slides are stored in tap water (with Complete protease inhibitor) to prevent desiccation. On the other hand for LCM with Arcturus PixCell II the slides have to be drained, dried, and desiccated after the 70% ethanol step according to the following protocol (*see* **Note 4**):

1. 90% Ethanol (30 s).
2. 100% Ethanol (30 s).

3. Xylene (30 s).
4. Xylene (30 s).

### 3.3. Microdissection

Lysis buffer for immediate SELDI application is a 1% solution of Nonidet NP40 in PBS (*see* **Note 5**).

### 3.3.1. Manual Microdissection

The hematoxylin-stained slide is removed from the storage water and processed immediately. Under the stereomicroscope with overall 40× magnification the ROIs are microdissected using careful scraping movements with a sterile needle (*see* **Note 6**). With this needle the small tissue pieces are transferred into a 200-µL PCR tube, which contains 12 µL lysis buffer (*see* **Note 7**). Samples are then stored at –80°C.

### 3.3.2. LCM with PixCell II, Arcturus

#### 3.3.2.1. DESIGN OF PIXCELL II-LCM DEVICE

1. Inverted microscope (Olympus IX50, modified by Arcturus).
2. X-Y stage with manual joystick control.
3. Pulsed IR laser, solid-state laser diode (wavelength 810 nm, power output max 100 mW) with three manual interchangeable predefined laser spot sizes (7.5, 15, 30 µm).
4. LCM unit with cap insertion tool.
5. Interface for complete control over laser functions (pulse duration, output power).
6. Color CCD camera (0.5" chip), attached to microscope.
7. PC for software-driven control and performance of the LCM.

#### 3.3.2.2. PRINCIPLE OF LCM WITH PIXCELL II

The pulsed IR laser is attached and collimated with an inverted microscope, replacing its regular transmitted light source. The laser beam is focused by a three-step optic to a manual interchangeable spot size between 7.5 and 30 µm. A special cap (CapSure™ LCM Cap) is positioned directly onto the stained section. This cap is covered with a thin ethylene vinyl acetate polymer sheet with low melting temperature capabilities. The cap is designed to fit as a lid for a routine Eppendorf centrifuge cup. The ROIs are focused and irradiated with the low-power IR laser beam. Each pulse of the laser melts the thermoplastic film within the predefined laser spot size (7.5, 15, or 30 µm) and glues the tissue onto the film. When lifting the cap from the slide, the selected tissue remains attached and is so separated from surrounding connective tissue and captured for further analysis. The whole setup is software-controlled, allowing the adjustment of the laser parameters.

### 3.3.2.3. Performing LCM With PixCell II

The slide is put onto the precentered manual stage and is moved manually until the ROI is visible. The slide is then fixed to the stage by applying a vacuum. Subsequently fine movements to center the ROI and to perform the LCM movements are carried out using the X-Y stage. The CapSure caps are loaded into their dovetail assembly unit and one of them is picked up, moved to the work area on the slide, and lowered onto the section. The laser is turned on and the typical settings as needed to achieve melting (power: 40 mW; duration: 6.00 ms; repeat: 0.2 ms; spot size: 30 μm; target: 300 mV; current: 4.7/4.8 mA; temperature: 21.5°C) are adjusted. After location of the ROI the laser is fired, moved slightly and fired again until the whole work area is processed (*see* **Note 8**). When complete, the cap is removed from the cryostat section, transferred to an Eppendorf tube, and stored on dry ice or at –80°C (*see* **Note 9**).

### 3.3.3. Laser Microbeam Microdissection With PALM

#### 3.3.3.1. Design of the PALM-LMM Device

1. Inverted microscope (Axiovert 135, Zeiss).
2. PALM Robot-Stage (computer-controlled X-Y stage).
3. Pulsed air-cooled UV-nitrogen laser (wavelength 337 nm, pulse energy 270 nJ/pulse).
4. Interface for complete control over laser functions (continuously adjustable focus, pulse adjustment, output power).
5. LPC capture unit.
6. Color CCD camera (0.5" chip), attached to microscope.
7. PC for software-driven control and performance of the LMM.
8. Unit mounted on a vibration-isolated stone tabletop.

#### 3.3.3.2. Principle of LMM With PALM

A pulsed UV laser is attached and collimated with an inverted microscope at its port of the fluorescence excitation light source. The laser beam is focused using the original microscope's objective to a minimal beam spot size of less than 1 μm. The high photon density of the laser enables tissue dissection by means of locally restricted ablative photodecomposition without heating. After an internal defocusing of the laser the power is too low to cut tissue, but the pure pressure of photons is high enough to move ("shoot") small tissue pieces against gravity. This phenomenon (laser pressure catapulting [LPC]) is used for transferring ready-cut samples into a collecting cap. The whole setup is coupled with software (PALM RoboSoftware) combining drawing software and a control unit, which allows the use of different laser software features.

### 3.3.3.3. LMM With Manual Transfer

For this kind of LMM only the feature "CUT" is used. ROIs are encircled with one of the line-drawing tools. When using the "CUT" feature the laser cuts along predefined lines. The cutting itself is dependent on predefined laser focus and laser energy. The cut pieces are then manually picked up with a sterile needle and transferred into a 200-µL PCR tube containing 12 µL lysis buffer as above.

### 3.3.3.4. AutoLPC Into Buffer or Oil

The ROIs are encircled with a line drawing tool, and using the "autoLPC" feature the whole predefined area is covered by laser shots. The density of laser shots (number of shots per µm$^2$) can be predefined in the setup menu. Each shot transfers a small piece of tissue solely by photon pressure rather than by cutting. In order to harvest the material a capture cap is filled with 3 µL lysis buffer or mineral oil and is inserted into the LPC capture unit above the specimen.

### 3.3.3.5. LPC Into Oil/AdhesiveCap After LMM

For this feature the section has to be mounted onto a PEN-membrane-coated slide. After encircling, the "CUT" function is used to cut the outer border of the ROI, while leaving a tiny bridge between the ROI and the surrounding tissue, between the start and stop positions of the initial laser cut. A few single laser shots are then positioned at this catwalk for catapulting the whole area, with the membrane attached, into a prepared cap located in the capture unit. The capture caps are filled with 3 µL mineral oil or lysis buffer, respectively. In addition, special AdhesiveCaps (P.A.L.M.) with a self-adhesive surface can be used; therefore the usage of oil can be omitted.

### 3.3.3.6. Performing LMM With PALM

The slide is positioned in the holding device, specimen facing upward. With the graphic tools all the ROIs are marked. Depending on the type of application, a capture cap is prepared, inserted into the capture unit, and positioned just above the slide. The desired laser feature is chosen and applied. When the LCM is completed, the capture cap is removed from the capture unit, transferred to an Eppendorf cup, and stored on dry ice or at –80°C (*see* **Note 10**).

### *3.4. Sample Lysis*

The samples are put on wet ice and thawed. Over a period of 1 h they are vortexed several times and spun down to ensure maximum protein recovery. Pipetting the whole small volume several times is a good alternative but has to be undertaken carefully to minimize loss of material and volume (*see* **Note 11**). Finally the sample is spun down, to separate supernatant from the remaining tissue-debris (*see* **Note 12**).

### 3.5. SELDI Analysis

### 3.5.1. Design of the Ciphergen SELDI Device

1. Time-of-flight mass spectrometer (TOF MS), tube length 0.8 m.
2. Pulsed UV-nitrogen laser (337 nm, max 150 mJ).
3. High vacuum supply.
4. Chip loading interface.
5. PC for software-driven control of the laser parameters and performing the TOF MS.
6. Unit mounted in an under-the-bench steel case.

### 3.5.2. Principle of SELDI

Matrix-assisted laser desorption ionization (MALDI) mass spectrometry allows the analysis of high molecular ("biologic") molecules by mass spectrometry. MALDI deals with a "soft" UV laser irradiation following cocrystallization with an energy-absorbing molecule (EAM), called the matrix. Therefore the high-energy ionization with an electron beam is avoided. Combining MALDI with retentive surface chromatography of the initial carrier material itself gives the highly versatile surface-enhanced laser desorption ionization (SELDI) method. Miniaturizing the whole experiment down to 1 µL sample volume per experiment results in the protein chip arrays with chromatographic active surfaces, thus reproducibly capturing proteins according to their specific chemistry out of the whole lysate. Only a fraction of all proteins is analyzed by mass spectrometry, thus minimizing overloading effects as well as ion signal suppression and possibly enhancing proton ionization efficiency. The resulting peaks, according to protein-specific mass-over-charge values, represent a characteristic fingerprint of the original lysate and can be used successfully to spot disease-related protein patterns by means of heuristic self-learning pattern detection algorithms *(13)*.

### 3.5.3. Chip Preparation

According to the desired chip surface chemistry a different chip preparation protocol is required. All steps are performed in 50-mL tubes on an appropriate orbital shaker.

For WCX2:

1. Wash the chips in 10 m*M* HCl for 5 min.
2. Wash the chips two times with autoclaved water for 1 min each.
3. Wash the chips two times with 10 m*M* ammonium acetate for 5 min each.

For SAX2:

Wash the chips two times with 10 m$M$ ammonium acetate for 5 min each.

For IMAC3:

1. Wash the chips two times with 50 m$M$ nickel sulfate for 5 min each.
2. Wash the chips two times with autoclaved water for 1 min each.
3. Wash the chips two times with 0.1% Triton X100 in PBS for 5 min each.
4. Let all chips dry and place them into the bioprocessor. Allow the chips to dry for about 2 min, until no more liquid is connecting the single spots.

### 3.5.4. Sample Application

Apply 1 μL of sample onto each spot and attach the disposable cartridge and gasket. Cover the whole unit with an appropriate lid (e.g., plastic lid of a pipet tip tray). Place a precisely fitting damp paper tissue into the lid first. Set the whole unit onto the orbital shaker and let incubate for 20 min. Using the 8-channel pipettor, wash every well three times with PBS and three times with autoclaved water using 200 μL, each. Use a vacuum device (*see* **Note 13**) to dry each well completely before removing the cartridge and gasket from the bioprocessor. Allow the chips to air dry completely. In the meantime prepare the EAM solution (*see* **Note 14**):

1. 2500 μL autoclaved water.
2. Add 25 μL trifluoroacetic acid.
3. Vortex briefly.
4. Add 2475 μL acetonitrile.
5. Vortex briefly.

and the matrix solution (*see* **Note 15**):

1. Measure 7 mg of cinnamic acid.
2. Add 200 μL of EAM solution.
3. Vortex.
4. Centrifuge at 15,000$g$ for 1 min.
5. Supernatant is the readymade matrix solution.

Then apply 0.8 μL of matrix solution to each spot twice. Allow the spots to dry before performing the second round (*see* **Note 16**).

### 3.5.5. Performing SELDI Analysis

Insert the chip into the Proteinchip Biology System II, following the manufacturer's instructions. Using the controlling software, measure the chip. Instrument settings have to be optimized for each sample. A good starting point is the following spot protocol (screen shot):

1. Set high mass to 20,000 Dalton, optimized from 1000 Dalton to 20,000 Dalton.

2. Starting laser intensity 220.
3. Detector sensitivity 8.
4. Focus mass at 6000 Dalton.
5. Set SELDI acquisition parameters 20 delta to 5 transients per to 15 ending position to 80.
6. Set warming positions with 2 shots at intensity 155 and do not include warming shots.

Analysis of spectra is performed with the Ciphergen software features "Comparison Wizard" and "Biomarker Wizard" (*see* **Note 17**).

## 4. Notes

1. P.A.L.M. does offer pre-prepared slides. These slides are already covered with the plastic film described; therefore, handling is simplified. In addition the reproducibility is enhanced, as the tension of the sheet is crucial for successful microdissection.
2. Hematoxylin is the best staining technique for this purpose. As hematoxylin is a dye specific for chromatin it stains nuclei exclusively and no proteins. The last detail is important as a dye stains proteins by modification or attachment. In both cases the protein itself is altered or shows at least a different mass-over-charge value in the mass spectrum. For this purpose the routine counterstaining with eosin is omitted. We have tried several other microscopic dyes (e.g., methylene blue, toluidine blue) vs no staining at all (native tissue) (data not shown), but according to efficiency of protein recovery, protein integrity, and visibility during microdissection, hematoxylin is the first choice.
3. Even if a protease inhibitor is present, a time frame of 30 min for the overall procedure (first thawing of the cryostat section until final sample freezing) should not be exceeded in order to maintain sufficient protein integrity. Due to autolysis and protein degradation, peak intensities will decrease and new peaks (degradation products) will occur (data not shown).
4. Complete sample dehydration is mandatory to obtain good microdissection results for LCM, especially regarding sample transfer. For this reason the two steps with xylene follow 100% ethanol. To prevent the 100% ethanol stock bottle from absorbing moisture add a little dehydrated copper sulfate into the bottle. If the copper sulfate turns color from gray to blue the ethanol is no longer absolute.
5. The lysis buffer contains Nonidet NP40 (10% stock solution) and phosphate-buffered saline (PBS) 1:10, giving a final concentration of 1% Nonidet NP40 in PBS. The stock solution must be refrigerated. The ready-to-use dilution should be stored on ice until use and should be discarded after 12 h. Previous experiments have shown the 1% solution to be best for lysis. Ciphergen recommends a different lysing buffer containing 8 *M* urea. This particular buffer has given an insufficient lysing efficiency for our samples. In addition one has to work at room temperature, as the buffer crystallizes on wet ice.
6. A minimum of 15,000 cells is strongly recommended.

7. We have tried other cups prior to the ones now listed in the materials section. We once favored the use of 500-μL Sarstedt screw cups.

   However, cup geometry is not perfect for handling the smallest volumes (compare to **Note 12**) and the plastic material used by the manufacturer tends to gain electrostatic voltage, thus complicating sample transfer into the 12-μL buffer in the bottom of the cup. The 12-μL buffer acts as a collecting main (*see* **Note 12**), thus enabling an easier transfer from the dissecting needle into the cup. In addition the immediate sample lysis improves protein recovery and protein integrity as proven in own experiments (data not shown).

8. A minimum of 5000 shots is strongly recommended.

9. After removing the cap from the section not only the microdissected ("glued") areas remain, but in addition some unwanted material may stick to the transfer film. Getting rid of these cells is important for obtaining pure fractions. Thus we recommend pressing the transfer film side of the cap onto the gluing area of a regular Post-it note. The microdissected areas will remain unaffected, whereas all other material is removed.

10. For the biopsy material as described we recommend manual microdissection. Tissue architecture (papillary tumors, exophytic papillary tumors) with comparative large ROIs shows that a manual technique is best regarding labor time vs cell numbers dissected. In addition, the manual methods demonstrate the highest protein recovery during lysis. The PALM autoLPC function seems to suffer from material loss while catapulting into the collecting cap. Perhaps grounding of the cap (electrostatic voltage) can significantly improve transfer efficiency. P.A.L.M. has detected this problem already and using the company's pre-prepared slides (*see* **Note 1**) increases transfer efficiency. If using the special AdhesiveCaps, these problems are significantly minimized.

11. Some tissue material seems to be hard to lyse; thus vortexing alone is not successful. We then recommend rigid pipetting of the whole volume. It is important to avoid the buildup of foam, as this will lead to a loss of material. In addition we emphasize the use of "high recovery pipet tips" (e.g., sterile aerosol pipet tip, Labcon, San Rafael, CA).

12. Most of the microdissected sections seem to remain unaffected by lysing. This debris should not find its way onto the chip surface. Therefore centrifugation is mandatory. For this step it is necessary to use a really small cup with a spiky cone end, as there is very little supernatant, which needs to be transferred by a pipet tip afterward. This dictates the amount of collecting volume as a too little volume will result in an insufficient supernatant. Trying to reduce the number of cells, which are necessary for downstream application, is possible by reducing the volume of lysing buffer/collecting gain. But one reaches very soon the limits of proper manageability.

13. We connect a regular tubing to the vacuum faucet at the bench. We insert the large end of a 1000-μL pipet tip into the tubing. The small end is cut so that the resulting diameter is small enough to be inserted into a single well but too large to touch the spot surface.

14. The ready-made EAM solution can be stored for 1 wk at room temperature in the dark.
15. Store cinnamic acid in the dark at –20°C. The matrix solution must be used the same day.
16. After complete drying the chips can be measured. However it may be necessary to store them until final analysis. We recommend vacuum sealing into aluminized bags and adding a capsule with desiccant. Our own experience proved this storage to be good for at least 5 d. For longer storage until measurement or a repeated measurement, one might consider the reapplication of matrix solution. It should be borne in mind that the ratio of protein to matrix cannot be significantly altered without influencing the mass spectrum.
17. These tools are the next level of spectra analysis after observation with the naked eye. For a high-resolution approach and especially when analyzing large case numbers we strongly recommend the use of a more sophisticated detection algorithm.

## Acknowledgments

The authors would like to thank Drs. Lance A. Liotta and Emanuel F. Petricoin III for generously supporting this study.

## References

1. Venter, J. C., Adams, M. D., Myers, E. W., Li, P. W., Mural, R. J., SuHon, G. G., et al. (2001) The sequence of the human genome. *Science* **291,** 1304–1351.
2. McPherson, J. D., Marra, M., Hillier, L., Waterston, R. H., Chinwalla, A., et al. (2001) A physical map of the human genome. *Nature* **409,** 934–941.
3. Wasinger, V. C., Cordwell, S. J., Cerpa-Poljak, A., Yan, J.X., Gooley, A.A., Wilkins, M.R., et al. (1995) progress with gene-product mapping of the mollicutes: *mycoplasma genitalium*. *Electrophoresis* **16,** 1090–1094.
4. Krieg, R. C., Paweletz, C. P., Liotta, L. A., and Petricoin, E. F. (2002) Clinical proteomics for cancer biomarker discovery and therapeutic targeting technology. *Cancer Research and Treatment* **1,** 1–10.
5. Srinivasan, R. (1986) Ablation of polymers and biological tissue by ultraviolet lasers. *Science* **234,** 559–565.
6. Schutze, K. and Clement-Sengewald, A. (1994) Catch and move—cut or fuse. *Nature* **368,** 667–669.
7. Emmert-Buck, M. R., Bonner, R. F., Smith, P. D., Chuaqui, R. F., Zhuang, Z., Goldstein, S. R., et al. (1996) Laser capture microdissection. *Science* **274,** 998–1001.
8. Bonner, R. F., Emmert-Buck, M. R., Cole, K., Pohida, T., Chuaqui, R., Goldstein, S., and Liotta, L. A. (1997) Laser capture microdissection: molecular analysis of tissue. *Science* **278,** 1481–1483.
9. Zaak, D., Kriegmair, M., Stepp, H., Stepp, H., Baumgartner, R., Oberneder, R., et al. (2001) Endoscopic detection of transitional cell carcinoma with 5-amino-levulinic acid: results of 1012 fluorescence endoscopies. *Urology* **57,** 690–694.

10. Kriegmair, M., Baumgartner, R., Lumper, W., Waidelich, R., and Hofstetter, A. (1996) Early clinical experience with 5-aminolevulinic acid for the photodynamic therapy of superficial bladder cancer. *Br. J. Urol.* **77,** 667–671.

11. Messmann, H., Knüchel, R., Bäumler, W., Holstege, A., and Schölmerich, J. (1999) Endoscopic fluorescence detection of dysplasia in patients with Barrett's esophagus, ulcerative colitis, or adenomatous polyps after 5-aminolevulinic acid-induced protoporphyrin IX sensitization. *Gastrointest. Endosc.* **49,** 97–101.

12. Burck, H. C. (ed.) (1988) *Histologische Technik.* Georg Thieme Verlag, Stuttgart.

13. Romeis, B. and Böck, P. (ed.) (1989) *Mikroskopische Technik.* Urban & Schwarzenberg, München.

14. Petricoin, E. F., Ardekani, A. M., Hitt, B. A., Levine, P. J., Fusaro, V. A., Steinberg, S. M., et al. (2002) Use of proteomic patterns in serum to identify ovarian cancer. *Lancet* **359,** 572–577.

# V

## MICRODISSECTION AND MOLECULAR ANALYSIS OF MICROORGANISMS

# Genetic Analysis of HIV by *In Situ* PCR-Directed Laser Capture Microscopy of Infected Cells

## Daniele Marras

### Summary

Behind the exponential expansion of the acquired immunodeficiency syndrome (AIDS) epidemic, there is a continuous and progressive molecular evolution of human immunodeficiency virus (HIV)-1. In this regard, the molecular analysis of viral strains infecting several anatomic compartments in humans has become critical to understanding AIDS-related pathologies and to improving emerging therapeutic protocols. Laser capture microdissection provides outstanding results in the genetic analysis of HIV-1 variants detectable in AIDS patients. The ability of the instrument to microdissect infected cells from a heterogeneous tissue compartment allows the investigator to obtain critical information regarding the genetic nature of a specific viral strain. To perform laser capture microdissection with better accuracy, *a priori* detection techniques may provide useful information about HIV distribution in the tissue specimen. An *in situ* polymerase chain reaction (PCR) assay on a serial slide results in a detailed map of the viral infection specific for the case under analysis. The knowledge of HIV distribution in the tissue section is critical for improving the dissection of infected cells by laser capture microscopy. This chapter describes laser capture microdissection and *in situ* PCR and its role in the analysis of the genetic nature of HIV-1 variants and quasispecies.

**Key Words:** HIV-1 envelope; HIV-associated nephropathy; *in situ* PCR; laser capture microdissection.

## 1. Introduction

Great strides have been made in developing potent antiretroviral regimens that suppress human immunodeficiency virus-1 (HIV-1) replication in infected persons *(1)*. Despite these therapeutic advances, major obstacles remain in eradicating HIV-1. Reservoirs of HIV-1 have been identified that represent major impediments to eradication *(2)*. In the last decade the virus dramatically

From: *Methods in Molecular Biology, vol. 293: Laser Capture Microdissection: Methods and Protocols*
Edited by: G. I. Murray and S. Curran © Humana Press Inc., Totowa, NJ

expanded its spectrum of infection, revealing an increased ability to infect unexpected cell targets. Besides the already known reservoirs, such as resting T cells *(3)*, the virus has been recently detected in intestinal epithelial cells *(4)*, in the placental trophoblastic layer *(5)*, and in renal tubules *(6)*.

The analysis of mutated viral strains is critical to understanding the viral evolution and the nature of acquired immunodeficiency syndrome (AIDS)-related pathologies. The tandem *in situ* polymerase chain reaction—laser capture microdissection (PCR-LCM) represents a very elegant methodology to obtain infected cells from tissue specimens without inadvertent admixture of adjacent unwanted cells. The approach of using *in situ* PCR-guided LCM of infected cells, HIV-1 nucleic acid extraction, DNA amplification, and analysis is described. This protocol was created during experiments performed on HIV-1-infected renal tubular cells of HIV-1-associated nephropathy patients.

## 2. Materials

### 2.1. Tissue Preparation

1. 4% Formaldehyde, Polysciences.
2. 10X phosphate-buffered saline (PBS) (Fluka).
3. Coplin jars (PBi International).
4. 100% Ethanol (Sigma Aldrich).
5. Xylene (Sigma Aldrich).
6. Paraffin wax.
7. Microtome blades.
8. Plain uncoated glass slides (PBi International).
9. Poly-L-lysine coated glass slides (PBi International).
10. Hematoxylin (Sigma Aldrich).
11. Eosin (Sigma Aldrich).

### 2.2. In Situ PCR

1. Omnislide *in situ* PCR System (Hybaid).
2. Proteinase K (Sigma Aldrich).
3. 10 m$M$ dNTPs mix (Invitrogen).
4. Klenow enzyme (Invitrogen).
5. Taq (Invitrogen).
6. DIG dUTP (Roche Biochemicals).
7. FRAME-SEAL (MJ Research, Inc).
8. 20X SSC buffer (Sigma Aldrich).
9. Sheep serum (Serotec Inc).
10. Daimont pen (PBi International).
11. DIG DNA labeling and detection kit (Roche Biochemicals).
12. Levamisole (Sigma Aldrich).
13. V3 forward: 5'-TGT CCA AAG GTA TCC TTT GAG CCA ATT CC–3'.

14. V3 reverse: 5'-AGT AGA AAA ATT CCC CTC CAC AAT TAA–3'.
15. β Globin forward: 5'-ACA CAA CTG TGT TCA CTA GC–3'.
16. β Globin reverse: 5'-CAA CTT CAT CCA CGT TCA CC–3'.

## 2.3. Laser Capture Microdissection

1. LCM PixCell II (Arcturus Engineering, Mountain View, CA).
2. CapSure LCM caps (Arcturus Engineering).
3. PrepStrip tissue preparation strip (Arcturus).
4. 0.5-mL safe-lock tubes (Eppendorf).
5. DNeasy tissue kit (Qiagen).

## 2.4. PCR, DNA Cloning, and Sequencing

1. Elongase enzyme mix (Invitrogen).
2. ES1: 5'-GAC TAA TAG AAA GAG CAG AAG ACA GTG GCA–3'.
3. ES2: 5'-AGT GCT TCC TGC TGC TCC CAA GAA CCC–3'.
4. IN1: 5'-ATG AGA GTG AAG GAG AAA TAT CAG CAC–3'.
5. IN2: 5'- GAA CAA AGC TCC TAT TCC CAC TGC TCT–3'.
6. ES73 5'-GAC TAA TAG AAA GAG CAG AAG ACA GTG GCA-3'.
7. ES75 5'-AGT GCT TCC TGC TGC TCC CAA GAA CCC-3'.
8. IN 74 5'-ATG AGA GTG AAG GAG AAA TAT CAG CAC-3'.
9. IN 76 5'-GAA CAA AGC TCC TAT TCC CAC TGC TCT-3'.
10. V3env s 5'-TGT CCA AAG GTA TCC TTT GAG CCA ATT CC-3'.
11. V3env as 5'-AGT AGA AAA ATT CCC CTC CAC AAT TAA-3'.
12. PCR Optimizer Kit (Invitrogen).
13. QIAquick PCR purification kit (Qiagen).
14. PGEM-T Easy Vector System (Promega).
15. ElectroMAX DH10B competent cells (Invitrogen).

## 2.5. Solutions

1. 0.1 $M$ Tris-HCl, 50 m$M$ EDTA, pH 8.0.
2. 0.1 $M$ Glycine/PBS.
3. 12.11 g Tris-HCl, 5.84 g NaCl, 0.4 g $MgCl_2$, 30 g BSA, 1 L distilled water (store at 4°C).
4. 100 m$M$ Tris-HCl, pH 9.5, 100 m$M$ NaCl.

# 3. Methods
## 3.1. Tissue Preparation
### 3.1.1. Gentle Tissue Fixation

To preserve the nucleic acid population, harvest fresh tissue, fix immediately and proceed directly to dehydration and embedding *(7,8)* (*see* **Note 1**).

1. Fix tissue at 4°C. Fixation times varies depending on the size of the tissue: 1 mm$^3$ (biopsy sample): 20–30 min; 1 cm$^3$ (mouse sample/autopsies) 3–4 h; Rinse briefly in cold 1X PBS; Tissue can be stored for a few days at 4°C in 1X PBS.

2. For tissue dehydration, perform all steps at room temperature.
Place tissue pieces in tissue-processing cassettes.

    Graded ethanol series:    15 min each with gentle shaking:

| | |
|---|---|
| 30% Ethanol | 15 min |
| 50% Ethanol | 15 min—if necessary, stop here and store at 4°C overnight |
| 75% Ethanol | 15 min |
| 80% Ethanol | 15 min |
| 95% Ethanol | 2 × 15 min |
| 100% Ethanol | 2 × 15 min |
| Xylene 30 min | 2 × 15 min |

    (*See* **Note 2**.)

### 3.1.2. Paraffin Embedding

1. Dip cassettes into liquid paraffin (prewarmed at 60°C) to remove the excess xylene and then transfer to fresh paraffin.
2. Keep blocks in paraffin at 60°C, under vacuum, for 2 h (*see* **Note 3**).
3. Cast paraffin blocks.
4. After the second hour of infiltration, remove tissue from the cassette and place it on a heated surface. Fill casting tray with liquid paraffin, arrange tissue in tray, and place cassette on top of casting tray.
5. Do not attempt to remove blocks until completely cooled to –20°C. Blocks are then ready to use or can be stored at –80°C.

### 3.1.3. Tissue Section Cutting

1. Remove the block from the –80°C freezer and transfer it to a small container with dry ice.
2. Carefully clean the microtome workstation.
3. Wash the bath and fill it with an appropriate volume of acidified water (50 μL of HCl per liter).
4. Set the bath temperature between 38 and 45°C.
5. Insert a new blade in the microtome (*see* **Note 4**).
6. Attach the block to the chuck in the microtome. The cutting surface should be as parallel as possible to the blade (*see* **Note 5**).
7. Cut 5-μm sections and place them in the warm bath.
8. Transfer sections onto poly-L-lysine-coated glass slides for *in situ* PCR and onto plain uncoated glass slides for LCM (*see* **Note 6**).
9. Incubate prepared slides at 37°C for at least 2 h (*see* **Note 7**).
10. Slides can now be stored at –80°C.

### 3.1.4. Section Staining

This procedure is necessary for sections to be used for LCM. These sections are mounted on uncoated slides. Sections for *in situ* PCR remain unstained.

1. Take slides out of the –80°C freezer and keep at room temperature for 60 s.

2. Stain sections with hematoxylin and eosin (*see* **Notes 8–10**):

| | |
|---|---|
| Xylene | 8 min |
| 100% Ethanol | 30 s |
| 95% Ethanol | 30 s |
| 70% Ethanol | 30 s |
| Distilled water | 30 s |
| Hematoxylin | 30 s |
| Distilled water | 30 s |
| Bluing solution | 30 s |
| 70% Ethanol | 30 s |
| 95% Ethanol | 30 s |
| Eosin | 1 min |
| 70% Ethanol | 30 s |
| 95% Ethanol | 30 s |
| 100% Ethanol | $2 \times 30$ s |
| Xylene | 5 min |
| Air-dry | 20 min |

### 3.2. In Situ *PCR*

For all sample types and all target sequences, the following steps are necessary for any PCR hybridization procedure *(6,9)*. The purpose of the following protocol is to obtain a detailed map of the HIV-1 infection specific to the specimen being analyzed. In this regard, the PCR is performed using primers for the hypervariable region V3 within the gp120 envelope protein. The resulting amplicons reveal the localization of the virus (**Fig. 1**).

### 3.2.1. Preparation of Tissue for In Situ *PCR*

1. Take slides out of the –80°C freezer and keep at room temperature for 60 s.
2. Dip slides sequentially in:

| | |
|---|---|
| Xylene | 30 min |
| 100% Ethanol | $2 \times 5$ min |
| 95% Ethanol | 5 min |
| PBS | 15 min |

### 3.2.2. Proteinase K Treatment

This is the most critical step in the entire procedure.

1. Prepare proteinase K solution by dissolving 10 μg of Proteinase K per mL of solution 1.
2. Gently cover the tissue section with the proteinase K solution and incubate at 37°C for 15 min (*see* **Note 11**).
3. Fill a slide cassette with solution 2 and dip slides for 5 min.
4. Wash in PBS for 5 min (three times).

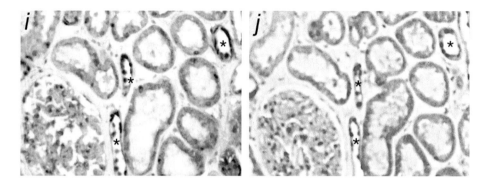

Fig. 1. *In situ* PCRs on a renal specimen for HIV-1 nucleic acid detection. To examine the virus distribution specific for the specimen being analyzed, it is useful to perform distinct *in situ* PCR experiments on serial slides using two primer sets specific for different viral sequences. In panel (i) the experiment was performed with primers specific for the V3 domain within HIV-1 gp120, whereas panel (j) reports the result obtained with primers for the viral circular forms. The similarity of the signal distribution in both conditions (highlighted with ✱) supports the efficiency of the technique.

### 3.2.3. Klenow Digestion

This step (optional but recomended) allows DNA repair after the proteinase K digestion. This step will increase the efficiency of DNA amplification.

1. For each slide prepare the following solution: 5 µL Buffer M; 0.5 µL 1 m$M$ dNTPs; 1 µL Klenow; and 43.5 µL Distilled water (*see* **Note 12**).
2. Pipet 50 µL directly onto the tissue section and incubate 30 min at 37°C.
3. Briefly dip slides in distilled water (*see* **Note 13**).

### 3.2.4. PCR Amplification Protocol

1. Prepare the following PCR Reaction mix:
   - 10 µL 10X Buffer
   - 6 µL 25 m$M$ MgCl$_2$
   - 0.5 µL 3 m$M$ DIG dUTP (final conc. 0.03 m$M$)
   - 2 µL 10 m$M$ dNTPs
   - 2 µL 100 ng/mL V3 forward primer
   - 2 µL 100 ng/mL V3 reverse primer
   - 1 µL Taq
   - 25 µL Distilled water

   Total: 50 µL

2. Set the plate temperature at 70°C and place the reaction mix tube on the plate.
3. Dry the glass surface around the tissue.

4. Fasten the adhesive-backed frame to the slide around the specimen area.
5. Place the slide on the 70°C heated plate.
6. Remove the upper layer covering the frame and pipet the reaction mix in one corner of the framed surface.
7. Carefully place the cover slip on the sticky frame starting from the corner where the mix was pipetted and then over the entire surface (*see* **Note 14**).
8. Run the following PCR cycle:

| 95°C | 2 min | × 1 |
| 95°C | 1 min | |
| 68°C | 30 s | × 35 |
| 72°C | 1 min | |
| 72°C | 5 min | × 1 |

## *3.2.5. Immunostaining*

1. To remove the unincorporated reagents, wash slides in 0.1X SSC for 20 min at 45°C. Repeat the wash twice in fresh 0.1X SSC.
2. Quickly rinse slides in solution 3.
3. To block nonspecific binding sites prepare an aliquot of solution 3 with 2% sheep serum.
4. Add 100 µL of the blocking solution to the tissue section and incubate for 2 h at 4°C (*see* **Note 15**).
5. Prepare 100 µL of solution 3 with 1% sheep serum and 1/1000 anti-DIG.
6. Using paper, gently adsorb the blocking solution from the incubation area and replace it with the anti-DIG Solution.
7. Incubate 4 h at room temperature.
8. Abundantly wash slides in solution 3: perform four washes for 10 min shaking at room temperature.
9. Incubate slides in solution 4: 10 min at room temperature.
10. Apply to the section the following detection solution: in 10 mL of Solution 4: 1 m*M* levamisole; 45 µL NBT; 35 µL BCIP; incubate in the dark for 15 to 25 min.
11. Check the tissue with a microscope; when the signals are clear stop the reaction by dipping slides in distilled water (*see* **Notes 16** and **17**).
12. Place permanent cover slips on the tissue using glycerin/PBS 1:1. To improve LCM accuracy, it is suggested to photograph the areas of the immunostained slide that show HIV-1 distribution.

## *3.3. Laser Capture Microdissection and HIV-1 Nucleic Acid Purification*

In the effort to characterize HIV-1 strains infecting a specific anatomic compartment, LCM allows an accurate dissection of single cell types without inadvertent admixture of adjacent cells *(11)*.

The new approach of using *in situ* PCR-guided LCM of infected renal tubular cells on a renal biopsy is described *(6)*.

### 3.3.1. Laser Capture Microdissection

1. Stain slides per instructions listed in **Subheading 3.1.3.**; let slides air-dry for 10 to 15 min (*see* **Note 18**).
2. While slides are air-drying, prepare the LCM facility: clean the microscope platform with fresh alcohol; turn on the instrument; prepare multiple series of LCM caps.
3. Analyze the specimen morphology at different magnifications to reproduce the pattern of the infection emphasized by *in situ* PCR in the serial section.
4. Keeping in consideration the cell type size, always try to work with the smallest laser beam spot size (7–30 μm).
5. Set the power between 50 and 80 mW.
6. Clean the tissue section using the Prepstrip provided by Arcturus.
7. Place the cap on the tissue (*see* **Note 19**).
8. Pulse laser at target cells following the map provided by the *in situ* PCR as much as possible (*see* **Note 20**).
9. Remove the cap with the dissected cells and place it in a 0.5-mL Perkin Elmer tube (*see* **Notes 21** and **22**).
10. Place the cap-tube complex on ice.

### 3.3.2. Nucleic Acid Purification

DNA from microdissected cells is extracted following the protocol of the QIAamp Tissue Kit.

1. Gently remove the cap from the 0.5-mL tube and pipet in the same tube with proteinase K and lysis solution as suggested in the user's manual. Then place the cap back in the tube and turn the complex upside down, making sure that the cap's inner surface is completely covered by the lysis mix.
2. Follow the kit's instructions for DNA recovery.
3. At the end of the purification process, nucleic acids are eluted in distilled water: perfom two elutions with 15 μL of distilled water (*see* **Note 23**).

Another possibility is to elute the DNA with a larger amount of distilled water (two elutions with 50 μL of distilled water) and to lyophilize the eluted solution. The PCR mix can be used to directly resuspend the lyophilized DNA.

## 3.4. PCR, DNA Cloning, and Sequencing

### 3.4.1. Standard PCR for HIV-1 Envelope DNA

In the effort to reveal the presence of HIV-1 in the microdissected samples, PCR allows the detection of even a few copies of the viral DNA. The subsequent cloning and sequencing of amplicons leads to important information about the viral strain under analysis.

Two PCR protocols for HIV-1 envelope are described here: a nested PCR for the entire viral envelope gp120, amplicon length 1.5 Kb, and a PCR for the V3 domain within the gp120, amplicon length 150 bp (**Fig. 2**).

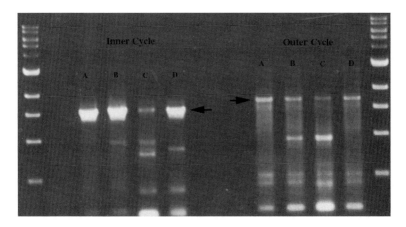

Fig. 2. Nested PCR for entire HIV-1 gp120 on DNA purified from microdissected renal tubules. Arrows indicate that both outer and inner cycles result in a DNA amplification in all experiments (lanes A, B, C, and D). In particular the inner cycle resulting amplicons are 1.5 Kb as expected. 1-Kb markers were used.

1. For the gp120 nested PCR, prepare the following PCR mix:
   - 5 μL    Buffer A
   - 5 μL    Buffer B
   - 4 μL    2.5 m*M* dNTPs
   - 2.5 μL   100 ng/mL ES73
   - 2.5 μL   100 ng/mL ES75
   - 2 μL    Elongase
   - 30 μL   Eluted DNA
2. Transfer the 50 μL PCR mix into a 0.5-mL PCR tube and add one drop of mineral oil.
3. Spin briefly.
4. Perform the following PCR cycle:

   | | | |
   |---|---|---|
   | 92°C | 1 min | |
   | 55°C | 3 min | × 35 |
   | 72°C | 3 min | |
   | 72°C | 10 min | × 1 |

5. Transfer 5 μL of the first round of PCR to the second-round PCR reaction mix. This solution is identical to the previous one but it contains the inner primers pair: 5 μL 100 ng/mL ES74; 5 μL 100 ng/mL ES76.
6. Perform the second-round PCR following the same PCR profile.
7. For the V3 loop PCR prepare the following PCR mix:
   - 10 μL   5X buffer C
   - 2.5 μL   100 ng/mL env s
   - 2.5 μL   100 ng/mL env as
   - 4 μL    2.5 m*M* dNTPs
   - 1 μL    Taq
   - 25 μL   Eluted DNA

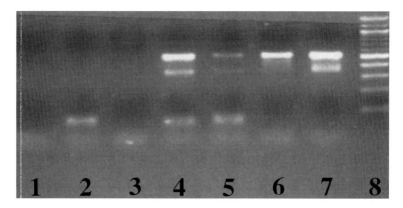

Fig. 3. PCR for HIV-1 V3 region on DNA purified from microdissected renal tubules. The gel shows 500-bp amplicons (lanes 4, 5, 6, and 7) as the result of the presence of viral DNA in distinct microdissected tubules. The absence of HIV-1 in other distinct purified DNA results in no amplification (lanes 1, 2, and 3). 100-bp markers were used (lane 8).

8. Follow directions listed in **steps 2** and **3**.
9. Perform the following PCR cycle (*see* **Note 24**):

| | | |
|---|---|---|
| 94°C | 1 min | |
| 68°C | 30 s | × 10 |
| 72°C | 1 min | |
| 94°C | 1 min | |
| 68°C | 30 s | × 35 |
| 72°C | 1 min extended 10 s/cycle | |
| 72°C | 3 min | × 1 |

10. Run an agarose gel with 5 μL of the PCR solution.

### 3.4.2. Amplicon Cloning in pGEM-T Easy Vector

The insertion of the PCR fragment in a cloning vector is necessary to sequence the amplified viral envelope. In the assessment of the resulting sequence it is critical to exclude a contamination event and more importantly to identify the viral sequence as unique and to obtain information about the viral tropism and nature.

1. Purify the PCR fragment using the QIAquick Purification Kit. Follow the manufacturer's instructions.
2. Quantify the purified fragment and calculate its concentration at 260 nm.
3. Briefly centrifuge the pGEM-T vector and the fragment tubes to collect contents at the bottom of the tube.

4. Set up the ligation reaction as described in the pGEM-T Easy Vector manual.
5. Proceed with transformation in DH10B competent cells.
6. Pick colonies for minipreps.

### 3.4.3. DNA Sequencing

1. Select those minipreps that released the insert after enzymatic digestion.
2. Proceed with the sequencing process using an automatic DNA sequencer (model ABI-PRISM).

## 4. Notes

1. To aid tissue penetration for the *in situ* PCR procedure and to optimize extraction of nucleic acids, keep formaldehyde concentration in the fixative low and keep fixation times at a minimum.
2. The protocol given above is for small tissue sections (1 cm$^3$ block), larger pieces of tissue will require longer incubation times at each step.
3. Excessive incubation at 60°C can damage the tissue.
4. To avoid cross contamination, a new blade needs to be used each time a new case is sectioned.
5. Allow the block to warm up slightly. If the block is too cold during cutting, section quality may decrease.
6. In general, few slides are prepared for *in situ* PCR, whereas more sections are dedicated to LCM. However, performing *in situ* PCR on a new tissue specimen requires several tests to design the optimal experimental conditions for the specific tissue or cell type.
7. This step results in a more homogeneous adherence of the tissue section to the glass support. A longer incubation does not affect the following procedures.
8. Always use fresh alcohol.
9. Hematoxylin and eosin can be used three to four times. It may be useful to extend the incubation times if the solutions have already been used.
10. Bluing solution: 500 µL ammonium hydroxide in 250 mL of distilled water.
11. The reported amount of proteinase K represents the optimal working condition for a 5-µm-thick kidney specimen. Performing *in situ* PCR on a naive specimen requires a test run to determine the best conditions. For this purpose, prepare 6 to 10 extra slides of the specific tissue and prepare an equal number of serial dilutions of proteinase K solutions. Set standard temperature and time of digestion, (e.g., 15 min at 37°C). Then perform the standard protocol for *in situ* PCR and observe which conditions preserve the morphology of the tissue while allowing the penetration of the reagents within the cells for the DNA amplification. A detailed methodology about the technique is published by Martinez et al. *(10)*.
12. 50 µL are generally enough to completely cover a biopsy section. Working with larger sections (autopsies, mouse tissues) requires more abundant enzymatic solution volumes (100 µL).
13. It is recomended to prepare the PCR reaction mix during the Klenow incubation to reduce the time lapse between the two steps.

14. This is a very delicate step: it is very important that the tissue section is centered in the frame and that no bubbles form during the positioning of the cover slip. Denaturation steps during the PCR will increase the temperature and therefore result in an enlarging of the inner bubbles with resulting leakages. It is recomended to practice using empty slides and 5% glycerol water.

15. This step can be extended overnight at 4°C. To avoid evaporation, it is recomended to place slides in a 10-cm culture dish with a moist paper towel underneath. Then place the dish at 4°C. It is also suggested to use a Daimont pen to mark the incubation area around the section.

16. As a positive control, it is recomended to dedicate one slide to a β-globin amplification. In this case, use the corresponding primers and perform the following PCR cycle:

    92°C    2 min    × 1
    92°C    1 min
    65°C    2 min    × 30
    65°C    3 min    × 1

17. As a negative control, several possibilities exist. It is recomended to perform at least two negative controls. One control consists of preparing the PCR reation mix without the DIG-dUTP. The second test is a PCR without primers.

18. It is critical that specimens remain in a dry/not humid environment before LCM to avoid low dissection efficiency.

19. Always try to avoid working near edges of the specimen. These areas are often weakly attached to the glass support and therefore may result in an aspecific dissection.

20. If the infection appears diffuse, it is suggested to collect on the same cap several cells or clusters, otherwise it is safer to stop with few cells and collect sequential caps in the same lysis mix.

21. It is highly recomended to use one tissue section to determine if any HIV-1 can be amplified. Simply scratch the whole section with a razor blade and follow lysis conditions as explained in **Subheading 3.3.2.** Then perform a standard PCR for HIV-1 on the purified total DNA. A negative result in this PCR would make useless any additional procedures, as no virus is detectable.

22. During the entire set of experiments, the risk of plasmid contamination is a constant possibility. Therefore, the LCM facility must be kept absolutely clean and no PCR tools or other molecular biology processes involving plasmids must occur nearby. The best setting consists of a room dedicated to LCM with a small hood specifically used for tissue processing, post-LCM lysis, and PCR preparation. A dedicated set of pipets is also recommended.

23. The HIV-1 DNA is most likely only a minimal percentage of the total DNA. Therefore it is suggested to use the entire volume for one of the following PCRs.

24. The PCR optimization kit has been used to verify the best pH and magnesium concentration for this specific PCR. Buffer C gave the best result in the amplification process.

## References

1. Autran, B., Carcelain, G., Li, T. S., Blanc, C., Mathez, D., Tubiana, R., et al. (1997) Positive effects of combined antiretroviral therapy on CD4+ T cell homeostasis and function in advanced HIV disease. *Science* **277**, 112–116.
2. Chun, T. W., Stuyver, L., Mizell, S. B., Ehler, L. A., Mican, J. A., Baseler, M., et al. (1997) Presence of an inducible HIV-1 latent reservoir during highly active antiretroviral therapy. *Proc. Natl. Acad. Sci. USA* **4**, 13,193–13197.
3. Bukrinsky, M. I., Stanwick, T. L., Dempsey, M. P., and Stevenson, M. (1991) Quiescent T lymphocytes as an inducible virus reservoir in HIV-1 infection. *Science* **254**, 423–427.
4. Meng, G., Wei, X., Wu, X., Sellers, M. T., Decker, J. M., Moldoveanu, Z., et al. (2002) Primary intestinal epithelial cells selectively transfer R5 HIV-1 to CCR5+ cells. *Nat. Med.* **8**, 150–156.
5. Lagaye, S., Derrien, M., Menu, E., Coito, C., Tresoldi, E., Mauclere, P., et al. (2001) Cell-to-cell contact results in a selective translocation of maternal human immunodeficiency virus type 1 quasispecies across a trophoblastic barrier by both transcytosis and infection. *J. Virol.* **75**, 4780–4791.
6. Marras, D., Bruggeman, L. A., Gao, F., Tanji, N., Mansukhani, M. M., Cara, A., et al. (2002) Replication and compartmentalization of HIV-1 in kidney epithelium of patients with HIV-associated nephropathy. *Nat. Med.* **8**, 522–526.
7. Tanji, N., Ross, M. D., Cara, A., Markowitz, G. S., Klotman, P. E., and D'Agati, V. D. (2001). Effect of tissue processing on the ability to recover nucleic acid from specific renal tissue compartment. *Exp. Nephrol.* **9**, 229–234.
8. Bruggeman, L. A., Ross, M. D., Tanji, N., Cara, A., Dikman, S., Gordon, R. E., et al. (2000) Renal epithelium is a previously unrecognized site of HIV-1 infection. *J. Am. Soc. Nephrol.* **11**, 2079–2087.
9. Bagasra,O. and Hansen, J. (1997) *In Situ PCR Techniques*, Wiley-Liss, New York.
10. Martinez, A., Miller, M. J., Quinn, K., Unsworth, E. J., Ebina, M., and Cuttitta, F. (1995) Non radioactive localization of nucleic acids by direct in situ PCR and *in situ* RT-PCR in paraffin embedded sections. *J. Histochem. Cytochem.* **43**, 739–747.
11. Emmert-Buck, M. R., Bonner, R. F., Smith, P. D., Chuaqui, R. F., Zhuang, Z., Goldstein, S. R., et al. (1996) Laser capture microdissection. *Science* **274**, 998–1001.

# 22

# Use of Laser Capture Microdissection Together With *In Situ* Hybridization and Real-Time PCR to Study Distribution of Latent Herpes Simplex Virus Genomes in Mouse Trigeminal Ganglia

Xiao-Ping Chen, Marina Mata, and David J. Fink

### Summary

The herpes simplex virus (HSV) frequently establishes a latent state in neurons, which can then be reactivated from the infected neurons. Quantifying the single-cell viral load is essential for understanding latency and reactivation of this virus. In this chapter the methods of laser capture microdissection and quantitative real-time polymerase chain reaction with *in situ* hybridization have been combined to determine the HSV copy number per neuron in latently infected trigeminal ganglia. The distribution of latent herpes simplex genomes at the individual cell level has been deteremined and the relationship of the number of latent genomes to the expression of latency-associated transcripts established.

**Key Words:** LCM; PCR; *in situ* hybridization; herpes simplex virus; trigeminal.

## 1. Introduction

Herpes simplex virus (HSV) characteristically establishes a lifelong latent state in neurons. The existence of latent genomes in sensory ganglia was first inferred from the ability of virus to be reactivated from infected ganglia that no longer harbor infectious virus, and subsequently confirmed on the cellular level by the detection of latency-associated transcripts (LATs), abundant non-polyadenylated RNA transcripts found in the nucleus of neurons of latently infected individuals *(1–4)*. There have been conflicting reports regarding the role of LAT expression in the establishment of the latent state, or reactivation

From: *Methods in Molecular Biology, vol. 293: Laser Capture Microdissection: Methods and Protocols*
Edited by: G. I. Murray and S. Curran © Humana Press Inc., Totowa, NJ

of virus from latency *(5–8)*. The ability to quantify the viral burden at the level of single cells is essential for understanding many aspects of latency and reactivation *(9)*.

The estimated total number of latent genomes per ganglion reported in the literature varies widely, in part depending on the method used to calculate that number, with estimates ranging from 3500 genomes per ganglion determined by semiquantitative Southern blot *(10)* or semiquantitative polymerase chain reaction (PCR) *(11)*, to $10^5$ genomes determined by other PCR-based methods *(12,13)*. Combining *in situ* hybridization (ISH) with PCR techniques, and making the assumption that only cells that are LAT-positive by ISH contain HSV genomes, Hill et al. *(14)* calculated the presence of 17–34 genomes per LAT-positive cell *(14)* in latently infected rabbit trigeminal ganglia (TG). In a more comprehensive analysis, using semiquantitative PCR to determine the number of genomes in neurons dissociated from latently infected fixed TG, Sawtell *(9)* determined that 20% of the neurons in a TG infected by the *Syn*17 strain of HSV contained HSV genomes, and that the number of latent genomes ranged from less than 10 to more than 1000 per neuron *(9)*. TG infected with KOS-strain HSV contained substantially fewer latent genomes per cell than TG infected by *Syn*17 *(15)*.

LATs are detected by ISH in only 0.1% to 3% of the neurons in latently infected ganglia *(16)*, but more sensitive *in situ* PCR techniques demonstrate that latent genomes are present in neurons that are LAT-positive by ISH, as well as LAT-negative by ISH *(13,17,18)*. The relationship between the number of latent genomes in an individual neuron to the expression of LATs detectable by ISH has not previously been investigated. One possibility is that expression of LATs to a level detectable by ISH is a function of the number of latent HSV genomes in the cell. According to this hypothesis, LATs would be detected when the number of latent genomes exceeded a threshold value. The alternative possibility is that LAT expression from latent HSV genomes depends on the presence of cell-specific factors within the host cell. According to this alternate hypothesis, the number of latent HSV genomes in LAT-positive and LAT-negative cells would display a similar distribution. In order to test the two hypotheses, we combined the methods of laser capture microdissection (LCM) and quantitative real-time PCR with ISH to determine the number of genomes per neuron in latently infected ganglia, in order to establish the distribution of latent genomes in TG at the single-cell level, and to examine the relationship of the number of latent genomes to the expression of LATs detectable by ISH.

From a methodologic perspective, these experiments illustrate the utility of LCM in defining parameters of viral infection of tissue on a single-cell basis. The details of the method employed are described below.

## 2. Materials

### 2.1. Virus

The laboratory strain of wild-type HSV-1, KOS, propagated and prepared from KOS-infected African green monkey kidney (vero) cells, was used in these experiments.

### 2.2. In Situ *Hybridization*

1. Precoated slides from Surgical Path (Richmond, IL).
2. 1X Standard saline citrate (SSC): 150 m$M$ NaCl and 15 m$M$ sodium citrate.
3. Hybridization buffer: 50% formamide, 5X SSC, and 40 µg/mL salmon sperm DNA.
4. Genius buffer III: 1X detection buffer (Roche), 50 m$M$ MgCl$_2$, 1 m$M$ levamisole (Sigma, St. Louis, MO). Levamisole was added before using.
5. The colorimetric substrate was made fresh by adding 87.5 µL nitroblue tetrazolium (Roche) and 87.5 µL of 5-bromo-4-chloro-3-indolylphosphatetoluidium (Roche) to 25 mL Genius buffer III.

### 2.3. PCR Analysis

1. PCR lysis buffer was made fresh before using. The lysis buffer contained 1X PCR buffer without MgCl$_2$ (Perkin-Elmer, Mountain View, CA), 1% Tween-20, and 0.4 mg/mL proteinase K.
2. Real-time PCR mixture for detection of the HSV-1 UL44 gene contained 900 n$M$ of primer (forward primer 5' GAT GCC GGT TTT GGA ATT C 3'; reverse primer 5' CCC ATG GAG TAA CGC CAT ATC T 3'), 250 n$M$ probe (5' FAM-ACC CGC ATG GAG TTG CGC CTC –TAMRA 3') (Synthegen, Houston, TX), and 1X TaqMan universal mix (Applied Biosystems, Foster City, CA).
3. 0.5 $M$ EDTA, pH 8.0.
4. 4 $M$ Lithium Chloride.
5. Standard sodium citrate chloride (SSC).
6. Phosphate-buffered saline.
7. Thermoplastic film-coated PCR caps (Capsure™, Arcturus, Mountain View, CA).
8. Gene AMP 5700 Sequence Detector (Applied Biosystems, Foster City, CA).

## 3. Methods

### 3.1. Animal Model

To establish herpes simplex virus latency in mouse trigeminal ganglia, 4–8-wk-old female Balb/c mice were inoculated with wild-type HSV-1 virus, KOS. $5 \times 10^6$ p.f.u. of virus in 5 µL of PBS was administered to each eye by corneal scarification under anesthesia. Thirty to forty days after infection, the animals were sacrificed and both trigeminal ganglia removed quickly and embedded in Cryo-Gel (Instrumedicis, Hackensack, NJ). Tissue blocks were stored at –70°C until used.

## *3.2. Probe Generation and* In Situ *Hybridization*

### *3.2.1. Generation of Probe*

1. The antisense riboprobe against latency-associated transcripts (LAT) was generated from a plasmid containing the BamHI fragment of HSV *(19)*.
2. The plasmid was linearized with EcoRI and transcribed in vitro with T7 RNA polymerase using digoxigenin-UTP to label the transcript.
3. Linearized template DNA (1 µg) was incubated with 4 µL DIG-nick translation mix in a final volume of 20 µL (the volume was filled with distilled $H_2O$) at 15°C for 90 min, and the reaction paused by placing on ice.
4. The distribution of labeled products was then checked on a 1.5% agarose gel. The ideal product size is between 200 bp and 600 bp, and the incubation time should be adjusted as needed to achieve the proper product size.
5. Once the reaction was completed, 1 µL 0.5 *M* EDTA (pH 8.0) was added to the reaction tube (for 1 µg DNA) and the reaction mixture heated to 65°C for 10 min to stop the reaction.
6. Carrier DNA (100 µg) from sheared salmon egg DNA (SSD) was added to 1 µg of template DNA.
7. After precipitation with 0.1 volume of 4 *M* LiCl and 2.5–3.0 vol of ice-chilled absolute ethanol, labeled probes were resuspended in hybridization buffer at about 100 ng/µL.

### *3.2.2.* In Situ *Hybridization*

1. Cryostat sections (10 µm) were sectioned from the snap-frozen trigeminal ganglia tissue blocks and mounted on precoated slides from Surgical Path (*see* Note 1).
2. Tissues were fixed with 4% paraformaldehyde for 30 min at room temperature.
3. For wax sections, sections were dewaxed by incubating slides at 60°C for 30 min, followed by two washes in xylene for 5 min, two washes of 100% ethanol for 2 min, two washes of 95% ethanol for 2 min, one wash with water for 2 min and the final two washes of PBS for 5 min.
4. Dewaxed sections were postfixed with 4% paraformaldehyde for 10 min at room temperature.
5. Fixed tissues were washed twice with PBS containing 0.1% active DEPC for 5 min each to protect RNA from degradation.
6. After equilibration with 5X SSC for 5 min at room temperature, tissues were hybridized with 7.5 ng of digoxigenin-labeled antisense probe in 15 µL hybridization buffer.
7. Sections were covered with cover slips, sealed with rubber cement, and incubated for 1 h or overnight at 56°C in a humidified chamber.
8. Sections were washed with 2X SSC for 5 min at RT, followed by 10 min in 2X SSC prewarmed to 65°C, by 10 min in 0.1X SSC prewarmed to 65°C, and by 2 min in PBS at room temperature.
9. After blocking with PBS containing 5% normal goat serum for 15 min at room temperature, sections were incubated for another 2 h at room temperature or 4°C

overnight with alkaline phosphatase conjugated anti-digoxigenin antibody (1:5000 dilution in blocking solution).

10. After three washes in PBS for 2 min each, sections were equilibrated with Genius buffer III for 5 min at room temperature.

11. Final detection was made by incubating slides with color substrate containing 5-bromo-4-chloro-3-indolylphosphatetoluidium-nitroblue tetrazolium (BCIP-NBT) for about 30 min or the time needed.

12. Sections were washed with tap water for 5 min, after which they were ready for laser capture.

## 3.3. Microdissection of Virus-Infected Cells

### 3.3.1. Dehydration of Slides

The dehydration step is critically important for the laser capture experiment *(19)*. However, slides must not be allowed to dry prior to the final dehydration step (*see* **Note 2**). Sections that had undergone ISH were dehydrated by two changes of 95% ethanol for 2 min, two changes of 100% ethanol for 2 min and two changes of xylene for 5 min before being air-dried.

### 3.3.2. Laser Capture Microdissection

1. The PixCell LCM™ (Arcturus) system was used to view and capture individual LAT-positive or LAT-negative neurons from the slides.

2. Individual neurons were captured with either the 15-μm or 30-μm laser beam (depending on the size of the cell) to thermoplastic film-coated PCR caps (**Fig. 1**).

3. The amplitude applied ranged between 30 mW and 50 mW for 44–66 ms as suggested by Arcturus (PixCell Laser Microdissection System Instruction Manual, Arcturus, 2002, p 16).

4. The cap with captured tissue on it was placed in a microcentrifuge tube for further DNA analysis (*see* **Notes 3** and **4**).

## 3.4. Real-Time PCR Analysis

1. Single LAT-positive or -negative neurons captured on the cap were dissolved in 20 μL of PCR lysis buffer at 37°C overnight (*see* **Note 5**).

2. Proteinase K was heat-inactivated in boiling water for 10 min before the PCR reaction was performed.

3. Lysate (10 μL) was added to 40 μL real-time PCR mixture to amplify the HSV-1 UL44 gene sequence in order to quantify HSV-1 genomes present in each LAT-positive or LAT-negative neuron.

4. The real-time PCR reaction included 2 min incubation at 50°C and 10 min of incubation at 95°C for denaturation and 50 cycles of PCR (95°C for 15s and 60°C for 1 min) in a GenAmp 5700 Sequence Detector.

5. Each PCR run contained two blank controls, serially diluted highly purified KOS DNA as standard, and the experimental samples run in duplicate.

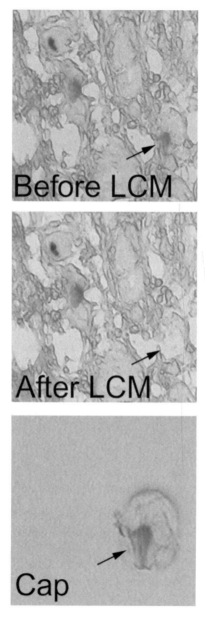

Fig. 1. LCM of individual neurons for real-time PCR. A single LAT-positive neu-
ron is shown before LCM (arrow, top) and the same section after LCM (arrow, middle).
The lifted cell on the cap is shown at the bottom (arrow, higher magnification). The
ISH signal for LAT appears blue in this micrograph of ISH performed with a
digoxigenin-labeled probe, and two additional LAT-positive neurons are seen.
(Reprinted with permission from *J. Neurovirol.* **8,** 204–210.)

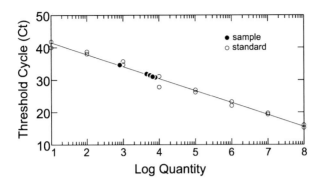

Fig. 2. Real-time PCR quantitation of HSV genomes. Purified HSV-1 DNA was amplified by PCR of UL44; the threshold cycle (Ct) is plotted against the amount of input DNA. (Reprinted with permission from *J. Neurovirol.* **8,** 204–210.)

6. The number of genomes in each sample was determined by comparing the threshold cycle (Ct) value to a standard curve obtained using KOS DNA.
7. The Ct value was defined, as the amplification cycle required observing a significant difference between emission intensity of the reporter dye in the sample and the emission intensity of the controls without DNA.
8. A standard curve was generated using known amounts of HSV KOS DNA purified by three rounds of phenol-chloroform extraction and ethanol precipitation (*see* **Note 6**).
9. The DNA amount was determined by UV spectrophotometer, and the DNA was serially diluted in nuclease-free water such that 1 µL of sample contained $10^7$, $10^6$, $10^5$, $10^4$, $10^3$, $10,^2$ or 10 copies of HSV-1 DNA.
10. Mouse genomic DNA (100 ng, Clontech, Palo Alto, CA) was added into each sample to mimic the DNA complexity of the virus-containing cell sample.
11. Real-time PCR amplification of HSV-1 DNA was linear over a range from 10 to $10^8$ copies of HSV DNA (**Fig. 2**).

## 4. Notes

1. Precoated slides from Surgical Path gave the authors the best results in successfully lifting the tissues.
2. The sections should not be dried completely during the process of *in situ* hybridization. This has proven to be crucial to lift tissues successfully in LCM step (Dr. Kening Wang, NIH, personal communication).
3. The upgraded PixCell™ system is beneficial for viewing dehydrated and unmounted tissue. This is particularly important in finding positive cells with weak signal (as described in detail on the Arcturus web site, http://www.arctur.com/products/pixcell_iie_upgrade.htm).
4. A single laser firing is usually sufficient to obtain complete capture of tissue. Occasionally, multiple hits may be required to lift a cell with larger amount of BCIP-NBT precipitates.

5. The minimum volume needed to cover the surface of cap is 20 µL. One should stop pushing the cap into the microcentrifuge tube at first resistance. Otherwise, leakage of solution may occur.

6. The copy numbers exist in highly purified KOS DNA were calculated based on the knowledge that HSV-1 is about 152,000 bp long and the molecular weight of one base pair is around 664 Da.

## References

1. Krause, P. R., Croen, K. D., Straus, S. E., and J. M. Ostrove. (1988) Detection and preliminary characterization of herpes simplex virus type 1 transcripts in latently infected human trigeminal ganglia. *J. Virol.* **62,** 4819–4823.

2. Rock, D. L., Nesbern, A. B., Ghiasi, H., Ong, J., Lewis, T. L., Lokensgard J. R., and Wechsler, S. (1987) Detection of latency-related viral RNAs in trigeminal ganglia of rabbits latently infected with herpes simplex virus type 1. *J. Virol.* **61,** 3820–3826.

3. Steiner, I., Spivack, J. G., O'Boyle, D. R., Lavi, E., and Fraser, N. W. (1988) Latent herpes simplex virus type 1 transcription in human trigeminal ganglia. *J. Virol.* **62,** 3493–3496.

4. Stevens, J. G., Haarr, L., Porter, D. D., Cook, M. L., and Wagner, E. K. (1988) Prominence of the herpes simplex virus latency-associated transcript in trigeminal ganglia from seropositive humans. *J. Infect. Dis.* **158,** 117–123.

5. Bergstrom, T. and E. Lycke. (1990) Neuroinvasion by herpes simplex virus. An in vitro model for characterization of neurovirulent strains. *J. Gen. Virol.* **71,** 405–410.

6. Hill, J. M., Sedarati, F., Javier, R. T., Wagner, E. K., and Stevens, J. G. (1990) Herpes Simplex Virus latent phase transcription facilitates in vivo reactivation. *Virology* **174,** 117–125.

7. Leib, D. A., Bogard, C. L., and Kosz-Vnenchak, M. (1989) A deletion mutant of the latency-associated transcript of herpes simplex virus type 1 reactivates from the latent state with reduced frequency. *J. Virol.* **63,** 2893–2900.

8. Trousdale, M. D., Steiner, I., Spivack, J. G., Deshmane, S. L., Brown, S. M., MacLean, A. R., et al. (1991) In vivo and in vitro reactivation impairment of a herpes simplex virus type 1 latency-associated transcript variant in a rabbit eye model. *J. Virol.* **65,** 6989–6993.

9. Sawtell, N. M. (1997) Comprehensive quantification of herpes simplex virus latency at the single-cell level. *J. Virol.* **71,** 5423–5431.

10. Sedarati, F., Izumi, K. M., Wagner, E. K., and Stevens, J. G. (1989) Herpes simplex virus type 1 latency-associated transcription plays no role in establishment or maintenance of a latent infection in murine sensory neurons. *J. Virol.* **63,** 4455–4458.

11. Rodahl, E., and Stevens, J. G. (1992) Differential accumulation of herpes simplex virus type 1 latency-associated transcripts in sensory and autonomic ganglia. *Virology* **189,** 385–388.

12. Katz, J. P., Bodin, E. T., and Coen, D. M. (1990) Quantitative polymerase chain reaction analysis of herpes simplex virus dna in ganglia of mice infected with replication-incompetent mutants. *J. Virol.* **64,** 4288–4295.

13. Ramakrishnan, R., Levine, M., and Fink, D. J. (1994) PCR-based analysis of herpes simplex virus type 1 latency in the rat trigeminal ganglion established with a ribonucleotide reductase- deficient mutant. *J. Virol.* **68,** 7083–7091.

14. Hill, J. M., Gebhardt, B. M., Wen, R., Bouterie, A. M., Thompson, H. W., O'Callaghan, R. J., et al. (1996) Quantitation of herpes simplex virus type 1 DNA and latency-associated transcripts in rabbit trigeminal ganglia demonstrates a stable reservoir of viral nucleic acids during latency. *J. Virol.* **70,** 3137–3141.

15. Sawtell, N. M., Poon, D. K., Tansky, C. S., and Thompson, R. L. (1998) The latent herpes simplex virus type 1 genome copy number in individual neurons is virus strain specific and correlates with reactivation. *J. Virol.* **72,** 5343–5350.

16. Roizman, B, and Sears, A. (1996) Herpes simplex viruses and their replication, in Fields, B. N., Knipe, D. M., Howley, P. M., Chanock, R. M., Hirsch, M. S., Melnick, J. L., et al. (eds.), *Field Virology*, 3rd ed., vol. 2. Lippincott-Raven, Philadelphia, pp. 2231–2295.

17. Mehta, A., Maggioncalda, J., Bagasra, Thikkavarapu, S., Saikumari, P., Valyi-Nagy, T., et al. (1995) *In situ* DNA PCR and RNA hybridization detection of herpes simplex virus sequences in trigeminal ganglia of latently infected mice. *Virology* **206,** 633–640.

18. Ramakrishnan, R., Poliani, P. L., Levine, M., Glorioso J. C., and Fink, D. J. (1996) Detection of herpes simplex virus type 1 latency-associated transcript expression in trigeminal ganglia by in situ reverse transcriptase PCR. *J. Virol.* **70,** 6519–6523.

19. Gordon, Y. J., Johnson, B., Romanowski, E., and Araullo-Cruz, T. (1988) RNA complementary to herpes simplex virus type I ICP0 gene demonstrated in neurons of human trigeminal ganglia. *J. Virol.* **62,** 1832–1835.

# 23

# Laser Capture Microdissection and PCR for Analysis of Human Papilloma Virus Infection

## Kheng Chew, Patrick H. Rooney, Margaret E. Cruickshank, and Graeme I. Murray

### Summary

Human papilloma virus (HPV) infection is considered one of the main factors involved in the pathogenesis of endocervical adenocarcinoma. However, the cellular location of HPV in this type of tumor is controversial. We have developed a method to determine the presence of HPV type 16 in endocervical cancer cells using laser capture microdissection followed by DNA extraction and qualitative polymerase chain reaction. Our results show that HPV type 16 is present in endocervical adenocarcinoma cells.

**Key Words:** Human papilloma virus; endocervical adenocarcinoma; laser capture microdissection.

## 1. Introduction

Human papilloma virus (HPV) infection is one of the main factors implicated in the pathogenesis of adenocarcinoma of the cervix (*1–3*). HPV infection is often found in the tumor. The antecedent cervical cytology and the HPV subtypes most commonly associated with cervical adenocarcinoma are types 16 and 18 (*4*). The reported prevalence of HPV infection in cervical adenocarcinoma ranges from 25 to 48%. This variation can be explained by the specific patient cohort sampled and variations in the method employed to detect individual HPV subtypes.

The incidence of concomitant squamous cervical intraepithelial neoplasia in endocervical adenocarcinoma is also well recognized and has been quoted to be as high as 40% (*5*). High-grade squamous cervical intraepithelial neoplasia is strongly correlated with HPV infection (*6*). However, it appears that cervical

From: *Methods in Molecular Biology, vol. 293: Laser Capture Microdissection: Methods and Protocols*
Edited by: G. I. Murray and S. Curran © Humana Press Inc., Totowa, NJ

screening may be a less-sensitive tool for detection of glandular as opposed to squamous intraepithelial neoplasia.

To date, most studies into the relationship between HPV infection and endocervical adenocarcinoma have been based on polymerase chain reaction (PCR) analysis of DNA extracted from whole tissue sections *(7)*. These sections will often contain squamous epithelium, which will frequently be infected with HPV. To establish the true prevalence of HPV infection in endocervical adenocarcinoma it is necessary to specifically analyze endocervical adenocarcinoma tumor cells.

The laser capture microdissection system developed at the National Institutes of Health and Arcturus Engineering enables the selective sampling of cells of interest from both archival and fresh-frozen histological sections under direct microscopic control *(8,9)*. The LCM system has the dual advantages of (1) allowing the analysis of specific cell types and (2) reducing the dilution of the target DNA by ensuring that DNA is extracted only from target cells. In this chapter we describe a method to determine the presence of HPV-16 in endocervical adenocarcinoma using laser microdissected cells.

## 2. Materials

1. Formalin-fixed wax-embedded sections of endocervical adenocarcinoma.
2. Xylene.
3. Ethanol.
4. PixCell II Laser Capture Microdissecting System (Arcturus Engineering, Mountain View CA).
5. Toluidine blue.
6. Digestion buffer (2 mg/mL proteinase K, 0.05 $M$ Tris-EDTA buffer, pH 8.5, containing 0.5% Tween-20).
7. β-globin forward primer: 5' ACA CAA CTG TGT TCA CTA GC 3' (*see* **Note 1**).
8. β-globin reverse primer: 5' CAA CTT CAT CCA CGT TCA CC 3' (*see* **Note 1**).
9. HPV-16 forward primer: 5' TCA AAA GCC ACT GTG TCC TG 3' (*see* **Note 2**).
10. HPV-16 reverse primer: 5' CGT GTT CTT GAT GAT CTG CA 3' (*see* **Note 2**).
11. Magnesium chloride.
12. dNTPs (Applied Biosystems).
13. AmpliTaq Gold DNA Polymerase (Applied Biosystems).
14. Sterile water.
15. Hybaid PCR Express thermocycler (Hybaid).
16. Agarose gel.
17. Ethidium bromide.
18. Gel loading buffer (6X; 10% Ficoll 400, 025% bromophenol blue, 0.25% xylene cyanol FF, 0.4% orange G, 10 m$M$ Tris-HCl, pH 7.9, and 50 m$M$ EDTA).
19. Horizontal gel electrophoresis apparatus for agarose gels.
20. Ultraviolet light transilluminator.
21. Polaroid type 665 instant film.

## 3. Methods

### 3.1. DNA Extraction From Whole Tumor Sections

1. For wax-embedded tumor biopsies, five 5-μm sections of each biopsy were used (approximate surface area = 10 mm × 10 mm). The sections were placed in a 1.5-mL Eppendorf tube.
2. The sections were dewaxed in xylene for three 10-min periods.
3. The sections were then rehydrated in 100% ethanol for three 10-min periods to remove the xylene (*see* **Note 3**).
4. The specimens were then left to dry, to allow the ethanol to evaporate, a heating block set at 56°C for 10 min.
5. Sections were then digested in 50 μL of digestion buffer for 4 h at 55°C in a shaking heating block.
6. The proteinase K was inactivated by incubation at 98°C for 8 min.
7. Samples were centrifuged in a benchtop microfuge at 13,000$g$ for 15 s and the supernatants carefully transferred to fresh sterile 0.5-mL microfuge tubes, ready for subsequent PCR analysis.

### 3.2. DNA Extraction From Laser Microdissected Tumor Cells

1. Sections of 5-μm thickness were cut from the wax-embedded tissue blocks onto uncoated glass slides (*see* **Note 4**).
2. The sections were dewaxed in xylene for three 10-min periods.
3. The sections were then rehydrated in 100% ethanol for three 10-min periods to remove the xylene (*see* **Note 3**).
4. The sections were then stained by immersing the sections in 0.25% toluidine blue (pH 4.5) for 5 s at room temperature. Excess staining solution was removed by washing the sections briefly in 100% ethanol, and then the sections were dehydrated sequentially in 100% ethanol and xylene (*see* **Note 5**).
5. Endocervical adenocarcinoma cells were microdissected from the stained tumor sections using the PixCell II laser microdissection system. The following settings of the laser were used: laser spot diameter15 μm, laser power 100 mW, and the duration of laser pulse was 10 ms (*see* **Note 6**). Under direct microscopic control the tumor cells of interest were aligned with the laser beam and the laser was fired at the targeted cells, causing the polymer film mounted on a disposable cap (CapSure HS™) directly above the targeted cells to be focally melted and the targeted cells captured onto the polymer film.
6. Approximately 500 laser pulses per tumor sample were used to obtain sufficient cells (corresponding to approx 1000 cells) for analysis.
7. The polymer film with the microdissected cells was carefully removed from the plastic cap, placed in an 1.5-mL Eppendorf tube, and digested in 50 μL of digestion buffer for 4 h at 55°C in a shaking hot block. This was followed by an 8-min incubation at 98°C to inactivate the proteinase K. PCR amplification for β-globin was then performed to confirm an adequate amount of PCR-amplifiable DNA.

### 3.3. β-Globin PCR

1. The amount of each primer was 150 pmol in a total reaction volume of 50 µL containing 2 m$M$ MgCl$_2$, 200 m$M$ of each dNTP, 0.25 units of AmpliTaq Gold DNA polymerase, and 1 µL of template DNA (*see* **Note 7**). The reaction volume was made up to 50 µL with sterile water.
2. DNA extracted from a fresh unstained cervical smear and distilled water were used as positive and negative controls, respectively.
3. PCR was carried out using a PCR Express thermocycler with the following cycling conditions: Initial 15-min period at 94°C followed by 40 two-step cycles at 72°C for 45 s and 55°C for 45 s.
4. The PCR products were then analysed by agarose gel electrophoresis (*see* **Subheading 3.5.**)

### 3.4. HPV-16 PCR

1. Only those samples showing amplifiable DNA as assessed by qualitative β-globin PCR were analyzed for HPV-16.
2. The amount of each primer was 150 pmol in a total reaction volume of 50 µL containing 200 m$M$ of each dNTP, 0.25 units of AmpliTaq Gold DNA polymerase, and 1 µL of template DNA. The MgCl$_2$ concentration was 3.5 m$M$ (*see* **Note 7**). The reaction volume was made up to 50 µL with sterile water.
3. The positive control was 1 µL of DNA extracted (as described in **Subheading 3.1.**) from five 10-µm sections of a paraffin-embedded block of SiHa cells (*see* **Note 8**).
4. The negative control consisted of 1 µL of sterile water.
5. PCR was carried out using a Hybaid thermocycler with the following conditions: initial 15-min period at 94°C followed by 40 two-step cycles of 72°C for 45 s and 55°C for 45 s.
6. When the PCR was complete the PCR products were then analyzed by agarose gel electrophoresis (*see* **Subheading 3.5.**).

### 3.5. Agarose Gel Electrophoresis

1. Gel electrophoresis using 2% agarose gels containing ethidium bromide (0.5 mg/mL) was used to analyze all PCR products. Ten µL of PCR product and 5 µL of loading buffer were added to each well.
2. Electrophoresis conditions were set at 80 V for 2 h. When the electrophoresis was complete, gels were placed on an ultraviolet light transilluminator to visualize the PCR products, and the gels were photographed using Polaroid type 665 film with a 12-s exposure time to obtain a permanent record of each gel.
3. Positive PCR results were regarded as being distinct bands of the predicted size for each PCR product (*see* **Notes 1** and **2**).
4. With DNA extracted from whole tumor sections, 33 cases were positive by PCR for β-globin. PCR was then carried out for HPV-16 in the 33 samples that were positive for β-globin; eight of those samples were positive for the HPV-16.

5. Using DNA extracted from laser microdissected cells, 52 of the 55 cases were positive for β-globin. The 52 samples that were β-globin-positive were analyzed for the presence of HPV-16 and 21 of the samples were HPV-16 positive.

## 4. Notes

1. The presence of amplifiable DNA in each sample is assessed using β-globin as an internal control *(10)*. The β-globin primers produce an amplicon of 110 base pairs. Formalin fixation and wax embedding result in marked DNA fragmentation; thus, only short fragments (up to 200–300 bp) of DNA are usually amplifiable. Therefore it is important to use primers designed to amplify a short segment of DNA.
2. The primer sequences for HPV-16 were derived from the L1 region of the HPV-16 genome, as sequence variations in this region characterize individual HPV subtypes. The HPV primers produce a 119-bp amplicon.
3. It is essential to completely remove all traces of xylene from the samples prior to proteolytic digestion, as xylene prevents the action of proteinase K.
4. It is important to use noncoated slides to ensure satisfactory microdissection. Slides coated with adhesive, e.g., poly-L-lysine or aminoproprylethoxy silane, prevent the satisfactory transfer of tissue from the glass slide to the plastic cap.
5. The stained tissue sections require complete dehydration prior to laser microdissection. Even a trace of moisture appears to inhibit the successful capture of microdissected cells.
6. The optimum settings of the laser have to be determined by the individual user and for their particular application.
7. The optimum magnesium chloride concentration for each PCR must be determined by the individual investigator.
8. The SiHa cell line is derived from a squamous carcinoma of the cervix and contains two copies of HPV-16 per cell.

## Acknowledgments

This research was supported by a grant from endowment funds of Grampian University Hospital NHS Trust.

## References

1. Pirog , E. C., Kleter, B., Olgac, S., Bobkiewicz. P., Lindeman. J., Quint, W. G., et al. (2000) Prevalence of human papillomavirus DNA in different histological subtypes of cervical adenocarcinoma. *Am. J. Pathol.* **157,** 1055–1062.
2. Parazzini, F. and La Vecchia, C. (1990) Epidemiology of adenocarcinoma of the cervix. *Gynecol. Oncol.* **39,** 40–46.
3. Ursin, G., Pike, M. C., Preston-Martin, S., d'Ablaing, G., and Peters, R. K. (1996) Sexual reproductive, and other risk factors for adenocarcinoma of the cervix: results from a population-based case-control study (California, United States). *Cancer Causes Control* **7,** 391–401.

4. Bosch, F. X., Manos, M. M., Munoz, N., Sherman, M., Jansen, A. M., Peto, J., et al. (1995) Prevalence of human papillomavirus in cervical cancer: a worldwide perspective. *J. Natl. Cancer Inst.* **87,** 796–802.

5. Maier, R. C. and Norris, H.J. (1980) Coexistence of cervical intraepithelial neoplasia with primary adenocarcinoma of the endocervix. *Obstet. Gynecol.* **56,** 361–364.

6. Bavin, P. J., Giles, J. A., Deery, A., Crow, J., Griffiths, P. D., Emery, V.C., and Walker, P. G. (1993) Use of semi-quantitative PCR for human papillomavirus DNA type 16 to identify women with high grade cervical disease in a population presenting with a mildly dyskaryotic smear report. *Br. J. Cancer* **67,** 602–605.

7. Ferguson, A. W., Svoboda-Newman, S. M., and Frank, T. S. (1998) Analysis of human papillomavirus infection and molecular alterations in the adenocarcinoma of the cervix. *Mod. Pathol.* **11,** 11–18.

8. Curran, S., McKay, J. A., McLeod, H. L., and Murray, G. I. (2000) Laser capture microscopy. *Mol. Pathol.* **53,** 64–68.

9. Curran, S. and Murray, G. I. (2002) Tissue microdissection and its applications in pathology. *Current Diagn. Pathol.* **8,** 183–192.

10. Cruickshank, M. E., Sharp, L., Chambers, G., Smart, L., and Murray G. (2002) Persistent infection with human papillomavirus following the successful treatment of high grade cervical intraepithelial neoplasia. *BJOG* **109,** 579–581.

# 24

## Laser Capture Microdissection of Hepatic Stages of the Human Parasite *Plasmodium falciparum* for Molecular Analysis

### Jean-Philippe Semblat, Olivier Silvie, Jean-François Franetich, and Dominique Mazier

### Summary

Despite the sequencing of parasite genomes and development of DNA microarray technology, gene profiling of parasites remains a difficult task. For example, transcriptome analysis cannot currently be applied to the hepatic stages of the malaria parasite *Plasmodium falciparum* due to difficulties in obtaining sufficient amounts of parasite material that lies among the large excess of host cell RNA. Here, we describe the isolation of *P. falciparum*-infected human hepatocytes by a laser capture microdissection approach. Reverse transcriptase polymerase chain reaction amplification of several *P. falciparum* transcripts demonstrates the high quality of the RNA recovered after microdissection. This approach should enable analysis of *P. falciparum* transcriptome during its hepatic development, a major step toward the identification of new therapeutic and vaccine targets.

**Key Words:** Gene expression; parasite; laser capture microdissection; *Plasmodium falciparum*.

## 1. Introduction

During the past few years, many parasite genomes have been completely sequenced and others are in progress. This sequence information, in combination with new large-scale genome analysis technologies, provides important tools for molecular studies such as gene expression profiling. However, in the field of parasitology, in the context of a complex background of host nucleic acids, such approaches remain difficult to perform because of the difficulty in isolating the parasite from its host to obtain parasite material pure enough to be

From: *Methods in Molecular Biology, vol. 293: Laser Capture Microdissection: Methods and Protocols*
Edited by: G. I. Murray and S. Curran © Humana Press Inc., Totowa, NJ

analyzed. This problem is strengthened when studying an intracellular para-site. For example, *Plasmodium falciparum*, the malaria parasite that kills more than one million children every year, invades and develops inside the hepato-cytes before setting cycles of erythrocytic development associated with the clinical symptoms. Hepatic stages are important targets for vaccine-induced protective immunity and prophylactic treatment *(1)*. However, little is known about the development of the parasites inside the liver. Indeed, molecular analy-sis, such as a DNA microarray approach, is hampered by the difficulty of iso-lating parasites from the host cells and by their low density either inside the liver or in hepatocyte cultures. Resulting preparations contain a very low pro-portion of nuclear material originated from the parasite. Therefore, it is neces-sary to enrich preparations for infected hepatocytes before molecular analysis.

Laser capture microdissection (LCM) technology appears to be one of the most appropriate approaches to isolate infected cells while saving nuclear ma-terial, notably RNA *(2,3)*. In parasitology research, few papers have used the LCM approach. One publication does concern the isolation of mouse brain vessels sections to detect gene expression induced by *P. berghei* infection *(4)*; another one demonstrates the feasibility of LCM on tissue sections from *P. yoelii* infected mouse liver for gene expression studies *(5)*. In contrast to rodent malaria, which can be investigated in vitro and in vivo, most of the studies concerning the human plasmodium, *P. falciparum*, hepatic stages need to be performed in cultures. We have recently applied LCM technology to cell cultures in order to isolate *P. falciparum*-infected hepatocytes, the aim being to obtain RNA preparations enriched in parasite material *(6)*.

## 2. Materials

1. HEPES buffer, pH 7.6: 8 g/L NaCl, 0.2 g/L KCl, 0.1 g/L Na$_2$HPO$_4$·12H$_2$O, 2.38 g/L HEPES.
2. Collagenase D (Roche, Meylan, France).
3. Percoll.
4. Collagen I (Beckton-Dickinson, Franklin Lakes, NJ).
5. Lab Tek glass slides (Nalge Nunc International, Cergy Pontoise, France).
6. Williams medium E (Life Technology, Cergy Pontoise, France).
7. Fetal calf serum (Life Technology).
8. L-glutamine (Life Technology).
9. Sodium pyruvate (Bio-Whittaker, Miami, FL).
10. Insulin (Sigma, Steinheim, Germany).
11. 5-Fluorocytosin (Roche).
12. Penicillin (Life Technology).
13. Streptomycin (Life Technology).
14. Dexamethasone.

15. Methanol.
16. Mouse antibodies directed against the *P. falciparum* heat-shock protein 70 (gift from D. Mattei, Institut Pasteur, France).
17. FITC-conjugated goat anti-mouse immunoglobulin (Sigma).
18. RNase inhibitor (Promega, Madison, WI, USA).
19. Ethanol.
20. Xylene.
21. PixCell II LCM system equipped with a fluorescent microscope (Arcturus Engineering, Mountain View, CA).
22. Cap (CapSure™ TF-100).
23. Sterile 0.5 mL microcentrifuge tubes.
24. Micro RNA isolation kit (Stratagene, La Jolla, CA).
25. Sensiscript Reverse Transcriptase kit (Qiagen).
26. Fast-start Taq polymerase (Roche).
27. dNTPs.
28. Oligonucleotide primers.
29. Agarose and DNA electroporation equipment.

## 3. Methods

The following methods describe (1) hepatocyte infection, (2) parasite labeling and microdissection, and (3) RNA extraction and reverse transcriptase polymerase chain reaction (RT-PCR).

### 3.1. Hepatocyte Infection

Descriptions of (1) the isolation and culture of human hepatocytes and (2) the infection with *P. falciparum* sporozoites are given below.

### 3.1.1. Human Hepatocyte Culture

1. Primary cultures of human hepatocytes are isolated from liver segments using the two-step enzymatic perfusion technique *(7)*. The hepatic fragments are successively perfused with HEPES buffer to eliminate all blood and with 0.05% collagenase D in HEPES buffer with 0.75 mg/mL $CaCl_2$ to dissociate the hepatocytes.
2. Viable cells are isolated by centrifugation at 800$g$ on a 36% Percoll phase. Cells are seeded at a density of $1.4 \times 10^5$ per $cm^2$ on a 16-chamber Lab-Tek glass slide (*see* **Note 1**) coated with collagen I (2 h with 100 µL/well of a 50 µg/mL solution in 0.02 $N$ acetic acid) and incubated at 37°C in 4% $CO_2$ atmosphere.
3. Hepatocytes are cultivated in Williams medium E with 10% fetal calf serum, 2 m$M$ L-glutamine, 1 m$M$ sodium pyruvate, 10 mg/L insulin, 2.5 µg/mL 5-fluorocytosin, 200 U/mL penicillin, and 200 µg/mL streptomycin. After adherence of the cells, culture medium is replaced by fresh medium supplemented with $10^{-7}$ $M$ dexamethasone.

### 3.1.2. P. falciparum Sporozoite Isolation and Hepatocyte Infection

1. *Anopheles stephensi* adult females were infected with the NF 54 strain of *P. falciparum*, using a membrane-based feeder system *(8)*.
2. After 14–21 d, mosquitoes are killed and their salivary glands are aseptically dissected and disrupted by trituration in a glass tissue grinder. Sporozoites are counted in a KovaSlide® chamber and diluted in culture medium.
3. Hepatocytes are inoculated with $2 \times 10^5$ sporozoites per well and incubated at 37°C in 4% $CO_2$ atmosphere for 3 h and washed three times in complete medium. Finally, fresh complete medium is added and renewed every day *(9)*.

## 3.2. Parasite Labeling and Microdissection

Infected hepatocytes are (1) labeled with an antibody directed against the heat-shock protein 70 (HSP-70) *(10)*, and (2) microdissected by laser capture.

### 3.2.1. Parasite Labeling

1. At appropriate times after inoculation, rinse the cultures three times in 1X PBS. Fix for 5 min in methanol and wash again three times in 1X PBS.
2. Incubate for 30 min at 37°C with a monoclonal antibody, directed against *P. falciparum* HSP-70, diluted 1/500 in PBS and containing 400 U/mL of RNase inhibitor (*see* **Note 2**).
3. Wash three times in PBS.
4. Incubate in a FITC-conjugated goat anti-mouse immunoglobulin, diluted 1/100 in PBS with RNase inhibitor, for 20 min at 37°C.
5. Wash three times in PBS.

### 3.2.2. Laser Capture Microdissection

To perform an efficient microdissection, it is essential to have the driest sample possible. Cultures are dehydrated through a freshly prepared series of increasing concentrations of ethanol.

1. Incubate first with 70% ethanol for 30 s, then incubate two times in 95% ethanol for 1 min each and two times for 1 min in 100% ethanol. Clear slides in xylene by incubating two times 10 min, and finally air-dry in the hood.
2. Without delay, proceed to microdissection using the PixCell II LCM system. Set up the microdissector to 90 mW of laser power and 15 μm diameter laser beam (*see* **Note 3**).
3. Put down the cap on the slide that has been inserted in the microdissector. Infected hepatocytes are distinguished from uninfected ones by the fluorescence of the parasites. Ensuring the attachment of the parasite to the membrane of the cap will require several laser pulses (*see* **Note 3**).
4. When all fluorescent cells have been captured, lift up the cap and control for the absence of the targeted cells. Place a sterile slide on the microscope and put the cap back down on this slide to verify the presence of the fluorescent hepatocytes on the cap.

5. Lift up the cap again and place into a 0.5-mL Eppendorf microcentrifuge tube for subsequent RNA extraction.

## 3.3. RNA Extraction and RT-PCR

In order to verify the integrity of the RNA after microdissection, we extract the RNA and perform a RT-PCR using primers designed from sequences of genes known to be expressed during the liver stage: *lsa*-1 *(11)*, *lsa*-3 *(12)*, and *hsp*-70 *(10)*. The RNA extraction should be performed directly in the 0.5-mL Eppendorf tube immediately after the microdissection to avoid any RNA degradation.

### 3.3.1. RNA Extraction

The Micro RNA isolation kit used is designed for a small number of cells (small amount of RNA). The protocol is conducted according to manufacturer's recommendations with column centrifugation. A DNase treatment (DNase I at 37°C for 15 min) is included during the extraction. RNA is eluted with 30 µL of preheated (65°C) elution buffer and stored at –80°C. Filter tips are used for the extraction and all Eppendorf tubes are autoclaved prior to use.

### 3.3.2. Reverse Transcription

1. To make the first-strand cDNA, use 12 µL of total RNA and mix with 1 µM oligo dT, 0.5 mM each dNTP, 10 U RNase inhibitor, 1 µL of Sensiscript reverse transcriptase in 1X buffer in 20 µL final volume. Perform a negative control reaction with the same volume of RNA using the same reaction mixture lacking the reverse transcriptase enzyme.
2. Incubate the reaction at 37°C for 1 h.
3. Inactivate the enzyme by incubating at 93°C for 5 min.

### 3.3.3. PCR Amplification

1. Take 2.5 µL of first-strand cDNA and mix with 0.4 µM of each primer, 1 unit of Fast Start Taq polymerase, 200 µM each of dNTP in a 2 mM MgCl$_2$ 1X PCR buffer for a total volume of 25 µL.
2. Amplify using the following PCR conditions: a first denaturation (and activation of the Taq polymerase) step of 4 min at 95°C preceding 40 cycles of amplification composed of a denaturation step at 95°C for 45 s, an annealing step at 45°C for 45 s, and an extension step at 60°C for 1 min. The PCR amplification is terminated by a final extension step at 60°C for 10 min.
3. If one round of amplification is not enough to detect the PCR products, amplify 1 µL of each PCR product using 0.625 U Taq polymerase in the same mixture and the same amplification conditions as described above, except with an increase in the annealing temperature to 50°C.
4. Stain the resulting products with ethidium bromide and run on a 1% agarose gel in 1X TAE buffer.

## 4. Notes

1. The diameter of the wells of the 16-chamber Lab-Tek glass slides used is the same as the diameter of the cap, so all the selected cells from one well are microdissected without moving the cap, thus avoiding any contaminations.

2. Hepatocytes are particularly RNase-rich. So it is very important to be careful throughout the experiment to avoid these RNases (similarly for parasite RNase). All steps preceding the microdissection itself should be performed under sterile conditions with sterile reagents. The remaining critical step is the parasite labeling, which is performed at 37°C. In order to avoid RNA degradation during the incubation at 37°C, add RNase inhibitor in a concentration of 400 U/mL to the primary and secondary antibodies. The labeling time can be reduced to 20 min for the first antibody, and to 15 min for the second one. Shorter labeling times have been tested, as described in Fend et al. *(13)*, but parasites were not bright enough to be detected during the microdissection. This labeling step should be avoided when transgenic fluorescent sporozoites are used *(14)*.

3. The main difficulty in performing microdissection with hepatocyte cultures is due to their tight attachment to the slide. It is therefore more difficult to capture the cells compared to microdissection of a tissue section, and a high laser power is required for the microdissection. Even so, the bottom membrane of the targeted cell is still attached to the slide and, consequently, the intracellular parasite is not easily captured and sometimes still remains on the slide after the cap is lifted. So, several laser shots (between 3 and 6) are necessary to capture only one parasite. This can explain why several noninfected hepatocytes may be captured along with the infected one. In this type of experiment, targeting an intracellular parasite in hepatocyte cultures, obtaining pure material is nearly impossible. Nevertheless, if the objective is to perform gene expression analysis, even microarray analysis, absolutely pure material is not required and experiments can be successful if the proportion of parasite RNA among the hepatocyte RNA is sufficient.

4. The use of GFP-*P. falciparum* infected human hepatocytes could be helpful to solve most of the encountered problems.

## References

1. Hoffman, S. L. and Doolan, D. L. (2000) Malaria vaccines-targeting infected hepatocytes. *Nat. Med.* **6,** 1218–1219.
2. Emmert-Buck, M. R., Bonner, R. F., Smith, P. D., Chuaqui, R. F., Zhuang, Z., Goldstein, S. R., et al. (1996) Laser capture microdissection. *Science* **274,** 998–1001.
3. Fend, F., Specht, K., Kremer, M., and Quintanilla-Martinez, L. (2002) Laser capture microdissection in pathology. *Methods Enzymol.* **356,** 196–206.
4. Ball, H. J., McParland, B., Driussi, C., and Hunt, N. H. (2002) Isolating vessels from the mouse brain for gene expression analysis using laser capture microdissection. *Brain Res. Protoc.* **9,** 206–213.

5. Sacci, J. B., Jr., Aguiar, J. C., Lau, A. O., and Hoffman, S. L. (2002) Laser capture microdissection and molecular analysis of Plasmodium yoelii liver-stage parasites. *Mol. Biochem. Parasitol.* **119**, 285–289.

6. Semblat, J. P., Silvie, O., Franetich, J. F., Hannoun, L., Eling, W., and Mazier, D. (2002) Laser capture microdissection of *Plasmodium falciparum* liver stages for mRNA analysis. *Mol. Biochem. Parasitol.* **121**, 179–183.

7. Guguen-Guillouzo, C., Campion, J. P., Brissot, P., Glaise, D., Launois, B., Bourel, M., and Guillouzo, A. (1982) High yield preparation of isolated human adult hepatocytes by enzymatic perfusion of the liver. *Cell Biol. Int. Rep.* **6**, 625–628.

8. Ponnudurai, T., Meuwissen, J. H., Leeuwenberg, A. D., Verhave, J. P., and Lensen, A. H. (1982) The production of mature gametocytes of *Plasmodium falciparum* in continuous cultures of different isolates infective to mosquitoes. *Trans. R. Soc. Trop. Med. Hyg.* **76**, 242–250.

9. Mazier, D., Beaudoin, R. L., Mellouk, S., Druilhe, P., Texier, B., Trosper, J., et al. (1985) Complete development of hepatic stages of *Plasmodium falciparum in vitro*. *Science* **227**, 440–442.

10. Renia, L., Mattei, D., Goma, J., Pied, S., Dubois, P., Miltgen, F., et al. (1990) A malaria heat-shock-like determinant expressed on the infected hepatocyte surface is the target of antibody-dependent cell-mediated cytotoxic mechanisms by non-parenchymal liver cells. *Eur. J. Immunol.* **20**, 1445–1449.

11. Guerin-Marchand, C., Druilhe, P., Galey, B., Londono, A., Patarapotikul, J., Beaudoin, R. L., et al. (1987) A liver-stage-specific antigen of *Plasmodium falciparum* characterized by gene cloning. *Nature* **329**, 164–167.

12. Daubersies, P., Thomas, A. W., Millet, P., Brahimi, K., Langermans, J. A., Ollomo, B., et al. (2000) Protection against *Plasmodium falciparum* malaria in chimpanzees by immunization with the conserved pre-erythrocytic liver-stage antigen 3. *Nat. Med.* **6**, 1258–1263.

13. Fend, F., Emmert-Buck, M. R., Chuaqui, R., Cole, K., Lee, J., Liotta, L. A., and Raffeld, M. (1999) Immuno-LCM: laser capture microdissection of immuno-stained frozen sections for mRNA analysis. *Am. J. Pathol.* **154**, 61–66.

14. Natarajan, R., Thathy, V., Mota, M. M., Hafalla, J. C. R., Menard, R., and Vernick, K. D. (2001) Fluorescent *Plasmodium berghei* sporozoites and pre-erythrocytic stages: a new tool to study mosquito and mammalian host interactions with malaria parasites. *Cell Microbiol.* **3**, 371–379.

# Index

## A

AIDS, *see* Human immunodeficiency virus

Allelotyping,
  microdissection advantages, 69, 70
  ovarian cancer,
    data acquisition and analysis, 74
    DNA extraction, 72, 73, 76
    materials, 70, 71
    microdissection,
      Leica LMD system, 72
      PixCell II LCM system, 72, 76
    polymerase chain reaction, 74
    tissue preparation, 71, 72, 75
    whole-genome amplification, 74
Atherosclerosis, *see* Macrophage foam cells

## B

Bladder cancer,
  genetic aberration detection,
    fluorescence *in situ* hybridization,
      denaturing of DNA, 85
      microdissected formalin-fixed, wax-embedded tissue, 86
      overview, 84, 85
      pretreatment, 85
      probe preparation, 85
      washing, 85, 91
    gene sequencing, 88, 89, 92
    laser microdissection,
      fresh-frozen material, 83, 91
      wax-embedded tissue, 83
    loss of heterozygosity analysis,
      microsatellite amplification, 87

microsatellite marker selection, 86, 91
      overview, 86
      polyacrylamide gel electrophoresis, 87
      scoring, 88
      silver staining, 87
      thermogradient polymerase chain reaction, 86, 87, 91, 92
    materials, 80–82
    whole-genome amplification, 83, 84
  proteomics analysis,
    biopsy acquisition, 257
    laser capture microdissection with PixCell II system, 259, 260, 265
    laser microbeam microdissection with PALM system, 260, 261, 265
    lysis of samples, 261, 262
    manual microdissection, 259, 264, 265
    materials, 256, 257
    overview, 255, 256
    sample preparation,
      sectioning, 257
      staining, 258, 259, 264
      tissue preparation, 258, 264
    SELDI mass spectrometry,
      chip preparation, 262, 263
      instrumentation, 262
      principles, 262
      sample application, 263, 265, 266

spectra analysis, 263, 264, 266
staging, 79, 80
Breast cancer,
    laser microdissection for
        microsatellite instability
        analysis,
        detection, 99–101
        DNA isolation, 95, 98, 100
        improved primer extension
            preamplification polymerase
            chain reaction, 98, 99, 101
        materials, 94, 95
        overview, 93, 94
    microsatellite instability, 94

**C**

Cervical cancer, *see* Human papilloma
    virus
CGH, *see* Comparative genomic
    hybridization
Colorectal cancer,
    laser microdissection for
        microsatellite instability
        analysis,
        detection, 99–101
        DNA isolation, 95, 98, 100
        improved primer extension
            preamplification polymerase
            chain reaction, 98, 99, 101
        materials, 94, 95
        overview, 93, 94
    microsatellite instability, 94
    multiplex real-time PCR of laser
        microdissected tissue,
        advantages, 27, 28
        colon tumor specimen
            preparation, 30, 35
        DNA extraction, 31
        gene copy number calculation, 28,
            29, 33–36
        laser microdissection, 31
        materials, 29, 30
        primer,

concentration optimization, 32,
    33, 35, 36
design, 31, 32
probe,
    concentration optimization, 33,
        36
    design, 31
proteomics,
    laser capture microdissection,
        246, 248, 251
    materials, 246, 247
    solubilization of tissue samples,
        248, 249
    two-dimensional gel
        electrophoresis,
        analysis, 250
        band excision and mass
            spectrometry analysis of
            proteins, 250–252
        denaturing gel electrophoresis,
            249
        isoelectric focusing, 249
        sample preparation, 249
        staining, 250
Comparative genomic hybridization
    (CGH),
    advantages, 41
    laser capture microdissected
        specimens,
        array-based hybridization, 42, 51,
            52
        degenerate oligonucleotide
            primed polymerase chain
            reaction,
            DNA labeling, 48, 53
            whole-genome amplification,
                41, 44–46, 52
        DNA extraction and
            quantification, 44, 52
        hybridization, washing and
            detection, 50, 51, 53
        labeled DNA analysis, 49
        materials, 41, 42

metaphase slide and probe
preparation, 50
microdissection, 43, 44
nick translation for DNA labeling,
48, 49, 53
probe precipitation, 49, 50
tissue preparation,
formalin-fixed, paraffin-
embedded tissue, 43
frozen tissue, 43
principles, 39–41

**D**

Degenerate oligonucleotide primed
polymerase chain reaction, *see*
Polymerase chain reaction
DNA microarray,
complementary DNA synthesis and
purification,
first round, 196, 197, 204
second round, 197–199
gene expression profiling, 187, 188
in vitro transcription and
purification,
first round, 197, 204
second round, 199, 204, 205
materials, 188, 189
RNA,
isolation, 191, 192, 203, 204
quality assessment, 192, 204
target labeling for GeneChip
analysis, 193, 195, 196, 204
tissue sectioning and staining, 189,
190, 201, 203

**F**

FISH, *see* Fluorescence *in situ*
hybridization
Fluorescence *in situ* hybridization
(FISH), bladder cancer,
denaturing of DNA, 85
microdissected formalin-fixed, wax-
embedded tissue, 86

overview, 84, 85
pretreatment, 85
probe preparation, 85
washing, 85, 91

**G**

GeneChip, *see* DNA microarray

**H**

Herpes simplex virus (HSV),
*in situ* hybridization of latency-
associated transcripts,
hybridization, 288, 289
laser capture microdissection of
virus-infected cells, 289, 291
materials, 287
mouse model, 287
overview, 286
probe generation, 288
real-time polymerase chain
reaction, 289, 291, 292
latency, 285, 286
HIV, *see* Human immunodeficiency
virus
HPV, *see* Human papilloma virus
HSV, *see* Herpes simplex virus
Human immunodeficiency virus (HIV),
cell specificity, 272
*in situ* polymerase chain reaction–
laser capture microdissection,
DNA cloning and sequencing,
280, 281
*in situ* polymerase chain reaction,
amplification, 276, 277, 282
immunostaining, 277, 282
Klenow incubation, 276, 281
proteinase K treatment, 275,
281
tissue preparation, 275
laser capture microdissection,
277, 279, 282
materials, 272, 273
nucleic acid purification, 278, 282

standard polymerase chain
   reaction, 278–280
tissue preparation,
   gentle fixation, 273, 274, 281
   paraffin embedding, 274
   sectioning, 274, 281
   staining, 274, 275, 281
Human papilloma virus (HPV),
   cervical cancer association, 295, 296
   laser capture microdissection–
      polymerase chain reaction
      analysis,
   DNA extraction,
      microdissected samples, 297,
         299
      tumor sections, 297, 299
   materials, 296, 299
   polymerase chain reaction,
      agarose gel electrophoresis of
         amplification products, 298,
         299
      γ-globin, 298
      HPV-16, 298

**I**

Immunohistochemistry,
   formalin-fixed, paraffin-embedded
      tissues, 176, 177
   macrophage foam cells, 224, 225,
      229
   methacarn fixated tissue, 16
   microdissection of membrane-
      mounted native tissue
      combination,
      formalin-fixed, paraffin-
         embedded tissue,
         epitope retrieval, 145, 147
         immunostaining and
            visualization, 145, 147
         wax removal and peroxidase
            blocking, 142, 147
      frozen sections, 145, 146
   materials, 140, 141

pituitary cells following laser
      capture microdissection,
      laser capture microdissection, 236
   materials, 234
   tissue preparation, 234–236, 238
transplanted organ tissue, 118, 121
Improved primer extension
      preamplification polymerase
      chain reaction, *see* Whole-
      genome amplification
*In situ* hybridization (ISH),
   herpes simplex virus latency-
      associated transcripts,
      hybridization, 288, 289
      laser capture microdissection of
         virus-infected cells, 289, 291
      materials, 287
      mouse model, 287
      overview, 286
      probe generation, 288
      real-time polymerase chain
         reaction, 289, 291, 292
   microdissection of membrane-
      mounted native tissue
      combination,
      hybridization and visualization,
         146–148
      materials, 141, 142
      slide preparation and sectioning,
         146
      tissue digestion and fixation, 146–
         148
*In situ* polymerase chain reaction, *see*
      Polymerase chain reaction
ISH, *see In situ* hybridization

**K**

Ki-*ras*, detection with laser capture
      microdissection and
      polymerase chain reaction,
      imaging, 63, 65, 66
      lysis and digestion, 61, 65, 66
      materials, 58–60

microdissection, 60, 61, 65
overview, 57, 58
polymerase chain reaction, 61, 66
single-strand conformational
 polymorphism analysis, 61, 63
tissue sectioning and staining, 60, 65
Kidney RNA analysis, *see* RNA
 analysis

**L**

Laser capture microdissection (LCM),
 automation, 7
 historical perspective, 4, 5
 principles, 5, 6, 128, 133
Laser microbeam microdissection
 (LMM), *see also*
 Microdissection of membrane-
 mounted native tissue; PALM
 Robot-Microbeam system,
 automation, 7
 instruments, 133, 134
 principles, 6, 128, 133
LCM, *see* Laser capture
 microdissection
LMM, *see* Laser microbeam
 microdissection
LOH, *see* Loss of heterozygosity
Loss of heterozygosity (LOH),
 bladder cancer analysis,
 microsatellite amplification, 87
 microsatellite marker selection,
 86, 91
 overview, 86
 polyacrylamide gel
 electrophoresis, 87
 scoring, 88
 silver staining, 87
 thermogradient polymerase chain
 reaction, 86, 87, 91, 92
 breast/colon cancer microsatellite
 instability analysis,
 detection, 99–101
 DNA isolation, 95, 98, 100

improved primer extension
 preamplification polymerase
 chain reaction, 98, 99, 101
 materials, 94, 95
 overview, 93, 94
microdissection advantages in
 analysis, 69, 70
ovarian cancer analysis,
 allelotyping,
 data acquisition and analysis,
 74
 polymerase chain reaction, 74
 data acquisition and analysis, 74
 DNA extraction, 72, 73, 76
 materials, 70, 71
 microdissection,
 Leica LMD system, 72
 PixCell II LCM system, 72, 76
 polymerase chain reaction, 74
 tissue preparation, 71, 72, 75
 whole-genome amplification, 74
 purity requirements, 93, 94
Macrophage foam cells,
 atherosclerosis role, 221
 RNA analysis in atherosclerosis,
 CD68 immunostaining, 224, 225,
 229
 laser capture microdissection, 225
 materials, 222–224
 overview, 221, 222
 reverse transcriptase-polymerase
 chain reaction, 227, 229
 RNA extraction, 225, 227, 229
 tissue processing, 224

**M**

Malaria, *see Plasmodium falciparum*
Mass spectrometry (MS),
 bladder cancer proteomics using
 SELDI mass spectrometry,
 chip preparation, 262, 263
 instrumentation, 262
 principles, 262

sample application, 263, 265, 266
spectra analysis, 263, 264, 266
colorectal cancer protein
identification from gels, 250–252
Messenger RNA, *see* RNA analysis
Methacarn fixation,
advantages, 11, 12
genomic DNA analysis in wax-embedded tissues,
DNA extraction from microdissected cells, 19, 22
fixation and wax embedding, 13–15, 21
materials, 12, 13
microdissection with PALM Robot-Microbeam system, 18, 19, 22
polymerase chain reaction,
nested PCR, 20
single-step PCR, 20, 23
tissue staining,
cresyl violet, 15, 16
effects on extracted DNA yield, 16–18, 22
hematoxylin and eosin, 16
immunostaining, 16
Microarray, *see* DNA microarray
Microdissection of membrane-mounted native tissue (MOMeNT),
disposable needles,
laser microbeam microdissection, 131, 132, 136
membrane-mounted glass slide preparation, 130, 134, 135
morphology improvements, 131, 135
DNA extraction, 132, 133, 136
immunohistochemistry combination,
formalin-fixed, paraffin-embedded tissue,
epitope retrieval, 145, 147
immunostaining and visualization, 145, 147

wax removal and peroxidase blocking, 142, 147
frozen sections, 145, 146
materials, 140, 141
*in situ* hybridization combination,
hybridization and visualization, 146–148
materials, 141, 142
slide preparation and sectioning, 146
tissue digestion and fixation, 146–148
materials, 128–130, 133, 134
principles, 128, 139, 140
RNA extraction, 133, 136
single-step collection, 131, 132, 134–136
tissue sectioning,
formalin-fixed, paraffin-embedded tissue, 131, 135, 136
frozen sections, 131, 135, 136
Microsatellite analysis, *see* Loss of heterozygosity
MOMeNT, *see* Microdissection of membrane-mounted native tissue
MS, *see* Mass spectrometry
Multiplex PCR, *see* Polymerase chain reaction

**N**

Nested PCR, *see* Polymerase chain reaction; Reverse transcriptase-polymerase chain reaction

**O**

Organ transplantation,
*in situ* microchimerism, 113
sex mismatches, 114
short tandem repeat analysis of recipient cells in transplanted organs,
contamination prevention, 116, 117

DNA isolation, 119, 120
foil-coated slide preparation, 117, 120
laser capture microdissection, 118, 119, 121
materials, 115, 116
overview, 114, 115
polymerase chain reaction and analysis, 120, 122
specimen preparation, 117
staining,
    immunohistochemical staining, 118, 121
    methylene blue, 118

## P

p53,
    genetic analysis of single cells from PALM system, 160, 162, 163
    mutation detection with laser capture microdissection and polymerase chain reaction,
        imaging, 63, 65, 66
        lysis and digestion, 61, 65, 66
        materials, 58–60
        microdissection, 60, 61, 65
        overview, 57, 58
        polymerase chain reaction, 61, 66
        single-strand conformational polymorphism analysis, 61, 63
        tissue sectioning and staining, 60, 65
PALM Robot-Microbeam system,
    bladder cancer microdissection for proteomics analysis, 260, 261, 265
    complementary DNA synthesis, 158, 164
    genetic analysis of single cells, 160, 162, 163
    genomic DNA analysis in wax-embedded methacarn-fixated tissues, 18, 19, 22
    laser microdissection, 153–155, 164

materials, 152, 153
microchip gel electrophoresis, 158, 163, 164
principles, 151, 152
reverse transcriptase-polymerase chain reaction, 159
RNA extraction, 158, 164
short tandem repeat analysis of recipient cells in transplanted organs, 118, 119, 121
staining of sections, 153, 163
tissue section preparation, 153, 163
PCR, *see* Polymerase chain reaction
Pituitary cells,
    DNA extraction following laser capture microdissection, 236, 237
    hormones, 233
    immunohistochemistry with laser capture microdissection,
        laser capture microdissection, 236
        materials, 234
        tissue preparation, 234–236, 238
    RNA analysis,
        materials, 234
        reverse transcriptase-polymerase chain reaction, 237, 238
        RNA extraction, 237, 238
PixCell II system,
    allelotyping, 72, 76
    bladder cancer proteomics analysis, 259, 260, 265
*Plasmodium falciparum,*
    genome sequencing, 301
    laser capture microdissection of liver infection,
        hepatocyte culture and infection, 303, 304, 306
        materials, 302, 303
        microdissection, 304–306
        parasite labeling, 304, 306
        rationale, 302
        RNA analysis,
            reverse transcriptase-polymerase chain reaction, 305

RNA extraction, 305
malaria mortality, 302
Polymerase chain reaction (PCR),
    allelotyping, 74
    bladder cancer loss of heterozygosity
        analysis, 86, 87, 91, 92
    degenerate oligonucleotide primed
        polymerase chain reaction for
        comparative genomic
        hybridization,
        DNA labeling, 48, 53
        whole-genome amplification, 41,
            44–46, 52
    genomic DNA analysis in wax-
        embedded methacarn-fixated
        tissues,
        nested PCR, 20
        single-step PCR, 20, 23
    herpes simplex virus latency-
        associated transcripts, 289,
        291, 292
    human immunodeficiency virus *in
        situ* polymerase chain
        reaction–laser capture
        microdissection,
        DNA cloning and sequencing,
            280, 281
        *in situ* polymerase chain reaction,
            amplification, 276, 277, 282
            immunostaining, 277, 282
            Klenow incubation, 276, 281
            proteinase K treatment, 275, 281
            tissue preparation, 275
        laser capture microdissection,
            277, 279, 282
        materials, 272, 273
        nucleic acid purification, 278, 282
        standard polymerase chain
            reaction, 278–280
        tissue preparation,
            gentle fixation, 273, 274, 281
            paraffin embedding, 274
            sectioning, 274, 281
            staining, 274, 275, 281

human papilloma virus,
    agarose gel electrophoresis of
        amplification products, 298,
        299
    β-globin, 298
    HPV-16, 298
improved primer extension
    preamplification polymerase
    chain reaction, *see* Whole-
    genome amplification
multiplex real-time PCR of laser
    microdissected tissue,
    advantages, 27, 28
    colon tumor specimen
        preparation, 30, 35
    DNA extraction, 31
    gene copy number calculation, 28,
        29, 33–36
    laser microdissection, 31
    materials, 29, 30
    primer,
        concentration optimization, 32,
            33, 35, 36
        design, 31, 32
    probe,
        design, 31
        concentration optimization, 33,
            36
mutation detection in cancer, *see*
    Ki-*ras*; *p53*
*p53* genetic analysis of single cells
    from PALM system, 160, 162,
    163
RNA, *see* Reverse transcriptase-
    polymerase chain
short tandem repeat analysis of
    recipient cells in transplanted
    organs, 120, 122
Proteomics,
    bladder cancer,
        biopsy acquisition, 257
        laser capture microdissection with
            PixCell II system, 259, 260,
            265

laser microbeam microdissection
with PALM system, 260, 261,
265
lysis of samples, 261, 262
manual microdissection, 259, 264,
265
materials, 256, 257
overview, 255, 256
sample preparation,
sectioning, 257
staining, 258, 259, 264
tissue preparation, 258, 264
SELDI mass spectrometry,
chip preparation, 262, 263
instrumentation, 262
principles, 262
sample application, 263, 265,
266
spectra analysis, 263, 264,
266
colorectal cancer,
laser capture microdissection,
246, 248, 251
materials, 246, 247
solubilization of tissue samples,
248, 249
two-dimensional gel
electrophoresis,
analysis, 250
band excision and mass
spectrometry analysis of
proteins, 250–252
denaturing gel electrophoresis,
249
isoelectric focusing, 249
sample preparation, 249
staining, 250
overview, 245, 246

**R**

Ras, *see* Ki-*ras*
RET/PTC oncogene, *see* Thyroid
cancer

Reverse transcriptase-polymerase chain
reaction,
macrophage foam cells, 227, 229
microdissected sample RNA,
complementary DNA synthesis,
178, 182
qualitative messenger RNA
analysis, 178, 179, 182
quantitative messenger RNA
analysis, 179, 180
RNA preamplification, 180, 181
PALM system samples, 159
pituitary cells, 237, 238
*Plasmodium falciparum*, 305
renal proximal tubule RNA analysis,
complementary DNA synthesis,
213, 215, 218
real-time polymerase chain
reaction, 215, 218
RET/PTC oncogene rearrangements,
laser capture microdissection,
105–107
materials, 105
nested reverse transcriptase-
polymerase chain reaction,
107–109
overview, 104, 105
RNA extraction, 107, 109
RNA analysis, *see* also DNA
microarray; Reverse
transcriptase-polymerase chain
reaction,
amplification, *see* Reverse
transcriptase-polymerase chain
reaction
extraction after microdissection of
membrane-mounted native
tissue, 133, 136
extraction after PALM Robot-
Microbeam system
microdissection, 158, 164
formalin-fixed, paraffin-embedded
tissue,
immunohistochemical staining,
176, 177

RNA preparation,
  higher cell number, 177, 178,
    181
  low cell number, 177
  sectioning, 175
  staining, 175, 176
frozen tissue analysis,
  immunohistochemical staining,
    172, 173, 181
  materials, 169–171
  microdissection, 173
  RNA preparation,
    DNase digestion, 175
    higher cell number, 174, 175
    low cell number, 173, 174
  sectioning, 171, 181
  staining, 172
macrophages, *see* Macrophage foam
    cells
pituitary cells, *see* Pituitary cells
*Plasmodium falciparum*,
  reverse transcriptase-polymerase
    chain reaction, 305
  RNA extraction, 305
renal proximal tubules,
  humans,
    laser capture microdissection,
      216, 218
    RNA extraction and
      amplification, 216, 218
    tissue processing, 215, 218
  kidney structure, 209
  materials, 210–212
  messenger RNA content, 210
  mouse,
    complementary DNA
      synthesis, 213, 215, 218
    laser capture microdissection,
      213, 216
    real-time polymerase chain
      reaction, 215, 218
    RNA extraction, 213, 218
    tissue processing, 212, 213,
      216

  tissue preservation, 209, 210
RET/PTC oncogene rearrangements,
    107–109
reverse transcriptase-polymerase
    chain reaction of
    microdissected sample RNA,
    complementary DNA synthesis,
      178, 182
    qualitative messenger RNA
      analysis, 178, 179, 182
    quantitative messenger RNA
      analysis, 179, 180
    RNA preamplification, 180, 181
  tissue processing optimization, 167–
    169

**S**

SELDI mass spectrometry, *see* Mass
    spectrometry
Short tandem repeat (STR), analysis of
    recipient cells in transplanted
    organs,
    contamination prevention, 116, 117
    DNA isolation, 119, 120
    foil-coated slide preparation, 117, 120
    laser capture microdissection, 118,
      119, 121
    materials, 115, 116
    overview, 114, 115
    polymerase chain reaction and
      analysis, 120, 122
    specimen preparation, 117
    staining,
      immunohistochemical staining,
        118, 121
      methylene blue, 118
Single-strand conformational
    polymorphism (SSCP),
    Ki-*ras*/*p53* mutation detection with
      laser capture microdissection
      and polymerase chain reaction,
      imaging, 63, 65, 66
      lysis and digestion, 61, 65, 66
      materials, 58–60

microdissection, 60, 61, 65
overview, 57, 58
polymerase chain reaction, 61, 66
single-strand conformational
polymorphism analysis, 61, 63
tissue sectioning and staining, 60,
65
principles, 58
SSCP, *see* Single-strand
conformational polymorphism
STR, *see* Short tandem repeat

**T**
Thyroid cancer,
progression, 103
RET/PTC oncogene,
mutations, 104
reverse transcriptase-polymerase
chain reaction of activation,
laser capture microdissection,
105–107
materials, 105
nested reverse transcriptase-
polymerase chain reaction
of rearrangements, 107–109
overview, 104, 105
RNA extraction, 107, 109
Two-dimensional gel electrophoresis,
*see* Proteomics

**W**
WGA, *see* Whole-genome
amplification
Whole-genome amplification (WGA),
degenerate oligonucleotide primed
polymerase chain reaction for
comparative genomic
hybridization, 41, 44–46, 52
improved primer extension
preamplification polymerase
chain reaction,
bladder cancer, 83, 84
breast/colon cancer microsatellite
instability analysis, 98, 99, 101
ovarian cancer, 74